Studies in Foundations and Combinatorics

ADVANCES IN MATHEMATICS
SUPPLEMENTARY STUDIES, VOLUME 1

ADVANCES IN Mathematics
SUPPLEMENTARY STUDIES

Edited by Gian-Carlo Rota

Editorial Board:

Michael F. Atiyah
Lipman Bers
Raoul Bott
Felix Browder
A. P. Calderón
S. S. Chern
J. Dieudonné
J. L. Doob
Samuel Eilenberg
Paul Erdös
Adriano Garsia
Marshall Hall, Jr.
Lars Hörmander
Konrad Jacobs
Nathan Jacobson
Mark Kac
Richard V. Kadison
Shizuo Kakutani
Samuel Karlin
Donald Knuth
K. Kodaira
J. J. Kohn
Bertram Kostant
Peter D. Lax
C. C. Lin
John Milnor
Calvin C. Moore
D. S. Ornstein
Claudio Procesi
Gerald E. Sacks
M. Schutzenberger
J. T. Schwartz
I. M. Singer
D. C. Spencer
Guido Stampacchia
Oscar Zariski

Studies in Foundations and Combinatorics

ADVANCES IN MATHEMATICS
SUPPLEMENTARY STUDIES, VOLUME 1

EDITED BY

Gian-Carlo Rota

Department of Mathematics
Massachusetts Institute of Technology
Cambridge, Massachusetts

With the Editorial Board
of *Advances in Mathematics*

ACADEMIC PRESS New York San Francisco London 1978
A Subsidiary of Harcourt Brace Jovanovich, Publishers

ACADEMIC PRESS, INC.
111 Fifth Avenue, New York, New York 10003

United Kingdom Edition published by
ACADEMIC PRESS, INC. (LONDON) LTD.
24/28 Oval Road, London NW1 7DX

Library of Congress Cataloging in Publication Data
Main entry under title:

Studies in foundations and combinatorics.

(Advances in mathematics : Supplementary studies ; v. 1)
Includes bibliographies.
1. Combinatorial analysis--Addresses, essays, lectures. 2. Logic, Symbolic and mathematical--Addresses, essays, lectures. I. Rota, Gian Carlo, Date II. Series.
QA164.S85 511'.6 78-12921
ISBN 0-12-599101-0

PRINTED IN THE UNITED STATES OF AMERICA

Contents

List of Contributors ix
Preface xi

Functional Completeness and Stone Duality 1

J. Lambek and B. A. Rattray

References 9

Linear Order in Lattices: A Constructive Study

Newcomb Greenleaf

Introduction 11
1. The Category of Sets and Apartness 12
2. Order 13
3. Kripke Models 14
4. Linear Order 16
5. Linear Order in Lattices 19
6. Ordered Algebraic Structures 21
7. Linear Order in L-Groups 23
8. Linear Order in L-Rings 24
9. Linear Order in L-Fields 27

References 29

Better-Quasi-Orderings and a Class of Trees

Richard Laver

1. Better-Quasi-Orderings 32
2. A Class of Trees 41

References 48

Three Cryptoisomorphism Theorems

G. A. Edgar

1. Introduction 49
2. Preliminaries 50
3. Topology 51
4. Pretopology 54
5. Pseudotopology 58
References 59

Topological Duality for Prevarieties of Universal Algebras

Brian A. Davey

1. Preliminaries 62
2. Dualities via Structured Compact Spaces 64
3. Dualities via Compact Topological Partial Algebras 74
4. Examples 81
References 97

Up–Down and Down–Up Partitions

L. Carlitz

1. Introduction 101
2. Notation 104
3. Down–Up Partitions 105
4. An Identity 108
5. Up–Down Partitions 111
6. Corollaries 114
7. Down–Up (k, t)-Partitions 114
8. Continuation 117
9. Up–Down (k, t)-Partitions 120
10. Corollaries 123
11. Partitions with Other Patterns 123
12. Some Special Cases 127
References 129

Plane Partitions (I): The MacMahon Conjecture

George E. Andrews

1. Introduction 131
2. Summation Lemmas 133
3. Determinants and Matrices 140
4. MacMahon's Conjecture 148
5. Conclusion 149
References 149

Graph Theory in Statistical Physics

F. Y. Wu

1. Introduction 151
2. The Whitney Rank Function 152
3. The Potts Model 154
4. The Vertex Model 158
5. Equivalence of the Whitney Polynomial with a Vertex Model 161
References 164

Secondary Structure of Single-Stranded Nucleic Acids

Michael S. Waterman

1. Introduction 167
2. The Graph Theory of Secondary Structure 168
3. First Order Single Loop Secondary Structures 179
4. The Hairpin Matrix 189
5. First Order Secondary Structures 191
6. Second Order Secondary Structures 195
7. *N*th Order Secondary Structures 205
8. Conclusion 208
References 211

Limits of Zeros of Recursively Defined Families of Polynomials

S. Beraha, J. Kahane, and N. J. Weiss

1. Introduction 213
2. Statement of the Main Result 214

3. Preliminaries 215
4. Proof of the Theorem 219
5. The Nondegeneracy Conditions 228
6. An Application 231
References 232

Time-Varying Linear Discrete-Time Systems: Realization Theory

Bostwick F. Wyman

1. Introduction 233
2. Generalized Difference Systems 233
3. Posets, Incidence Algebras, and Dynamical Modules 238
4. Linear Systems on Locally Finite Posets 241
5. Difference Equations on Locally Finite Posets 245
6. The Incidence Algebra of the Discrete Line 247
7. Linear Systems on the Discrete Line 249
8. Difference Equations, Adjoints, and Realizations 251
9. Some Examples 256
References 257

Combinatorial Problems on Subsets and Their Intersections 259

M. Deza, P. Erdos, and N. M. Singhi

References 265

List of Contributors

Numbers in parentheses indicate the pages on which the authors' contributions begin.

GEORGE E. ANDREWS (131), Mathematics Research Center, University of Wisconsin, Madison, Wisconsin 16802

S. BERAHA (213), Department of Mathematics, Queens College, City University of New York, Flushing, New York 11367

L. CARLITZ (101), Department of Mathematics, Duke University, Durham, North Carolina 27706

BRIAN A. DAVEY (61), Department of Mathematics, La Trobe University, Bundoora, Victoria, Australia

M. DEZA (259), Centre National des Recherches Scientifiques, Paris, France

G. A. EDGAR (49), Department of Mathematics, Ohio State University, Columbus, Ohio 43210

P. ERDOS (259), The Hungarian Academy of Sciences, Budapest, Hungary

NEWCOMB GREENLEAF (11), Science Department, Naropa Institute, Boulder, Colorado 80302

J. KAHANE (213), Department of Mathematics, Queens College, City University of New York, Flushing, New York 11367

J. LAMBEK (1), Department of Mathematics, McGill University, Montreal, Canada

RICHARD LAVER (31), Department of Mathematics, University of Colorado, Boulder, Colorado 80309

B. A. RATTRAY (1), Department of Mathematics, McGill University, Montreal, Canada

N. M. SINGHI (259), School of Mathematics, Tata Institute of Fundamental Research, Colaba, Bombay, India

MICHAEL S. WATERMAN (167), Los Alamos Scientific Laboratory, Los Alamos, New Mexico 87545

N. J. WEISS (213), Department of Mathematics, Queens College, City University of New York, Flushing, New York 11367

F. Y. WU (151), Department of Physics, Northeastern University, Boston, Massachusetts 02115

BOSTWICK F. WYMAN (233), Mathematics Department, Ohio State University, Columbus, Ohio 43210

Preface

The supplementary volumes of the journal *Advances in Mathematics* are issued from time to time to facilitate publication of papers already accepted for publication in the journal. The volumes will deal in general—but not always—with papers on related subjects, such as algebra, topology, foundations, etc., and are available individually and independently of the journal.

STUDIES IN FOUNDATIONS AND COMBINATORICS
ADVANCES IN MATHEMATICS SUPPLEMENTARY STUDIES, VOL. 1

Functional Completeness and Stone Duality

J. LAMBEK AND B. A. RATTRAY

Department of Mathematics
McGill University
Montreal, Canada

Every student of logic knows that all truth functions

$$\{\text{true, false}\}^n \to \{\text{true, false}\}$$

can be expressed in terms of "and," "or," and "not." It is also well known that these logical connectives can be expressed in terms of the Boolean ring operations. This property of the two element ring has been called *functional completeness.* We shall show that it lies at the bottom of the Stone duality between Boolean rings and Boolean spaces and obtain a similar duality for any functionally complete algebra.

While in the process of writing up this generalization of Stone duality, we learned that it had already been discovered by Hu [3]. Our methods, however, are quite different, as we place the theorem in the context of a general program of studying dualities induced by adjoint functors.

A Boolean ring may be described as a ring that is isomorphic to a subring of a power of $\mathbb{Z}/2\mathbb{Z}$. In the generalization of Stone duality, we replace the category of rings by any category $\mathscr{A}$ that is algebraic in the sense of Lawvere [6]. In the terminology of universal algebra, $\mathscr{A}$ is a Birkhoff variety or an equational class of algebras. Like Lawvere and Birkhoff, we use *operation* for what might be called an *abstract operation* as distinguished from its particular realization in a given algebra.

Whatever approach one takes to algebraic categories, an n-ary operation gives rise to (and may be identified with) a natural transformation $\omega: V^n \to V$, where $V: \mathscr{A} \to \text{Sets}$ is the *underlying set* functor. (Here n is a nonnegative integer and $V^n(A)$ means $(V(A))^n$.) The naturality expresses the fact that morphisms in $\mathscr{A}$ (homomorphisms) preserve all operations of $\mathscr{A}$. In fact, a homomorphism $A \to B$ is precisely a function $V(A) \to V(B)$ that preserves all operations of $\mathscr{A}$.

It is well known that $\mathscr{A}$ has powers. If A is an algebra in $\mathscr{A}$ and X is any set, there is an algebra A^X whose underlying set is

$$(V(A))^X = \text{Sets}(X, V(A)).$$

ISBN 0-12-599101-0

An operation ω operates on this set of functions pointwise, that is,

$$\omega_{A^X}:(\mathrm{Sets}(X, V(A)))^N \to \mathrm{Sets}(X, V(A))$$

is given by

$$(\omega_{A^X}(f_1, \ldots, f_n))(x) = \omega_A(f_1(x), \ldots, f_n(x)).$$

Thus $\omega_{A^X}(f_1, \ldots, f_n)$ is the composite:

$$X \xrightarrow[\langle f_1, \ldots, f_n \rangle]{} (V(A))^n \xrightarrow[\omega_A]{} V(A).$$

An algebra I of $\mathscr{A}$ is called *functionally complete* [2] if $V(I)$ is finite and has at least two elements and, for every nonnegative integer n, every function $(V(I))^n \to V(I)$ is ω_I for some operation ω.

For example, a finite field F is functionally complete in the category of F-algebras and a finite prime field $\mathbb{Z}/p\mathbb{Z}$ is functionally complete in the category of rings. These examples will be discussed later in more detail.

THEOREM 1. *Let I be a functionally complete algebra in an algebraic category $\mathscr{A}$. Let $\mathscr{A}_I$ be the full subcategory of $\mathscr{A}$ whose objects are the subalgebras of powers of I. Then:*

$\mathscr{A}_I$ is dual to the category $\mathscr{B}$ of Boolean spaces, that is, there is an equivalence $(\mathscr{A}_I)^{\mathrm{op}} \underset{F}{\overset{U}{\rightleftarrows}} \mathscr{B}$;

$U(A)$ is $\mathscr{A}(A, I)$ topologized as a subspace of $(I_{\mathrm{D}})^{V(A)}$, I_{D} being $V(I)$ equipped with the discrete topology;

$F(B)$ is $\mathscr{B}(B, I_{\mathrm{D}})$, that is, the set of continuous functions $B \to I_{\mathrm{D}}$, regarded as a subalgebra of I^B;

F is left adjoint to U with adjunction morphisms:

$$\eta_B: B \to UF(B) = \mathscr{A}(\mathscr{B}(B, I_{\mathrm{D}}), I), \qquad \eta_B(b) = \hat{b}, \qquad \hat{b}(f) = f(b);$$
$$\epsilon_A: A \to FU(A) = \mathscr{B}(\mathscr{A}(A, I), I_{\mathrm{D}}), \qquad \epsilon_A(a) = \hat{a}, \qquad \hat{a}(u) = u(a);$$

η_B and ϵ_A are isomorphisms for all B in $\mathscr{B}$ and A in $\mathscr{A}$.

Remark. The functors U and F are obviously defined on larger categories. The resulting situation will be discussed in Theorem 2.

Proof. For any Boolean space B, $\mathscr{B}(B, I_{\mathrm{D}})$ is a subalgebra of $A = I^B$. For, if $f_1, \ldots, f_n \in \mathscr{B}(B, I_{\mathrm{D}})$ and ω is an n-ary operation of $\mathscr{A}$, then $\omega_A(f_1, \ldots, f_n)$ is the composite:

$$B \xrightarrow[\langle f_1, \ldots, f_n \rangle]{} (I_{\mathrm{D}})^n \xrightarrow[\omega_I]{} I_{\mathrm{D}}.$$

Now $\langle f_1, \ldots, f_n \rangle$ is continuous, and ω_I is continuous because $(I_D)^n$ is discrete. (It is essential here that all operations be finitary.) Thus $\omega_A(f_1, \ldots, f_n) \in \mathscr{B}(B, I_D)$.

For any $g \in \mathscr{B}(B, B')$, $\mathscr{B}(g, I_D): F(B') \to F(B)$ is easily seen to be an $\mathscr{A}$-homomorphism, which is taken as $F(g)$. Thus we have a functor $F: \mathscr{B} \to (\mathscr{A}_I)^{op}$.

We can similarly define $U: (\mathscr{A}_I)^{op} \to \text{Top}$ (the category of topological spaces) with $U(A)$ as described above and $U(f) = \mathscr{A}(f, I)$. We need only show that each $U(A)$ is a Boolean space, that is, compact Hausdorff and zero dimensional. Since I_D is compact and discrete, each closed subspace of a power of I_D is Boolean. Thus we need only show that $\mathscr{A}(A, I)$ is closed in $(I_D)^{V(A)}$.

For any n-ary operation ω of $\mathscr{A}$ and $a_1, \ldots, a_n \in A$, define $\varphi, \psi: (I_D)^{V(A)} \to I_D$ by

$$\varphi(t) = \omega_I(t(a_1), \ldots, t(a_n)),$$
$$\psi(t) = t(\omega_A(a_1, \ldots, a_n)).$$

Now ψ is continuous, because it is a projection. Also φ is continuous, because it is a composition:

$$(I_D)^{V(A)} \xrightarrow[\langle \hat{a}_1, \ldots, \hat{a}_n \rangle]{} (I_D)^n \xrightarrow[\omega_I]{} I_D,$$

where $\hat{a}_1, \ldots, \hat{a}_n$ are projections, so that $\langle \hat{a}_1, \ldots, \hat{a}_n \rangle$ is continuous, and ω_I is continuous because $(I_D)^n$ is discrete. Since I_D is Hausdorff, the subset

$$\{t \in (I_D)^{V(A)} | \omega_I(t(a_1), \ldots, t(a_n)) = t(\omega_A(a_1, \ldots, a_n))\}$$

is closed. Now $\mathscr{A}(A, I)$ is the intersection of all such closed subsets, for all operations ω and all $a_1, \ldots, a_n \in A$. Thus $\mathscr{A}(A, I)$ is closed in $(I_D)^{V(A)}$.

The adjointness of F and U, that is, the existence of a natural isomorphism

$$\mathscr{A}(A, \mathscr{B}(B, I_D)) \to \mathscr{B}(B, A(A, I)),$$

is easily seen. The morphisms η_B and ϵ_A given above correspond to the identity mappings of $\mathscr{B}(B, I_D)$ and $\mathscr{A}(A, I)$; hence they are the adjunctions.

We show next that η_B is always a homeomorphism. We recall that

$$\eta_B(b) = \hat{b}, \qquad \hat{b}(f) = f(b),$$

for all $b \in B$ and $f \in \mathscr{B}(B, I_D)$. Since B is zero dimensional, for any $b_1, b_2 \in B$ there is a closed and open set G containing b_1 but not b_2, I_D has two distinct points, which we may call 0 and 1, and the function $f: B \to I_D$ given by

$$f(b) = \begin{cases} 1 & \text{if} \quad b \in G \\ 0 & \text{if} \quad b \notin G \end{cases}$$

is continuous. Since $\hat{b}_1(f) \neq \hat{b}_2(F)$, η_B is one-to-one. Since B is compact and $UF(B)$ is Hausdorff, η_B induces a homeomorphism of B with its image and the image of η_B is closed in $UF(B)$. To complete the proof, we need only show that the image of η_B is dense in $UF(B)$.

Our proof depends on the fact that any compact topological space has a unique uniformity inducing the given topology, and that any continuous mapping between compact spaces is uniformly continuous. Thus, the category of Boolean spaces is isomorphic to a full subcategory of Uni, the category of uniform spaces. This isomorphism preserves products and subspaces, so the uniformity on $UF(B)$ is that of a subspace of the power $(I_{\mathrm{D}})^{F(B)}$, where I_D is $V(I)$ with the discrete uniformity. Thus a basis $\mathscr{W}$ of vicinities (of the diagonal) for $UF(B)$ is given by the sets

$$W_f = \{(s, t) \in UF(B)^2 \mid s(f_1) = t(f_1), \ldots, s(f_n) = t(f_n)\},$$

where $f_1, \ldots, f_n \in \mathrm{Uni}(B, I_{\mathrm{D}})$. To show that the image of η_B is dense in $UF(B)$ we must show that, for any $s \in UF(B)$ and any W_f in $\mathscr{W}$, there is an element b of B such that $(s, \hat{b}) \in W_f$.

For clarity, we consider first a W_f with $n = 1$, that is,

$$W_f = \{(s, t) \in UF(B)^2 \mid s(f) = t(f)\},$$

with $f \in \mathrm{Uni}(B, I_{\mathrm{D}})$. Let $e: V(I) \to V(I)$ be a retraction onto the image of f, that is, $e(x) = x$ if $x \in \operatorname{im} f$ and $e(x) \in \operatorname{im} f$ for all $x \in V(I)$. Thus $ef = f$. By functional completeness, $e = \omega_I$ for some operation $\omega: V \to V$. For any $s \in UF(B)$, we have a commutative diagram

$$\begin{array}{ccc} F(B) & \xrightarrow{\omega_{F(B)}} & F(B) \\ \Big\downarrow s & & \Big\downarrow s \\ I & \xrightarrow[\omega_I]{} & I \end{array}$$

and $f \in F(B)$, so

$$s(\omega_{F(B)}(f)) = \omega_I(s(f)).$$

Recalling that $\omega_{F(B)}(f) = \omega_I f$, we have

$$s(f) = s(ef) = s(\omega_I f) = s(\omega_{F(B)}(f)) = \omega_I(s(f)) = e(s(f)).$$

Thus $s(f) \in \operatorname{im} f$, that is,

$$s(f) = f(b) = \hat{b}(f)$$

for some $b \in B$, that is, $(s, \hat{b}) \in W_f$.

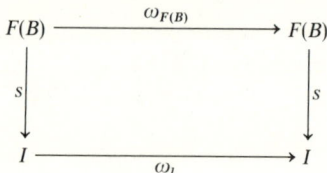

The general case is quite similar. We put $f = \langle f_1, \ldots, f_n \rangle : B \to I^n$ and let $e: V(I)^n \to V(I)^n$ be a retraction onto the image of f. For each projection $p_k: V(I)^n \to V(I)$, we may express $p_k e: V(I)^n \to V(I)$ as $(\omega_k)_I$, where ω_k is an operation of $\mathscr{A}$. Letting $\omega = \langle \omega_1, \ldots, \omega_n \rangle$, we have $e = \omega_I$ and proceed as above to show that

$$(s(f_1), \ldots, s(f_n)) = \omega_I(s(f))$$

is in the image of f, that is, of the form $(f_1(b), \ldots, f_n(b))$ for some $b \in B$.

Thus, η_B is always a homeomorphism. To discuss ϵ_A, we need the following lemmas.

LEMMA 1. *Let I be functionally complete and let A be a subalgebra of I^X. Then $A = \mathrm{Uni}(B, I_{\mathrm{D}})$, where B is the set X equipped with the weak uniformity $\mathscr{W}_A$ induced by A (and I_{D}), that is, the smallest uniformity on X such that all a in A become uniformly continuous functions $X \to I_{\mathrm{D}}$.*

Proof. Surely $A \subseteq \mathrm{Uni}(B, I_{\mathrm{D}})$. $\mathscr{W}_A$ has a basis consisting of the sets

$$V_a = \{(x, y) \in X^2 \mid a_1(x) = a_1(y), \ldots, a_n(x) = a_n(y)\},$$

where $a_1, \ldots, a_n \in A$. Putting

$$a(x) = (a_1(x), \ldots, a_n(x)),$$

we see that a is an element of $\mathrm{Uni}(B, (I_{\mathrm{D}})^n)$ and we may write

$$V_a = \{(x, y) \in X^2 \mid a(x) = a(y)\}.$$

If $f \in \mathrm{Uni}(B, I_{\mathrm{D}})$, then

$$\{(x, y) \mid f(x) = f(y)\} \supseteq V_a$$

for some a; hence

$$\forall_{x, y \in X}(a(x) = a(y) \Rightarrow f(x) = f(y)).$$

Thus, there is a function $\varphi: \mathrm{im}\, a \to V(I)$ such that $\varphi(a(x)) = f(x)$ for all x in X. Extend φ to $\psi: V(I)^n \to V(I)$; then $f = \psi a$. Since I is functionally complete, there is an operation $\omega: V^n \to V$ such that $\psi = \omega_I$. Then

$$f(x) = \psi(a_1(x), \ldots, a_n(x)) = \omega_I(a_1(x), \ldots, a_n(x)) = (\omega_A(a_1, \ldots, a_n))(x),$$

since A is a subalgebra of I^X. Therefore $f = \omega_A(a_1, \ldots, a_n)$ and so $f \in A$.

LEMMA 2. *Let I be a functionally complete algebra, $\mathscr{U}$ a compact uniformity on a set X and A a subalgebra of $\mathrm{Uni}((X, \mathscr{U}), I_{\mathrm{D}})$ that separates points. Then $A = \mathrm{Uni}((X, \mathscr{U}), I_{\mathrm{D}})$. In particular, if B is a Boolean space and A is a subalgebra of $F(B) = \mathscr{B}(B, I_{\mathrm{D}})$ that separates points, then $A = F(B)$.*

Proof. In view of Lemma 1, we need only show that $\mathscr{U}$ is $\mathscr{W}_A$, the weak uniformity induced by A. Clearly $\mathscr{W}_A \subseteq \mathscr{U}$. Since A separates points and I_D is separated, $(X, \mathscr{W}_A)$ is separated. Thus, the topologies induced by $\mathscr{W}_A$ and $\mathscr{U}$ are Hausdorff and compact respectively, and the first is contained in the second. Therefore they are the same compact topology, and so $\mathscr{W}_A = \mathscr{U}$.

To complete the proof of Theorem 1 it remains to show that ϵ_A is an isomorphism. We recall that

$$\epsilon_A(a) = \hat{a}, \qquad \hat{a}(u) = u(a),$$

for all $a \in A$ and $u \in \mathscr{A}(A, I)$. Since $\mathscr{A}$ is an algebraic category, any homomorphism that is one-to-one and onto is an isomorphism. We consider an A in $\mathscr{A}_I$, that is, a subalgebra of some power I^Y of I. If a_1 and a_2 are distinct elements of A, then there is some $y \in Y$ such that the homomorphism $u_y: A \to I$, defined by $u_y(a) = a(y)$, has $u_y(a_1) \neq u_y(a_2)$, that is, $\hat{a}_1(u_y) \neq \hat{a}_2(u_y)$. Thus ϵ_A is one-to-one. Let $\hat{A}$ be the image of ϵ_A; then $\hat{A}$ is a subalgebra of $FU(A)$, which separates points of $U(A)$, by an argument similar to that above. By Lemma 2, $\hat{A} = FU(A)$, and so ϵ_A is onto.

This completes the proof of Theorem 1.

It follows from the theorem that the categories $\mathscr{A}_I$ are equivalent for all functionally complete I. Thus, any categorical theorem about Boolean rings is true also for arbitrary $\mathscr{A}_I$. For example, every Boolean ring is contained is an injective (Dedekind complete) Boolean ring, so the same is true for every $\mathscr{A}_I$. Any categorical theorem about Boolean spaces can be similarly extended. For example, since every finite Boolean space is a coproduct of one-point spaces, it follows that every finite algebra in $\mathscr{A}_I$ is a power of I. (It is evident that an algebra A of $\mathscr{A}_I$ is finite if and only if $U(A)$ is finite.)

It is well known [2, p. 178] that, for any functionally complete algebra I, $\mathscr{A}_I$ is the equational closure of I. That is, an algebra A of $\mathscr{A}$ is in $\mathscr{A}_I$ if and only if every equation (between operations) satisfied identically in I is also satisfied identically in A.

EXAMPLE 1. For any prime p, it is well known that $I_p = \mathbb{Z}/p\mathbb{Z}$ is functionally complete in the category $\mathscr{A}$ of rings. For completeness, we give a simple proof of this by induction.

We claim that any function $V(I_p)^n \to V(I_p)$ can be realized as an operation. This is clearly so for $n = 0$, since all elements of I_p can be obtained by repeated addition from the nullary operation 1. Assume the result for n and consider a function $f: V(I_p)^{n+1} \to V(I_p)$. For each $i \in I_p$, the inductional assumption assures the existence of an operation $\omega_i: V^n \to V$ such that

$$(\omega_i)_I(x_1, \ldots, x_n) = f(x_1, \ldots, x_n, i).$$

But

$$f(x_1, \ldots, x_{n+1}) = \sum_i \delta(x_{n+1} - i)(\omega_i)_I(x_1, \ldots, x_n),$$

where

$$\delta(x) = \begin{cases} 1 & \text{if} \quad x = 0 \\ 0 & \text{if} \quad x \neq 0. \end{cases}$$

Thus f can be realized as ω_I by means of an operation ω provided δ can be so realized. But, by Fermat's little theorem,

$$\delta(x) = 1 - x^{p-1}.$$

It was shown in [8] that a ring is in $\mathscr{A}_{I_p}$ if and only if it is a p-ring, that is, satisfies the identities $x^p = x$, $px = 0$. We include a short proof of this here.

Surely any subring of $I_p{}^X$ is a p-ring. Conversely, given any p-ring A, we shall show that the canonical homomorphism $A \to I_p^{\mathscr{A}(A, I_p)}$ is one-to-one. Indeed, suppose $a \neq 0$ in A; then $1 \notin (1 - a^{p-1})A$, since $a(1 - a^{p-1}) = 0$. Pick a maximal ideal M containing $1 - a^{p-1}$; then A/M is a p-field. It cannot have more than p elements, since all elements satisfy the same equation of degree p. It has at least p elements, since p is its characteristic. Therefore, A/M is a field of p elements and so isomorphic to I_p. Let $f: A \to A/M \cong I_p$; then $f(a) \neq 0$, for otherwise $a \in M$ and so $1 = a^{p-1} + (1 - a^{p-1}) \in M$.

Example 2. Let I_q be the Galois field of $q = p^n$ elements. By much the same argument as in Example 1, we may show that an I_q-algebra is in $\mathscr{A}_{I_q}$ if and only if it satisfies the identity $x^q = x$, and that I_q is functionally complete in the category $\mathscr{A}$ of I_q-algebras. The only difference is that we now express the δ function $I_q \to I_q$ as

$$\delta(x) = (a_1 - x) \cdots (a_k - x)/a_1 \cdots a_k,$$

where $k = q - 1$ and $a_1, \ldots, a_k$ are the nonzero elements of I_q.

We note that the characterizations of $\mathscr{A}_I$ by identities given in Examples 1 and 2 demonstrate in these special cases the general theorem that $\mathscr{A}_I$ is the equational closure of I, for any functionally complete algebra I.

Before stating our next theorem, we recall that the category $\mathscr{B}$ of Boolean spaces can be considered as a full subcategory of either Top or Uni.

Theorem 2. *Let $\mathscr{A}$, I, $\mathscr{A}_I$ be as in Theorem 1. Then*
(a) there is an adjoint pair of functors

$$\mathscr{A}^{\text{op}} \underset{F}{\overset{U}{\rightleftarrows}} \text{Top}$$

with adjunctions η and ϵ defined as in Theorem 1, and

(b) *there is an adjoint pair of functors*

$$\mathscr{A}^{\mathrm{op}} \underset{F}{\overset{U}{\rightleftarrows}} \mathrm{Uni}$$

with adjunctions η and ϵ defined as in Theorem 1, except that I_{D} is $V(I)$ with the discrete uniformity and $F(B) = \mathrm{Uni}(B, I_{\mathrm{D}})$.

In both cases the triples $(FU, \epsilon, F\eta U)$ on $\mathscr{A}$ and $(UF, \eta, U\epsilon F)$ on Top *or* Uni *are idempotent, with fixed subcategories $\mathscr{A}_I$ and $\mathscr{B}$ respectively. Thus they determine reflectors $\mathscr{A} \to \mathscr{A}_I$ and* $\mathrm{Top} \to \mathscr{B}$ *or* $\mathrm{Uni} \to \mathscr{B}$ *with reflection maps ϵ_A and η_B respectively.*

Moreover, $\mathscr{A}_I$ is the limit closure of I in $\mathscr{A}$ and $\mathscr{B}$ is the limit closure of I_{D} in Top *or* Uni.

We recall from [4] that the *fixed subcategory* $\mathrm{Fix}(FU, \epsilon)$ consists of *all A* in $\mathscr{A}$ for which $\epsilon(A)$ is an isomorphism and also that the *limit closure* $\mathscr{L}(I)$ of I is the smallest full subcategory of $\mathscr{A}$ containing I and closed under limits.

Proof. We observe that, in both (a) and (b), every $FU(A)$ is in $\mathscr{A}_I$ and every $UF(B)$ is in $\mathscr{B}$. Now ϵ_A and η_B are isomorphisms for A in $\mathscr{A}_I$ and B in $\mathscr{B}$, by Theorem 1. Thus (UF, η) and (FU, ϵ) are idempotent and $\mathscr{A}_I \subseteq \mathrm{Fix}(FU, \epsilon)$ and $\mathscr{B} \subseteq \mathrm{Fix}(UF, \eta)$. Also, if ϵ_A is an isomorphism, then A is isomorphic to $FU(A)$, which is in $\mathscr{A}_I$; hence A is in $\mathscr{A}_I$. Similarly, if η_B is an isomorphism, then B is in $\mathscr{B}$.

We would like to show that $\mathscr{A}_I$ is the limit closure of I in $\mathscr{A}$. For completeness, we also sketch a proof that $\mathscr{B}$ is the limit closure of I_{D} in Top or Uni, although this is well known.

We consider the functors

$$\mathscr{A}^{\mathrm{op}} \underset{F}{\overset{U}{\rightleftarrows}} \mathrm{Top},$$

the case of Uni being similar. For any subspace E of a power $(I_{\mathrm{D}})^X$ of I_{D} in Top, with inclusion morphism i, the closure of E in $(I_{\mathrm{D}})^X$ is easily seen to be the joint equalizer of all pairs of morphisms $u, v\colon (I_{\mathrm{D}})^X \to I_{\mathrm{D}}$ such that $ui = vi$. That is, $\bar{E}$ is the equalizer of a pair of continuous maps $(I_{\mathrm{D}})^E \to (I_{\mathrm{D}})^F$ where F is the set of all real pairs u, v. Thus $\bar{E}$ is in $\mathscr{L}(I_{\mathrm{D}})$. Now it is clear that for any Boolean space B the obvious map $B \to {I_D}^{\mathrm{Top}(B, I_D)}$ is a homeomorphism with a closed subspace. Thus $\mathscr{B} \subset \mathscr{L}(I_{\mathrm{D}})$. But $\mathscr{B}$ is limit closed so $\mathscr{L}(I_{\mathrm{D}}) \subset \mathscr{B}$. Thus $\mathscr{B} = \mathscr{L}(I_{\mathrm{D}})$.

Since a mono $i\colon B \to B'$ in $\mathscr{B}$ is clearly a homeomorphism of B with a closed subspace of B', it is an equalizer of some pair of morphisms $B' \rightrightarrows (I_{\mathrm{D}})^X$. It follows that every epi in $\mathscr{B}$ is a surjection, and it is easily shown that every surjection in $\mathscr{B}$ is the coequalizer of its kernel pair. Thus, all epis in $\mathscr{B}$ are regular. By duality, all monos in $\mathscr{A}_I$ are regular. Now, for every A in $\mathscr{A}_I$, there is a mono $A \to I^X$; hence there is an equalizer diagram $A \to I^X \rightrightarrows I^Y$. Thus $\mathscr{A}_I$ is the limit closure of I in $\mathscr{A}$.

This completes the proof of Theorem 2.

We observe that I_{D} is easily proved to be injective in $\mathscr{B}$. Also I is injective in $\mathscr{A}_I$ since $I = F(1)$, where 1 is the one-point space, which is clearly projective in $\mathscr{B}$. Thus I_{D} and I are κ-injective in Top and $\mathscr{A}$ respectively, in the sense of our paper [4]. It follows that the functors $Q_{I_{\mathrm{D}}}$ and Q_I defined as in [4] are UF and FU respectively.

References

1. G. Birkhoff, "Lattice theory," Colloquium Publications XXV, *Amer. Math. Soc.*, Providence, Rhode Island, 1967.
2. G. Grätzer, "Universal Algebra," Van Nostrand Reinhold Princeton, New Jersey, 1968.
3. T.-K. Hu, Stone duality for primal algebra theory, *Math. Z.* **110** (1969), 180–198.
4. J. Lambek and B. A. Rattray, Localization at injectives in complete categories, *Proc. Amer. Math. Soc.* **41** (1973), 1–9.
5. J. Lambek and B. A. Rattray, Localization and duality in additive categories, *Houston J. Math.* **1** (1975), 87–100.
6. F. W. Lawvere, Functorial semantics of algebraic theories, *Proc. Nat. Acad. Sci. U.S.A.* **50** (1963), 869–872.
7. S. MacLane, "Categories for the Working Mathematician," Springer–Verlag, Berlin and New York, 1971.
8. N. H. McCoy and D. Montgomery, A representation of generalized Boolean rings, *Duke Math. J.* **3** (1937), 455–459.
9. M. H. Stone, The theory of representation of Boolean algebras, *Trans. Amer. Math. Soc.* **40** (1936), 37–111.
10. R. W. Stringall, The categories of p-rings are equivalent, *Proc. Amer. Math. Soc.* **29** (1971), 229–235.

AMS (MOS) 1970 subject classifications: 54H10, 18A40, 18C05, 08A25.

STUDIES IN FOUNDATIONS AND COMBINATORICS
ADVANCES IN MATHEMATICS SUPPLEMENTARY STUDIES, VOL. 1

Linear Order in Lattices: A Constructive Study

NEWCOMB GREENLEAF

Science Department
Naropa Institute
Boulder, Colorado

Contents

Introduction.
1. The Category of Sets with Apartness.
2. Order.
3. Kripke Models.
4. Linear Order.
5. Linear Order in Lattices.
6. Ordered Algebraic Structures.
7. Linear Order in L-Groups.
8. Linear Order in L-Rings.
9. Linear Order in L-Fields.

INTRODUCTION

This paper investigates the question, Are the real numbers (constructively) linearly ordered? The answer can be either yes or no, depending on the precise definition of linear order used, but we shall see that the former predominates. Indeed, Theorem 9.2 lists eighteen different ways in which the continuum can be considered linearly ordered. Rather than deal with the continuum, we have chosen to work axiomatically and to investigate the linear order concept in various settings, starting with sets and ending with lattice ordered fields, where, rather remarkably, the eighteen conditions mentioned above are all equivalent.

In §1 and §2 we give a new axiomatic foundation for the theory of order on sets with apartness (rather than equality) as the basic relation. In §4 we investigate the relations between various formulations of the linear order concept for sets, and show their inequivalence by Kripke models. In §5 we consider the linear order concept for lattices. In classical mathematics it is hardly worth mentioning that a linearly ordered set is a lattice, but the lattice structure of the continuum is constructively nontrivial, since it is not generally possible to identify $x \wedge y$ with either x or y ($x \wedge y$ exists since it can be defined termwise on Cauchy sequences of rationals). Theorem 5.1 shows that for lattices linear order is a natural, unified concept.

ISBN 0-12-599101-0

The rest of the paper deals with additional formulations of linear order that can be made when algebraic structure is put on a lattice. While the results here are considerably easier to prove in classical mathematics, they do not appear to have been fully recognized.

All of our reasoning is constructive (or intuitionistic). A good summary of this logic is found in [15]. Throughout we omit universal quantifiers, using the convention that free variables are implicitly universally quantified.

§1. The Category of Sets with Apartness

Our approach to set theory is similar to that of Bishop in [1] and [2], except that we take difference or apartness as the basic relation on a set rather than equality. To construct a *set* X it must be made precise what must be done to construct an element x of X, and what must be done to prove that two elements of X are apart. We denote this relation by $x \# y$, and it must further be proved that it is an *apartness relation*, that it is antireflexive, symmetric, and cotransitive:

$$\neg x \# x, \tag{A1}$$

$$x \# y \to y \# x, \tag{A2}$$

$$x \# y \to (x \# z \text{ or } z \# y). \tag{A3}$$

Equality in X is defined to be the negation of apartness:

$$x = y \leftrightarrow \neg x \# y.$$

It is immediate that this relation is an equivalence relation. Inequality, the negation of equality, is denoted by the symbol "$\neq$." Inequality is not generally an apartness relation, for instead of (A3), it satisfies the following weaker property

$$x \neq y \to \neg(x = z \text{ and } z = y).$$

A relation is called *stable* if it is equivalent to its double negation. Since equality is the negation of apartness, it is always stable. Apartness is not generally stable. The stability of apartness for the continuum is equivalent to the Markov principle (see §4). Apartness always implies inequality.

We have described the objects of our category; we now describe the morphisms or functions. By a *function* f from X to Y we mean a rule that routinely transforms elements of X into elements of Y, and that "pulls back" apartness:

$$f(x) \# f(x') \to x \# x'.$$

It is immediate that $x = x'$ implies that $f(x) = f(x')$.

Binary operations will be functions in the above sense. Two elements of a cartesian product are apart if their coordinates in some factor space are apart. This means that a binary operation "$\circ$" will satisfy

$$x \circ y \# z \circ w \rightarrow (x \# z \text{ or } y \# w) \tag{$*$}$$

and also the special cases

$$x \circ y \# x \circ w \rightarrow y \# w,$$
$$x \circ y \# z \circ y \rightarrow x \# z.$$

A set is called *discrete* if any two elements are either apart or equal (i.e., if the apartness relation is decidable). This term is used because it is on such sets that the discrete metric can be (everywhere) defined. The integers and the rationals are discrete, but the reals are not. Indeed, the discreteness of **R** is equivalent to the limited principle of omniscience of Bishop (see [1, p. 9] and §4).

We indicate briefly how apartness is defined for the continuum. Let (x_n) and (y_n) be Cauchy sequences of rational numbers. They are apart if there exist natural numbers M and N such that $n \geqslant N$ implies $|x_n - y_n| \geqslant 1/M$. In the proof that this relation satisfies (A3), essential use is made of the fact that the sequences are Cauchy.

Apartness for real numbers was first defined by Brouwer and was extensively used in analysis [3]. While the real numbers and indeed any metric space carry natural apartness relations, there are important sets that do not. Familiar examples are the power set of a nonempty set, and the quotient group **R**/**Q**. These are best considered in the category of sets having compatible equality and inequality relations, neither the negation of the other. The same holds for the "constructive Dedekind reals" of Staples [22], and the "fickle numbers" of Bishop [1]. Algebra in this category is studied in [17] and [23]. Brouwer also considered other notions of difference. These are compared in [18], and, in the context of category theory, in [11].

Brouwer introduced the symbol "$\#$" for apartness, while Bishop uses "$\#$."

§2. Order

Only slight modifications of the classical axioms are necessary to establish a constructive theory of order. Clearly we want an order relation $\leqslant$ to be reflexive, antisymmetric, and transitive:

$$x = y \leftrightarrow (x \leqslant y \text{ and } y \leqslant x) \tag{O1}$$
$$(x \leqslant y \text{ and } y \leqslant z) \rightarrow x \leqslant z. \tag{O2}$$

Strict order is defined in terms of order and apartness:

$$x < y \quad \text{means} \quad (x \leqslant y \text{ and } x \# y).$$

Surprisingly, it is not possible to deduce from (O1) and (O2) that strict order is transitive, so an extra axiom is necessary (this will be shown in the next section):

$$(x < y \text{ and } y < z) \rightarrow x < z. \tag{O3}$$

Using (A3) we easily obtain from (O3) the apparently stronger assertions

$$(x < y \text{ and } y \leqslant z) \rightarrow x < z,$$

$$(x \leqslant y \text{ and } y < z) \rightarrow x < z.$$

It is misleading to read $x \leqslant y$ as "x is less than or equal to y" since it does not imply ($x < y$ or $x = y$) (see §3). However, the following is easily proved.

PROPOSITION 2.1. (a) *If* $x \leqslant y$ *and* $\neg x < y$, *then* $x = y$. (b) *If* $x \leqslant y$, *then* $\neg\neg(x < y$ *or* $x = y)$, *and the latter is equivalent to* $\neg\neg x \leqslant y$.

Note that if $x = x'$ and $y = y'$, then $x \leqslant y$ is equivalent to $x' \leqslant y'$ by (O1) and (O2), so this need not be made an explicit assumption.

It is customary to define a lattice as an ordered set in which every pair of elements has a g.l.b. and a l.u.b. We need a slightly stronger definition—the g.l.b. and l.u.b. must be given by binary operations, so that we can use (∗). So by a *lattice* we mean a set L along with two binary operations $\wedge$ and $\vee$, which are associative and commutative and satisfy the "absorptive laws":

$$x \wedge (x \vee y) = x \vee (x \wedge y) = x.$$

We then define an order on L by taking $x \leqslant y$ to mean $x \wedge y = x$. The proof that this relation satisfies (O1) and (O2) is standard (see [6, pp. 63–64]). To prove (O3) note that if $x < y$ and $y < z$, then $x \wedge y = x$ and $z \wedge y = y$. Since $x \# y$, by (∗) we obtain $x \# z$. Since $x \leqslant z$ by (O2), we are done. It is also easily checked that $x \wedge y$ and $x \vee y$ are a g.l.b. and l.u.b. for x and y with respect to this order.

Previous constructive formulations of order have taken strict order as the basic relation. These researches go back to work of Brouwer, and are described in [4, 14, 15, 24].

§3. KRIPKE MODELS

We shall describe the very simple Kripke models that we use for counterexamples. For an excellent discussion of the general theory of Kripke models, see [21].

To construct a Kripke model one must give the set of nodes or "states of knowledge," the "domain of discourse" at each node, and the model structure. The set K of nodes is an ordered set. We denote its elements by Greek letters. In our examples K will always be one of the following two sets.

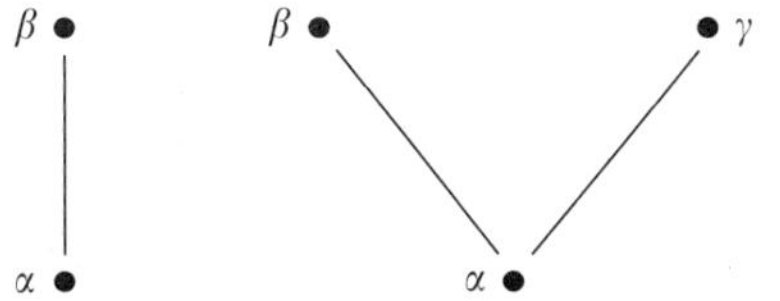

In general there is a domain of discourse D_λ for each $\lambda \in K$. But in our examples all D_λ will be equal, and this common domain of discourse D will always either be $\{a, b\}$ or $\{a, b, c\}$.

Finally, to give the model structure we must decide which atomic formulas are true or "forced" at λ for all $\lambda \in K$. If A is an atomic formula, we use the notation $\lambda \vdash A$ for forcing. The model structure must satisfy the following condition: if $\lambda \vdash A$ and $\lambda \leqslant \mu$, then $\mu \vdash A$. If the order in K is thought of as representing the passage of time, then this condition says that knowledge can only increase. All of our atomic formulas will be of the form $d \# e$ or $d \leqslant e$, but we shall also use the defined symbols "$=$" and "$<$." Once the atomic formulas have been forced, all else follows by an inductive schema:

$\lambda \vdash (A \text{ and } B)$	when	$\lambda \vdash A$ and $\lambda \vdash B$,
$\lambda \vdash (A \text{ or } B)$	when	$\lambda \vdash A$ or $\lambda \vdash B$,
$\lambda \vdash (A \rightarrow B)$	when	for all $\mu \geqslant \lambda$ $(\mu \vdash A \rightarrow \mu \vdash B)$,
$\lambda \vdash \neg A$	when	for all $\mu \geqslant \lambda$ $\mu \nVdash A$,
$\lambda \vdash \exists x\, P(x)$	when	$\exists d \in D_\lambda$ $(\lambda \vdash P(d))$,
$\lambda \vdash \forall x\, P(x)$	when	for all $\mu \geqslant \lambda\ \forall d \in D_\lambda$ $(\mu \vdash P(d))$.

From this it easily follows that $\lambda \vdash \neg\neg A$ when for all $\mu \geqslant \lambda$ there exists $\rho \geqslant \mu$ such that $\rho \vdash A$. In particular, $\lambda \vdash (d = e)$ when for all $\mu \geqslant \lambda$ $\mu \nVdash (d \# e)$, and $\lambda \vdash (d \neq e)$ when for all $\mu \geqslant \lambda$ there exists $\rho \geqslant \mu$ such that $\rho \vdash (d \# e)$.

We will describe Kripke models by listing the atomic formulas that are forced at each node. It will be clear whether $D = \{a, b\}$ or $D = \{a, b, c\}$. We shall not explicitly mention those atomic formulas that follow from the listed ones by (A2) and by (O1), (O2), and (O3).

Kripke models have the following significance for us. If we can construct a Kripke model in which $\lambda \vdash A$ and $\lambda \nVdash B$, then it is not possible to prove (constructively) that A implies B [21, p. 329].

Kripke models will be used extensively in the next section. For the moment we give two simple examples that concern the concepts of the previous two sections. The first example shows that an apartness relation need not be stable, and also that even if any two elements are either equal or not equal, we cannot conclude that a set is discrete.

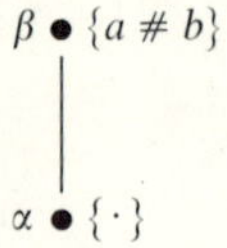

If we take equality as a primitive notion, then a similar example shows that equality need not be stable. It is also possible to construct a Kripke model for sets with equality in which equality is stable but is not the negation of any apartness relation (see [8]). However, this cannot be done when the set K of nodes is finite.

The second example shows the necessity for axiom (O3).

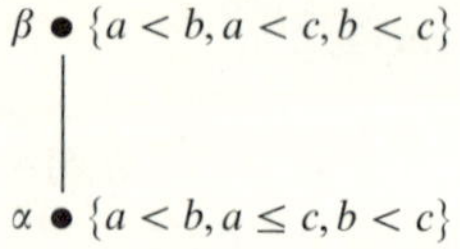

If we omit b from this example, we see that from $a \leqslant c$ we cannot deduce ($a < c$ or $a = c$).

Kripke models for the Brouwer theory of order are given in [16], while that theory is analyzed via the "topological interpretation" in [20].

§4. Linear Order

The term "linear order" will be reserved for that property that best describes our intuition of order in the continuum. Looking at the continuum in a naive geometric way, we might express its linearity as follows: if you can see a difference between x and y, then you can see that one is to the right and the other to the left. Hence we will say that a set is *linearly ordered* if it satisfies

$$x \# y \rightarrow (x < y \text{ or } y < x). \qquad \text{(LO)}$$

On the other hand, given two points on the continuum, we have no routine way of comparing them, so we cannot assert that *trichotomy* or *dichotomy* hold in $\mathbf{R}$:

$$x < y \text{ or } y < x \text{ or } x = y, \qquad \text{(T)}$$

$$x \leqslant y \text{ or } y \leqslant x. \qquad \text{(D)}$$

These three conditions are classically equivalent. The (constructive) relations between them are given in the following easily proved proposition.

PROPOSITION 4.1. *Trichotomy implies dichotomy, and the latter implies linear order. Trichotomy holds in an ordered set S if and only if S is discrete and linearly ordered.*

We now state some other conditions that seem to express the idea of linear order. The first is "constructive dichotomy," which is extremely useful in analysis as a substitute for the classical use of (D) [1, p. 22]:

$$x < y \to (x < z \text{ or } z < y). \tag{CD}$$

We next introduce three conditions that all express some version of "that which is not to the left must be to the right." The latter two are valid for the continuum.

$$\neg x \leqslant y \to y < x, \tag{S}$$

$$\neg x < y \to y \leqslant x, \tag{S_1}$$

$$\neg x \leqslant y \to y \leqslant x. \tag{S_2}$$

Since trichotomy and dichotomy are too strong to hold for the continuum, we also consider their double negations, which we take in the following weak form (making the "double negation shift" from $\neg\neg\forall$ to $\forall\neg\neg$ [25, p. 93]):

$$\neg(\neg x < y \text{ and } \neg y < x \text{ and } \neg x = y) \tag{T_n}$$

$$\neg(\neg x \leqslant y \text{ and } \neg y \leqslant x). \tag{D_n}$$

These conditions state that there do not exist "absolutely incomparable" elements. We can now state the main result of this section.

THEOREM 4.2. *The following implications link the conditions introduced above. No other implications are possible. (The dotted line encloses the conditions which hold for the continuum.)*

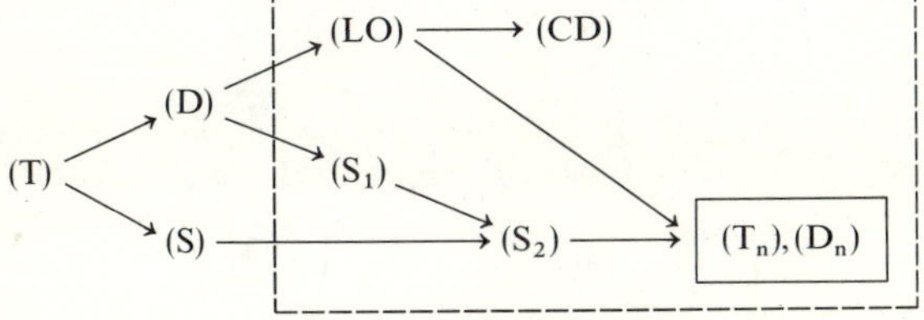

Proof. We have already remarked that the implications (T) → (D) → (LO) are trivial. So are $(S_1) \to (S_2) \to (D_n)$ and $(T) \to (S) \to (S_2)$.

(D) $\to$ (S_1). Assume $\neg x < y$. By (D) we have ($x \leqslant y$ or $y \leqslant x$). In the first case by Proposition 2.1 we have $x = y$ so in either case $y \leqslant x$.

(LO) $\to$ (CD). From $x < y$ we deduce $x \# y$ so by (A3) ($x \# z$ or $z \# y$). By (LO) we have ($x < z$ or $z < x$ or $z < y$ or $y < z$) and then ($x < z$ or $z < y$) follows from (O3).

(LO) $\to$ (T_n). Assume ($\neg x < y$ and $\neg y < x$ and $\neg x = y$). If $x \# y$ then by (LO) we have ($x < y$ or $y < x$), which is impossible. Hence $\neg x \# y$ or $x = y$, which is also contradictory.

(T_n) $\to$ (D_n) is an immediate consequence of (T) $\to$ (D).

(D_n) $\to$ (T_n). Assume (D_n) and ($\neg x < y$ and $\neg y < x$ and $\neg x = y$), and seek a contradiction. If $x \leqslant y$, then by Proposition 2.1 we have $x = y$, which is contradictory. Hence $\neg x \leqslant y$ and by symmetry $\neg y \leqslant x$, which yields a contradiction with (D_n).

To complete the proof we need to supply some examples. The first example is the four element lattice of subsets of $\{0, 1\}$. This ordered set satisfies (CD) but none of the other conditions, so even classically (CD) does not imply any other conditions. Since classically (T_n) implies (T), the remaining conditions are classically equivalent, and our other examples will all be Kripke models.

In the first Kripke model all conditions but (S) and (T) are forced at α. Hence (D) does not imply (S).

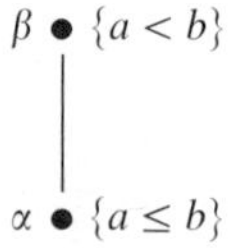

In the second Kripke model (CD) and the conditions that imply it are not forced at α, while all other conditions are. Hence neither (S) nor (S_1) implies (CD).

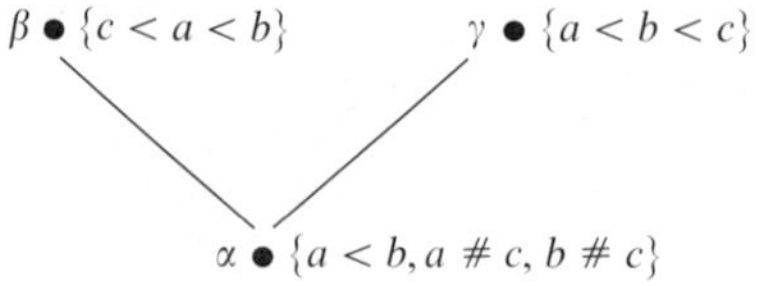

In the next model, (S_1) and the conditions that imply it are not forced at α, but all others are. Hence (S) does not imply (S_1).

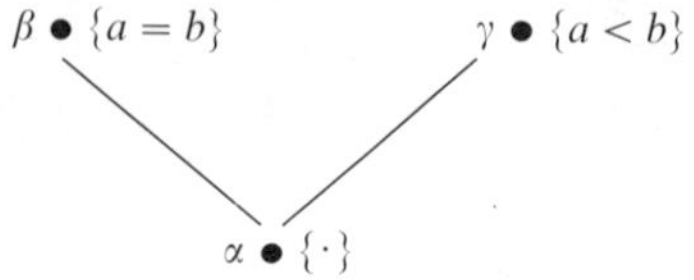

In the last model only (LO) and the conditions it implies are forced at α. Hence (LO) does not imply (S_2).

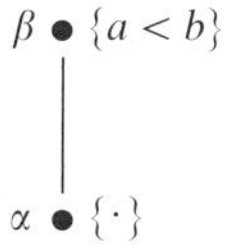

We will see in the next section that (LO), (S_1), and (S_2) are equivalent for lattices. Hence the last three examples do not support a lattice structure.

We close this section with a brief discussion of the relationship of the properties (T), (D), and (S) to the continuum. The validity of these properties for the continuum is equivalent to the limited principle of omniscience (LPO), the lesser limited principle of omniscience (LLPO), and the Markov principle (MP), respectively. The latter conditions concern elements of the Cantor space, infinite sequences of zeros and ones. (LPO) states that every such sequence is either identically zero or contains a one. (LLPO) states that if such a sequence contains at most one one, then either all its even index terms are zero, or all its odd index terms are zero. (MP) states that if it is contradictory that such a sequence is identically zero, then it contains a one.

Counterexamples to (LPO) and (LLPO) can be obtained by coding unsolved problems (such as the Goldbach conjecture, or perverse questions about the decimal expansion of π) into elements of the Cantor space. Such counterexamples have a "temporal" quality—the problem in question may be capable of solution. There are differences of opinion about the Markov principle. Brouwer claimed to have a temporal counterexample, but his construction is not widely accepted [5]. (MP) is accepted as an axiom by the Russian constructivists [19]. On the other hand it should not be regarded as a theorem, since from a proof that it is contradictory that a sequence is identically zero, there is no known general way to compute an index at which it is one, though in all known cases (except for Brouwer's problematic construction) this is possible.

§5. Linear Order in Lattices

One advantage of studying order in lattices is that since $x \leqslant y$ if and only if $x \wedge y = x$, the order relation is necessarily stable. Recall also that both $\wedge$ and $\vee$ are assumed to be binary operations. There are also new formulations of the linear order concept that can be made for lattices. The following condition states that the greatest lower bound of two elements is also an "infimum" in the sense that anything greater than it must be greater than one of the two elements:

$$x \wedge y < z \rightarrow (x < z \text{ or } y < z). \tag{L}$$

It is the following special case that will be central in our theory, and through which a great unification of the linear order concept will be achieved:

$$x \wedge y < y \rightarrow x < y. \tag{L_1}$$

We also introduce the condition

$$(z < x \text{ and } z < y) \rightarrow z < x \wedge y, \tag{L_2}$$

which will be useful in §8.

THEOREM 5.1. *In a lattice the following conditions are equivalent*:

$$x \# y \rightarrow (x < y \text{ or } y < x), \tag{LO}$$

$$\neg x < y \rightarrow y \leqslant x, \tag{S_1}$$

$$\neg x \leqslant y \rightarrow y \leqslant x, \tag{S_2}$$

$$\neg(\neg x < y \text{ and } \neg y < x \text{ and } \neg x = y), \tag{T_n}$$

$$\neg(\neg x \leqslant y \text{ and } \neg y \leqslant x), \tag{D_n}$$

$$x \wedge y < z \rightarrow (x < z \text{ or } y < z), \tag{L}$$

$$x \wedge y < y \rightarrow x < y, \tag{L_1}$$

$$(z < x \text{ and } z < y) \rightarrow z < x \wedge y. \tag{L_2}$$

Proof. By Theorem 4.2 we need only prove the implications in the following diagram.

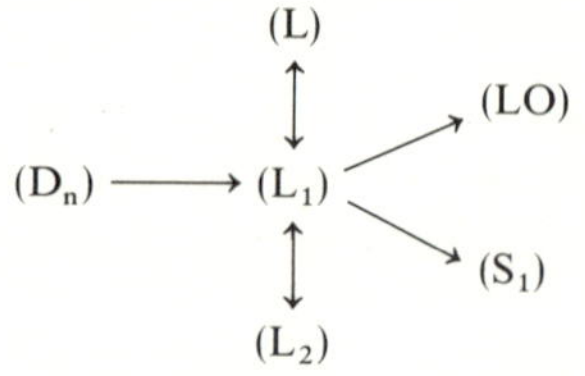

$(L_1) \rightarrow (S_1)$. Using the contrapositive of (L_1) and the assumption $\neg x < y$ we obtain $\neg x \wedge y < y$, and since $x \wedge y \leqslant y$ we obtain $x \wedge y = y$ by Proposition 2.1. Thus $y \leqslant x$.

$(L_1) \rightarrow (LO)$. From $x \# y$ we obtain by cotransitivity $(x \wedge y < x$ or $x \wedge y < y)$, and $(x < y$ or $y < x)$ follows by (L_1).

$(L_1) \rightarrow (L_2)$. If $(z < x$ and $z < y)$ then $z \leqslant x \wedge y$. Applying (A3) to $z \# y$, we obtain $(z \# x \wedge y$ or $x \wedge y \# y)$. In the first case $z < x \wedge y$. In the second, $x \wedge y < y$, so by (L_1) $x < y$ and $z < x = x \wedge y$.

$(L_2) \to (L_1)$. From $x \wedge y < y = y \wedge y$ we deduce by $(*)$ that $x \# y$. Hence we need to show that $x \leqslant y$, or $x \wedge y = x$. Assume $x \wedge y \# x$, which implies that $x \wedge y < x$. Then from (L_2) we have $x \wedge y < x \wedge y$, and this contradiction proves that $x \wedge y = x$.

$(L_1) \to (L)$. Assume that $x \wedge y < z$. We know that (L_1) implies (CD), so we can deduce $(x \wedge y < x$ or $x < z)$. In the first case by (L_1) $y = x \wedge y < z$ and in the second case we are already done.

$(L) \to (L_1)$. Assume $x \wedge y < y$. By (L) we have $(x < y$ or $y < y)$ and this implies $x < y$.

$(D_n) \to (L_1)$. Assume $x \wedge y < y$. If $y \leqslant x$ then $x \wedge y = y$, so $\neg y \leqslant x$. Then from (D_n) we get $\neg\neg x \leqslant y$, and since order is stable in a lattice $x \leqslant y$. From $x \wedge y < y = y \wedge y$ we obtain $x \# y$ by $(*)$, so $x < y$.

From this theorem we can deduce that the conditions (L), (L_1), and (L_2) are equivalent to their dual statements.

If real numbers are defined as Cauchy sequences, then the lattice operations can be defined termwise on the sequences. We are preparing an expository account of **R** as a lattice ordered field [12].

§6. Ordered Algebraic Structures

In this section we outline the results on lattice ordered algebraic structures that we shall need. Before doing this we make some remarks about the constructive theory of algebra on sets with apartness. Systematic study of algebra on sets with apartness was begun by Heyting in [13], a work of great significance that had the misfortune of appearing in Amsterdam in 1941, and is still little known. Heyting's axiom schemes can be simplified by making the assumption that all functions pull back apartness, so that $(*)$ holds for addition and multiplication.

By a *group* we shall always mean an additive abelian group, and our *rings* will be commutative with unit. It follows from $(*)$ that in a group

$$x + y \# 0 \to (x \# 0 \text{ or } y \# 0)$$

and in a ring

$$xy \# 0 \to (x \# 0 \text{ and } y \# 0).$$

A ring is an *integral domain* if $0 \# 1$ and

$$(x \# 0 \text{ and } y \# 0) \to xy \# 0.$$

This does not imply that if $xy = 0$ then either $x = 0$ or $y = 0$, but it does yield the cancellation law

$$(x \# 0 \text{ and } xy = xz) \to y = z.$$

A domain is a *field* if every element apart from zero has a multiplicative inverse. It is easily proved that the rule $x \mapsto x^{-1}$ is a function (i.e., that it pulls back apartness).

We now turn to the theory of lattice ordered groups and rings. A good reference for the classical theory, the elementary part of which is constructively valid with few changes, is [10]. By an *L-group* (lattice ordered group) we shall mean a set that is both a group and a lattice, where the following holds:

$$x \leqslant y \to x + z \leqslant y + z. \tag{OG}$$

It then follows from (∗) that $x < y$ implies $x + z < y + z$, and it follows from (O3) that if both x and y are strictly positive, so is their sum. In an *L-ring* (*L-domain*, *L-field*) order and multiplication are related by

$$(0 \leqslant x \text{ and } y \leqslant z) \to xy \leqslant xz. \tag{OR}$$

In a domain the cancellation law then yields:

$$(0 < x \text{ and } y < z) \to xy < xz.$$

PROPOSITION 6.1. *If x, y, z are elements of an L-group, then*

(a) $x \wedge y + z = (x + z) \wedge (y + z)$,
(b) $x + y = x \wedge y + x \vee y$,
(c) $(x \wedge y = 0$ *and* $x \wedge z = 0) \to x \wedge (y + z) = 0$,
(d) $x \wedge y = 0 \to mx \wedge ny = 0$ *for all* $m, n \in \mathbf{N}$.

If x is an element of an L-group, then its *positive part*, *negative part*, and *absolute value* are defined as usual:

$$x^+ = x \vee 0$$
$$x^- = (-x) \vee 0$$
$$|x| = x \vee (-x).$$

PROPOSITION 6.2. *Let x, y be elements of an L-group. Then*

(b) $x = x^+ - x^-$,
(b) $x^+ \wedge x^- = 0$,
(c) $(-x)^+ = x^-$ *and* $(-x)^- = x^+$,
(d) $|x| = x^+ + x^- = x^+ \vee x^-$,
(e) $|x + y| \leqslant |x| + |y|$
(f) $||x| - |y|| \leqslant |x - y|$.

From these two propositions we can deduce the following.

COROLLARY 6.3. *If m, n are natural numbers then* $mx^+ \wedge nx^- = 0$.

PROPOSITION 6.4. *Let* $x \wedge y = u$. *Then*

(a) $(x - y)^+ = x - u$,
(b) $(x - y)^- = y - u$,
(c) $|x - y| = x + y - 2u$.

The above proposition is of considerable use when $u = 0$. When we combine it with Proposition 6.1, we obtain the following result.

COROLLARY 6.5. *Let* x *be an element of an* L*-group and* n *be a positive integer. Then*

(a) $(nx)^+ = nx^+$ *and* $(nx)^- = nx^-$,
(b) $x \# 0 \rightarrow nx \# 0$,
(c) $0 \leqslant nx \rightarrow 0 \leqslant x$.

We will also need the following easily proved results about L-rings.

PROPOSITION 6.6. *If* x, y, z *are elements of an* L*-ring, then*

(a) $0 \leqslant x \rightarrow x(y \wedge z) \leqslant xy \wedge xz$,
(b) $|xy| \leqslant |x|\,|y|$.

§7. LINEAR ORDER IN L-GROUPS

We return to the concept of linear order and introduce three new formulations of it in the context of groups. The first two apply to arbitrary ordered (abelian) groups, the third to L-groups:

$$0 \leqslant x + y \rightarrow (0 \leqslant x \text{ or } 0 \leqslant y), \tag{G}$$

$$0 < x + y \rightarrow (0 < x \text{ or } 0 < y), \tag{G_1}$$

$$0 < x^+ \rightarrow 0 < x. \tag{LG}$$

While the conditions (G) and (G_1) look similar, the former is actually much stronger, as the following easy result shows.

PROPOSITION 7.1. *In an ordered group* G *the conditions* (G) *and* (D) *are equivalent, as are the conditions* (G_1) *and* (CD).

THEOREM 7.2. *In an* L*-group the following conditions are equivalent*:

(LO), (S_1), (S_2), (CD), (T_n), (D_n), (L), (L_1), (L_2), (G_1), (LG).

Proof. By Theorem 5.1 and Proposition 7.1, it will suffice to prove (G_1) implies (LG) and (LG) implies (LO).

$(G_1) \to (LG)$. If $0 < x^+$, then $0 < x^+ + x^-$, and the latter can be written as $(2x^+ - x^-) + (2x^- - x^+)$. Applying (G_1) we obtain $(2x^+ > x^-$ or $2x^- > x^+)$ which by Corollary 6.3 implies that $(x^- = 0$ or $x^+ = 0)$. The latter case is impossible, so $x = x^+ > 0$.

$(LG) \to (LO)$. If $x \# 0$, then since $x = x^+ - x^-$, we have by $(*)$ $(x^+ > 0$ or $x^- > 0)$. In the first case apply (LG) to x, in the second to $-x$.

While (G_1) is valid for the continuum, the similar condition (G) is not. Its use could be a source of nonconstructivity in analysis.

§8. Linear Order in L-Rings

When we consider L-rings (recall all rings are commutative with 1) we find many different forms of the linear order concept. The following two conditions make no reference to the lattice structure:

$$0 \leqslant x^2, \tag{R_1}$$

$$(0 < x \text{ and } 0 < xy) \to 0 < y. \tag{R_2}$$

PROPOSITION 8.1. *In an ordered ring,* ((LO) *and* (S_1)) *implies* (R_1) *and* (LO) *implies* (R_2).

Proof. Assume that (LO) and (S_1) hold, and that $x^2 < 0$. It follows from $(*)$ that $x \# 0$, so by (LO) $(x < 0$ or $0 < x)$. In either case $0 < x^2$ and this contradiction, with (S_1), shows that $0 \leqslant x^2$.

Now assume that (LO) holds, and that $0 < x$ and $0 < xy$. Since $xy \# x0$, by $(*)$ $y \# 0$, and by (LO) $(y < 0$ or $0 < y)$. The former case would imply $xy < 0$, which is not possible.

Turning to L-rings, we introduce the following conditions:

$$(x \wedge y = 0 \text{ and } 0 \leqslant z) \to x \wedge yz = 0, \tag{LR_1}$$

$$x \wedge y = 0 \to xy = 0, \tag{LR_2}$$

$$|xy| = |x|\,|y|, \tag{LR_3}$$

$$0 \leqslant z \to z(x \wedge y) = zx \wedge zy. \tag{LR_4}$$

These conditions are considerably weaker than linear order for L-rings (but become equivalent for L-fields). For example. (R_1) and (LR_1)–(LR_4) all hold for the ring of continuous real valued functions on a topological space. Condition (LR_1) is often used as the definition of an F-ring [10, p. 143].

THEOREM 8.2. *The following implications hold for an L-ring.*

$$\begin{array}{|l|}\hline (LO), (S_1), (S_2), (CD), (L), (L_1) \\ (L_2), (T_n), (D_n), (G_1), (LG) \\ \hline \end{array} \begin{array}{l} \longrightarrow (R_2) \\ \longrightarrow (LR_1) \end{array} \begin{array}{l} \nearrow \; (LR_2) \rightarrow (R_1) \\ \searrow \; \boxed{(LR_3), (LR_4)} \end{array}$$

Proof. To facilitate the proof we also consider the following special cases of (LR_3) and (LR_4):

$$0 \leqslant x \rightarrow |xy| = x|y|, \qquad (LR_3')$$

$$(0 \leqslant z \text{ and } x \wedge y = 0) \rightarrow zx \wedge zy = 0. \qquad (LR_4')$$

By Theorem 7.2 and Proposition 8.1 it suffices to prove the implications in the following diagram.

$$\begin{array}{ccccccc} (L_2) & \longrightarrow & (LR_1) & \longrightarrow & (LR_4') & \longrightarrow & (LR_4) \\ & & \downarrow & & \updownarrow & & \\ & & (LR_2) & & (LR_3') & \longrightarrow & (LR_3) \\ & & \downarrow & & & & \\ & & (R_1) & & & & \end{array}$$

$(L_2) \rightarrow (LR_1)$. Assume that $0 \leqslant z$ and $x \wedge y = 0$. Since $0 \leqslant x \wedge yz$, by Proposition 2.1 it suffices to prove $\neg 0 < x \wedge yz$. If $0 < x \wedge yz$, then $0 < x$ and $0 < yz$. Since $0 \leqslant y$, by (*) we have $0 < y$, and then (L_2) implies that $0 < x \wedge y$, which is a contradiction.

$(LR_1) \rightarrow (LR_2)$. Applying (LR_1) twice we obtain $0 = x \wedge y = xy \wedge y = xy \wedge xy = xy$.

$(LR_2) \rightarrow (R_1)$. Using $x^+ \wedge x^- = 0$ and (LR_2) we obtain

$$\begin{aligned} x^2 &= (x^+ - x^-)^2 \\ &= (x^+)^2 - 2x^+x^- + (x^-)^2 \\ &= (x^+)^2 + (x^-)^2 \\ &\geqslant 0. \end{aligned}$$

$(LR_1) \rightarrow (LR_4')$ is immediate.

$(LR_4') \rightarrow (LR_4)$. Assume $0 \leqslant z$ and set $u = x \wedge y$. Then we have

$$\begin{aligned} 0 &= (x - u) \wedge (y - u) \\ &= z(x - u) \wedge z(y - u) \\ &= (zx - zu) \wedge (zy - zu) \\ &= zx \wedge zy - zu. \end{aligned}$$

$(\mathrm{LR}_3') \to (\mathrm{LR}_3)$. Using (LR_3') and the fact that $||u| - |v|| \leqslant |u - v|$, we obtain

$$\begin{aligned} |x|\,|y| &= ||x|y| \\ &= ||x|y^+ - |x|y| \\ &= ||xy^+| - |xy^-|| \\ &\leqslant |xy^+ - xy^-| \\ &= |xy|. \end{aligned}$$

(LR_3) is now a consequence of the fact that always $|xy| \leqslant |x|\,|y|$.

$(\mathrm{LR}_3') \leftrightarrow (\mathrm{LR}_4')$. Assume $0 \leqslant z$ and $x \wedge y = 0$. If $v = zx \wedge zy$, then $|z(x - y)| = zx + zy - 2v$ (Proposition 6.4) while $z|x - y| = zx + zy$. The result follows by Corollary 6.5. This concludes the proof of the theorem.

We know of no further implications between individual conditions for L-rings, but we do have the following result.

PROPOSITION 8.3. *If an L-ring satisfies both* (R_1) *and* (LR_3), *then* (LR_2) *holds.*

Proof. Assume $x \wedge y = 0$. Consider $(x - y)^2 = x^2 + y^2 - 2xy$, and set $u = (x^2 + y^2) \wedge 2xy$. Then it follows from $(\mathbf{R}_1)$ that $u = 2xy$, and it follows from (LR_3), applied to $|(x - y)^2| = |x - y|^2$, that $u = 0$. Hence $xy = 0$.

If we assume that our ring is a domain we obtain some additional implications.

PROPOSITION 8.4. *In an L-domain* (LR_2) *implies* (LG), *and* (LR_3') *implies* (R_2).

Proof. $(\mathrm{LR}_2) \to (\mathrm{LG})$. From (LR_2) we obtain $x^+x^- = 0$. Hence if $x^+ \# 0$, then $x^- = 0$ and $0 < x$.

$(\mathrm{LR}_3') \to (\mathrm{R}_2)$. Assume $0 < x$ and $0 < xy$. Then $y \# 0$ and it suffices to show $y^- = 0$. By (LR_3') we have

$$\begin{aligned} x(y^+ - y^-) &= x|y| \\ &= |xy| \\ &= xy \\ &= x(y^+ - y^-). \end{aligned}$$

Hence $2xy^- = 0$ and $y^- = 0$.

This gives the following diagram of implications for L-domains.

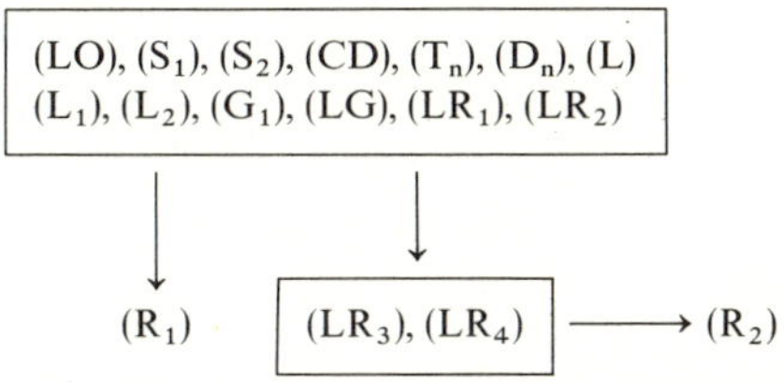

It follows from Propositions 8.3 and 8.4 that if an L-domain satisfies both (R_1) and (LR_3), then it is linearly ordered. We do not have examples of nonlinearly ordered L-domains that satisfy either of these. An example of a nonlinearly ordered L-domain can be obtained by taking the polynomial ring $\mathbf{R}[X]$, and defining the nonnegative elements to be those all of whose coefficients are nonnegative. This satisfies neither (R_1) nor (R_2).

§9. Linear Order in L-Fields

In the case of ordered fields we can introduce one last formulation of linear order that will bring our theory full circle:

$$0 < x \to 0 < x^{-1}. \tag{F}$$

If a field is trivially ordered, then (F) holds vacuously. If the order is nontrivial, then there exist positive elements, and it follows from (F) that $0 < 1$, and hence that $\{x \mid 0 < x\}$ is a subgroup of the multiplicative group of F. It easily follows from the axiom (OR) that it is an ordered group. The following result is easily proved.

Proposition 9.1. *In an ordered field* (R_2) *is equivalent to* (F), *and* (R_1) *implies* (F).

We are now in a position to prove our main result, which shows that for L-fields a remarkable variety of conditions are equivalent to linear order.

Theorem 9.2. *In an L-field the following conditions are equivalent*:

$$x \# y \to (x < y \text{ or } y < x), \tag{LO}$$

$$\neg x < y \to y \leqslant x, \tag{S$_1$}$$

$$\neg x \leqslant y \to y \leqslant x, \tag{S$_2$}$$

$$x < y \to (x < z \text{ or } z < y), \tag{CD}$$

$$\neg(\neg x < y \text{ and } \neg y < x \text{ and } \neg x = y), \tag{T$_n$}$$

$$\neg(\neg x \leqslant y \text{ and } \neg y \leqslant x), \tag{D_n}$$

$$x \wedge y < z \rightarrow (x < z \text{ or } y < z), \tag{L}$$

$$x \wedge y < y \rightarrow x < y, \tag{L_1}$$

$$(z < x \text{ and } z < y) \rightarrow z < x \wedge y, \tag{L_2}$$

$$0 < x + y \rightarrow (0 < x \text{ or } 0 < y), \tag{G_1}$$

$$0 < x^+ \rightarrow 0 < x, \tag{LG}$$

$$0 \leqslant x^2, \tag{R_1}$$

$$(0 < x \text{ and } 0 < xy) \rightarrow 0 < y, \tag{R_2}$$

$$(x \wedge y = 0 \text{ and } 0 \leqslant z) \rightarrow x \wedge yz = 0, \tag{LR_1}$$

$$x \wedge y = 0 \rightarrow xy = 0, \tag{LR_2}$$

$$|xy| = |x|\,|y|, \tag{LR_3}$$

$$0 \leqslant z \rightarrow z(x \wedge y) = zx \wedge zy, \tag{LR_4}$$

$$0 < x \rightarrow 0 < x^{-1}. \tag{F}$$

Proof. From Theorem 8.2, Proposition 8.4, and Proposition 9.1 it suffices to prove that (F) implies (LO). Our proof is a constructivisation of a similar proof of Dubois in [9]. Assume that $x \# 0$. Set $p = x^+$, $q = x^-$. Then $p \# 0$ or $q \# 0$. Let

$$t = p^2/(p + q).$$

Then

$$t - x = q^2/(p + q)$$

and

$$p - t = pq/(p + q).$$

Since each of these quantities is positive, we have $p = x \vee 0 \leqslant t$ and $t \leqslant p$, so $t = p$ and $pq = 0$. If $p \# 0$, then $x = p > 0$, while if $q \# 0$, then $x = -q < 0$. This completes the proof.

While the proof of Theorem 9.2 is easier in classical mathematics, its full force does not appear well known there. For instance, the question as to whether (R_1) implies linear order for L-fields is the second open problem in [7].

COROLLARY 9.3. *The set of strictly positive elements in a linearly ordered L-field is an L-group.*

Proof. We have already remarked that it is an ordered group. We need only to show that it is a lattice. This follows from (L_2).

COROLLARY 9.4. *Let D be an L-domain that is not linearly ordered, but satisfies* ($\mathbf{R}_1$) *or* ($\mathbf{R}_2$). *Then there is no extension of the order of D to a lattice order of its field of fractions satisfying the same condition.*

Proof. If this were possible, then the field of fractions, and hence D, would be linearly ordered.

EXAMPLES. The field of real numbers, and hence any L-subfield of $\mathbf{R}$, satisfies the eighteen equivalent conditions of Theorem 9.2. The rational numbers are discrete and so also satisfy (T), (D), and (S). The same is true of the real algebraic numbers (this was first proved by Vandiver in [26]). We give one example of an L-field that is not linearly ordered (more examples can be found in [7]). Namely, take the quadratic extension of $\mathbf{Q}$ generated by $\sqrt{2}$, and define $a + b\sqrt{2}$ to be positive when a and b are both positive.

In conclusion we mention some open questions:

1. L-rings. Do there exist additional implications among the conditions (LR_1), (LR_2), (LR_3), (R_1), and (R_2)? Note that rings of real valued functions satisfy all but the last.
2. L-domains. Do there exist nonlinearly ordered L-domains that satisfy (R_1) or (R_2)? If such domains do exist, can their order be extended to a lattice order of the field of fractions (note Corollary 9.4)?
3. L-fields. Does (D) imply (T) for L-fields? Since (S) is equivalent to the Markov principle, we know that (S) does not imply (D).

ACKNOWLEDGMENTS

Some of this material was developed in a course in constructive mathematics at The University of Texas. I would like to thank the students of that class, particularly Mary Parker, Kieth Topham, Rodney Long, and Pearl Olson, for their contributions. Our colleague Simon Bernau showed us many helpful tricks of the theory of lattice algebra. Finally I would like to thank Gabriel Stolzenberg, who first inspired me to learn to think constructively.

REFERENCES

1. E. BISHOP, "Foundations of Constructive Analysis," McGraw-Hill, New York, 1967.
2. E. BISHOP, "Schizophrenia in Contemporary Mathematics," *Amer. Math. Soc.*, Colloquium Lectures, Providence, Rhode Island, 1973.
3. L. E. J. BROUWER, Begründung der Mengelehre unabhängig vom logischen Satz vom ausgeschlossenen Dritten, I, II, *Verh. Nederl. Akad. Wetensch.* **12**(5) (1918), and **12**(7) (1919).

4. L. E. J. Brouwer, Virtuelle Ordnung und unerweiterbare Ordnung, *J. Reine Angew. Math.* **157** (1927), 255–257.
5. L. E. J. Brouwer, On order in the continuum, and the relation of truth to non-contradictority, *Indag. Math.* **13** (1951), 357–358.
6. P. M. Cohn, "Universal Algebra," Harper, New York, 1965.
7. P. Conrad and J. Dauns, An embedding theorem for lattice ordered fields, *Pacific J. Math.* **30** (1969), 385–398.
8. D. van Dalen and C. D. Gordon, Independence problems in subsystems of intuitionistic arithmetic, *Indag. Math.* **33** (1971), 448–456.
9. D. W. Dubois, On partly ordered fields, *Proc. Amer. Math. Soc.* **7** (1956), 918–930.
10. L. Fuchs, "Partially Ordered Algebraic Systems," Pergamon, Oxford, 1962.
11. N. Greenleaf, The categories of constructive mathematics, to appear.
12. N. Greenleaf, The real numbers as a lattice ordered field, to appear.
13. A. Heyting, Untersuchungen über intuitionistische Algebra, *Verh. Nederl. Akad. Wetensch.* **18** (1941), 3–36.
14. A. Heyting, Axiomatic method and intuitionism, *in* "Essays on the Foundations of Mathematics," Jerusalem, 1961, 237–247.
15. A. Heyting, "Intuitionism, an Introduction," North–Holland Publ., Amsterdam, 1972.
16. N. G. Hull, Counterexamples in intuitionistic analysis using Kripke's schema, *Z. Math. Logik Grundlagen Math.* **15** (1969), 241–246.
17. W. Julian, R. Mines, and F. Richman, Algebraic numbers, a constructive development *Pacific J. Math.* **74** (1978), 91–102.
18. B. van Rootselaar, On intuitionistic difference relations, *Indag. Math.* **22** (1960), 316–322, and **25** (1963), 132–133.
19. N. A. Sanin, "Constructive Real Numbers and Function Spaces," Translations of Mathematical Monographs, *Amer. Math. Soc.*, Providence, Rhode Island, 1968.
20. D. Scott, Extending the topological interpretation to intuitionistic analysis, *Compositio Math.* **20** (1968), 194–210.
21. C. A. Smorynski, Applications of Kripke models, *in* "Metamathematical Investigations of Intuitionistic Arithmetic and Analysis," Lecture Notes in Mathematics No. 344, Springer–Verlag, Berlin and New York, 1973, 324–391.
22. J. Staples, On constructive fields, *Proc. London Math. Soc.* **23** (1971), 753–768.
23. J. Staples, Axioms for constructive fields, *Bull. Austral. Math. Soc.* **8** (1973), 221–232.
24. A. S. Troelstra, "Principles of Intuitionism," Lecture Notes in Mathematics No. 95, Springer–Verlag, Berlin and New York, 1969.
25. A. S. Troelstra, "Metamathematical Investigation of Intuitionistic Arithmetic and Analysis," Lecture Notes in Mathematics No. 344, Springer–Verlag, Berlin and New York, 1973.
26. H. S. Vandiver, On the ordering of real algebraic numbers by constructive methods, *Annals of Math.* **37** (1936), 7–16.

AMS (MOS) 1970 subject classification: 02E06.

STUDIES IN FOUNDATIONS AND COMBINATORICS
ADVANCES IN MATHEMATICS SUPPLEMENTARY STUDIES, VOL. 1

Better-Quasi-Orderings and a Class of Trees

RICHARD LAVER

Department of Mathematics
University of Colorado
Boulder, Colorado

Section 1 explicates the theory of better-quasi-orderings (Nash-Williams [11, 12, 13]). In Section 2 we prove an extension of the Nash-Williams infinite tree theorem ([12]) to a class of trees that includes the countable trees and contains trees of all heights. Call a tree T scattered if the binary tree of height ω is not embeddable in T. Let $\mathcal{M}$ be the class of all T such that there are T_n $(n < \omega)$ with $T = \bigcup_{n<\omega} T_n$, each T_n downwards closed in T, each T_n scattered. The main result is that $\mathcal{M}$ is better-quasi-ordered under embeddability. The proof shows, more generally, that if Q is a better-quasi-ordering, then the class $\mathcal{M}_Q$ of all Q-labeled $\mathcal{M}$ trees is better-quasi-ordered.

In Section 1 we prove a fact that comes out of Nash-Williams' "fore-running" method of [12]. This result, Theorem 1.9, implies that a quasi-ordered set, which admits a ranked construction having certain properties, is better-quasi-ordered. We derive from this theorem another of Nash-Williams' results, the transfinite sequence theorem of [13] (Theorem 1.10 here), in an effective form for application in Section 2.

The notation $\mathcal{M}$ is also used in [8] for a similarly defined class of linear order types ("countable unions of scattered types"). I do not know how to prove the above theorem for the set of countable trees without essentially doing it for $\mathcal{M}$-trees—this is in contrast with the situation for order types in [8]. The question of whether the class of countable trees is better-quasi-ordered was raised by Galvin.

A word about the possibility of extending the theorem. Galvin had shown that the class of trees of height $\omega + 1$ is not well-quasi-ordered. Aronszajn trees (trees of height ω_1 with every level and path countable) are not in $\mathcal{M}$. They need not be well-quasi-ordered (for instance, they are not if Jensen's $\diamondsuit_{\omega_1}$ axiom holds), so an extension of the theorem to a class of trees that includes the Aronszajn trees would take the form of a relative consistency result.

On is the class of all ordinals. For κ a cardinal, $[X]^\kappa$ $([X]^{<\kappa})$ is the collection of subsets of X of power κ $(<\kappa)$. $\mathcal{P}(X)$ is the power set of X; more generally, $\mathcal{P}^\alpha(X)$, for $\alpha \in$ On, is defined by $\mathcal{P}^0(X) = X$, $\mathcal{P}^{\alpha+1}(X) = \mathcal{P}(\mathcal{P}^\alpha(X))$,

ISBN 0-12-599101-0

and for limit ordinals λ, $\mathscr{P}^{\lambda}(X) = \bigcup_{\beta<\lambda} \mathscr{P}^{\beta}(X)$. Set $(X)^Y = \{f:\text{dom}\, f = Y$ and $\text{rng} f \subseteq X\}$. If α is an ordinal, $(X)^{<\alpha} = \bigcup_{\beta<\alpha} (X)^{\beta}$.

The letter Q is reserved for quasi-ordered sets, i.e., sets on which a transitive, reflexive relation $\leqslant$ is defined. For $q_1, q_2 \in Q$, $q_1 < q_2$ means that $q_1 \leqslant q_2$ and $q_2 \nleqslant q_1$, and $q_1 \equiv q_2$ means that $q_1 \leqslant q_2$ and $q_2 \leqslant q_1$.

We define quasi-orders $\leqslant_{\text{m}}$, $\leqslant_1$ on $\mathscr{P}(Q)$. If $X, Y \subseteq Q$, then $X \leqslant_{\text{m}} Y$ means there is an $H:X \to Y$ with $q \leqslant H(q)$ for all $q \in X$, and $X \leqslant_1 Y$ holds if there is an H as above that is 1–1. $\bigcup_{\alpha\in\text{On}} (Q)^{\alpha}$ (the class of "transfinite Q-sequences") is quasi-ordered by the rule: $J \leqslant K$ if and only if there is a 1–1 increasing $H:\text{dom}\, J \to \text{dom}\, K$ with $J(\alpha) \leqslant K(H(\alpha))$ ($\alpha \in \text{dom}\, J$).

1. Better-Quasi-Orderings

If Q is quasi-ordered, call an $f:\omega \to Q$ *good* if for some $i < j < \omega$, $f(i) \leqslant f(j)$; f is *bad* otherwise. Q is *well-quasi-ordered* (wqo) just in case every $f:\omega \to Q$ is good.

We recall a few facts about wqo's. Q is *well founded* if there is no descending sequence $q_0 > q_1 > \cdots > q_n > \cdots$ of members of Q. An *antichain* is an $A \subseteq Q$ such that if $q, r \in A$ and $q \neq r$ then $q \nleqslant r$. In view of Ramsey's theorem, two equivalences to Q being wqo are: Q is well founded and has no infinite antichains; for every $f:\omega \to Q$ there is an infinite $M \subseteq \omega$ with $f(i) \leqslant f(j)$ ($i,j \in M$, $i \leqslant j$). Higman ([4]) proved that if Q is wqo then $(Q)^{<\omega}$ is wqo (as is $\{F \subseteq Q: F \text{ finite}\}$, under $\leqslant_{\text{m}}$, $\leqslant_1$). Kruskal ([6]) proved that if Q is wqo then the collection of finite Q trees is wqo under $\leqslant_{\text{m}}, \leqslant_1$.[1] Rado [14] showed that there is a wqo Q such that $\mathscr{P}(Q)$ is not wqo (under $\leqslant_{\text{m}}$ or $\leqslant_1$), nor is $(Q)^{\omega}$. Namely, Q is $\{\langle i, j\rangle : i < j < \omega\}$ ordered by $\langle i, j\rangle \leqslant \langle k, l\rangle$ if and only either $i = k$ and $j \leqslant l$ or $j < k$. An account of results of wqo theory is given in, e.g., [10, pp. 2–4].

Note that Higman's and Kruskal's results apply to spaces whose objects are finite. Consider the following spaces, though:

(a) the class of all trees having height $\leqslant \omega$,
(b) $\bigcup_{\alpha\in\text{On}} (\text{On})^{\alpha}$.

Nash-Williams ([12, 13]) proved that these spaces are wqo. The methods of wqo theory, though, did not suffice for the proofs. This turned out to be for the reason that the property of being wqo was too weak for inductive arguments about infinite objects to be carried out; one cannot pass from Q wqo to $\mathscr{P}(Q)$ wqo because of Rado's counterexample.

[1] Definitions about trees are in Section 2.

Nash-Williams discovered a stronger property than wqo, which is preserved in passage from Q to $\mathscr{P}(Q)$. He showed that the spaces (a) and (b) above in fact have this stronger property.

DEFINITIONS. If $s, t \subseteq \omega$, then $s \leqslant t$ $(s < t)$ means that s is a (proper) initial segment of t. Define $s \lhd t$ to hold if there is an $n > 0$ and $i_0 < \cdots < i_n < \omega$ such that for some $m < n$, $s = \{i_0, \ldots, i_m\}$ and $t = \{i_1, \ldots, i_n\}$. (Thus, e.g., $\{3\} \lhd \{5\}$, $\{3, 5, 6\} \lhd \{5, 6, 8, 9\}$, $\{3, 5, 6\} \not\lhd \{5, 6\}$.)

For $X \in [\omega]^{\aleph_0}$, a *barrier* on X is a set $B \subseteq [X]^{<\aleph_0}$ such that $\varnothing \notin B$ and

(a) for every $Y \in [X]^{\aleph_0}$ there is an $s \in B$ such that $s < Y$,
(b) if $s, t \in B$ and $s \neq t$ then $s \not< t$.

If B is a barrier, $f: B \to Q$ is *good* if there are $s, t \in B$, $s \lhd t$ and $f(s) \leqslant f(t)$, and f is *bad* otherwise; a quasi-ordered set Q is *better-quasi-ordered* (bqo) if, for every barrier B and every $f: B \to Q$, f is good.

We give some motivation for this definition.[2] We look for a condition, stronger than wqo (but nontrivial in that it at least includes finite and well-ordered sets and is preserved under finite products) that is preserved from Q to $\mathscr{P}(Q)$. The condition cannot be "Q and $\mathscr{P}(Q)$ are wqo" because of an (unpublished) example of Kruskal of a Q such that Q, $\mathscr{P}(Q)$ are wqo but $\mathscr{P}^2(Q)$ is not wqo. This example is like Rado's but on triples instead of pairs. Kruskal also found analogous examples with 2 replaced by any $n < \omega$.

Consider then $\bigcup_{\alpha \in \mathrm{On}} \mathscr{P}^\alpha(Q)$. To avoid a trivial notational problem, assume that $Q \cap \mathscr{P}^{\alpha+1}(Q) = \varnothing$, for all α. Define a quasi-order $\leqslant_{\mathrm{m}}$ on this space: if $X \in \mathscr{P}^\beta(Q)$, $Y \in \mathscr{P}^\gamma(Q)$ (the definition is by induction on β, γ) then $X \leqslant_{\mathrm{m}} Y$ if

(a) $\beta = 0, \gamma = 0$, and $X \leqslant Y$ as members of Q, or
(b) $\beta = 0, \gamma > 0$, and for some $Y' \in Y$, $X \leqslant_{\mathrm{m}} Y'$, or
(c) $\beta > 0, \gamma > 0$, and for some $H: X \to Y$, $X' \leqslant_{\mathrm{m}} H(X')$ for all $X' \in X$.

That $\leqslant_{\mathrm{m}}$ is a quasi-order may be seen by induction. The $\leqslant_1$ quasi-order is defined by the same scheme except that H must be 1–1.

The working definition of Q being bqo, given above, is equivalent to $\bigcup_{\alpha \in \mathrm{On}} \mathscr{P}^\alpha(Q)$ being wqo, and to $\bigcup_{\alpha < \omega_1} \mathscr{P}^\alpha(Q)$ being wqo, under $\leqslant_{\mathrm{m}}$ or $\leqslant_1$. We indicate without details how the consideration that $\bigcup_{\alpha \in \mathrm{On}} \mathscr{P}^\alpha(Q)$ be wqo under $\leqslant_{\mathrm{m}}$ could lead to the definition of bqo (see also Theorem 1.11). Suppose $f: \omega \to \bigcup_{\alpha \in \mathrm{On}} \mathscr{P}^\alpha(Q)$ is bad.

Let $A_0 = \{\{i\}: i < \omega\}$. Define f_0 on A_0 by $f_0(\{i\}) = f(i)$. Define inductively A_n, f_n $(n < \omega)$ with $A_n \subset [\omega]^{<\aleph_0}$, $f_n: A_n \to \bigcup_{\alpha \in \mathrm{On}} \mathscr{P}^\alpha(Q)$, so that the following

[2] We are partially repeating here some remarks in [8, p. 92].

hold. Let

$$C_n = \{s \in A_n : f_n(s) \in Q\},$$
$$D_n = \{s \in A_n : f_n(s) \notin Q \text{ and there is no } t \in A_n \text{ with } s \lhd t\},$$
$$E_n = \{\langle s, t\rangle \in A_n : f_n(s) \notin Q \quad \text{and } s \lhd t\}.$$

(a) $A_{n+1} = C_n \cup D_n \cup \{s \cup t : \langle s, t\rangle \in E_n\}$,
(b) $f_{n+1}(s) = f_n(s)$, all $s \in C_n$,
(c) if $s \in D_n$ then $s - \{\min s\} \in C_n$, $f_{n+1}(s) \in f_n(s)$,
(d) if $\langle s, t\rangle \in E_n$, then $f_{n+1}(s \cup t) \in f_n(s)$,
(e) if $s, t \in A_{n+1}$ and $s \lhd t$ then $f_{n+1}(s) \nleq_{\mathrm{m}} f_{n+1}(t)$.

Let $A = \bigcup_{n<\omega} \bigcap_{m>n} A_m$. It may be seen that $A = \bigcup_{n<\omega} C_n$, and that $\bigcup_{n<\omega} f_n \upharpoonright C_n$ is a function $f: A \to Q$ such that if $s, t \in A$ and $s \lhd t$ then $f(s) \nleq f(t)$.

Now A need not be a barrier, but A is a *block* in the terminology of [12], that is, A satisfies the definition of a barrier on ω with (b) replaced by: if $s, t \in B$ and $s \neq t$ then $s \not\subset t$. And every block contains a barrier ([12], Lemma 20), so for $B \subseteq A$ a barrier, $f \upharpoonright B$ is bad, showing that Q is not bqo. Our definition of bqo in terms of barriers, instead of blocks as in [12], is for convenience—blocks that aren't barriers are not encountered this way. The Rado and Kruskal examples cited above are wqo but not bqo. However, Nash-Williams has expressed the opinion that every wqo that "occurs in nature" is a bqo. From here through Theorem 1.9 is an account of methods and results in [12].

LEMMA 1.1. *If Q is* bqo *then Q is* wqo.

Proof. Since $m < n$ just in case $\{m\} \lhd \{n\}$, wqo is the special case of bqo with $B = \{\{m\} : m < \omega\}$.

We state some facts about $\lhd$ and barriers that will be used without comment below. Let B be a barrier on X. Then for each $Y \in [X]^{\aleph_0}$, the $b \in B$ with $b < Y$ is unique. We must have $X = \bigcup B$, namely, $\bigcup B \subseteq X$ by definition, and if $n \in X$ there is a nonempty $b \in B$ with $b < \{m \in X : m \geqslant n\}$. Let $s \in B$. Then there is a $t \in B$ with $s \lhd t$. Namely, given any $Y \in [X]^{\aleph_0}$ with $s < Y$, then any $t \in B$ with $t < (Y - \{\min Y\})$ must satisfy $\max x < \max t$ (lest $t \subset s$), whence $s \lhd t$.

LEMMA 1.2. *If Q is well ordered then Q is* bqo.

Proof. Given $f: B \to Q$. Pick $b_0, b_1, \ldots, b_n, \ldots$ from B with $b_i \lhd b_{i+1}$. Since Q is well ordered there is an i with $f(b_i) \leqslant f(b_{i+1})$.

Let B be a barrier on X. Define $B(2) = \{b_1 \cup b_2 : b_1, b_2 \in B \text{ and } b_1 \lhd b_2\}$. Note that if $s \in B(2)$ then the expression $s = b_1 \cup b_2$ as above is unique. We claim that $B(2)$ is a barrier on X. If $s, t \in B(2)$ and $s \subseteq t$ then $s = t$, namely, letting $s = b_1 \cup b_2$, $t = c_1 \cup c_2$, $b_1, b_2, c_1, c_2 \in B$, $b_1 \lhd b_2, c_1 \lhd c_2$, then $b_2 = s - \{\min s\} \subseteq t - \{\min t\} = c_2$, so $b_2 = c_2$, thus $\min s = \min t$, so $s = t$. If $Y \in [X]^{\aleph_0}$ there is an $s \in B(2)$ with $s < Y$. Namely, pick $b_1 \in B$, $b_1 < Y$, and $b_2 \in B$, $b_2 < Y - \{\min Y\}$; then $s = b_1 \cup b_2$ is as desired, showing that $B(2)$ is a barrier. Note that if $s, t \in B(2)$ and $s \lhd t$ then there are $b_1, b_2, b_3 \in B$ with $b_1 \lhd b_2 \lhd b_3$ so that $s = b_1 \cup b_2$ and $t = b_2 \cup b_3$.

LEMMA 1.3. *Let $\mathscr{P}(Q)$ be quasi-ordered under $\leqslant_{\mathrm{m}}$. If $f : B \to \mathscr{P}(Q)$ is bad then there is a bad $g : B(2) \to Q$ such that if $b_1, b_2 \in B$ and $b_1 \lhd b_2$ then $g(b_1 \cup b_2) \in f(b_1)$. Hence if Q is* bqo *then $\mathscr{P}(Q)$ is* bqo *under $\leqslant_{\mathrm{m}}$.*

Proof. Given $b_1, b_2 \in B$ with $b_1 \lhd b_2$, pick $g(b_1 \cup b_2) \in f(b_1)$ to satisfy $g(b_1 \cup b_2) \nleqslant q$, all $q \in f(b_2)$. Thus $g(b_1 \cup b_2) \nleqslant g(b_2 \cup b_3)$ for all $b_1, b_2, b_3 \in B$ with $b_1 \lhd b_2 \lhd b_3$, so g is bad.

If $D \subseteq [\omega]^{<\aleph_0}$ and $Y \in [\omega]^{\aleph_0}$ let $D \restriction Y = D \cap [Y]^{<\aleph_0}$. Let B be a barrier on X. Note that if $Y \in [X]^{\aleph_0}$ then $B \restriction Y$ is a barrier on Y.

The next result is the Nash-Williams partition theorem [11]. It plays a role in bqo theory analogous to that played by Ramsey's theorem in wqo theory.

THEOREM 1.4. *Suppose $D \subseteq [\omega]^{<\aleph_0}$ is such that $c \not\subset d$ for all $c, d \in D$. Then if $D = \bigcup_{i<n} D_i$ there is a $Y \in [\omega]^{\aleph_0}$ such that for some i, $D \quad Y \subseteq D_i$.*

We omit the well-known proof ([11, 3]). See [3, 15, 1] for generalizations. Noting that the theorem holds equally well replacing ω by any $X \in [\omega]^{\aleph_0}$, we get

COROLLARY 1.5. *If B is a barrier and $B = \bigcup_{i \leqslant n} B_i$ for some $n < \omega$, then some B_i contains a barrier.*

Since singletons are bqo, the following lemma shows that finite sets are bqo.

LEMMA 1.6. *Suppose $n < \omega$ and $Q = \bigcup_{i<n} Q_i$. Then if each Q_i is* bqo *then Q is* bqo.

Proof. Suppose $f : B \to Q$ were bad. By Corollary 1.5, for some i, $f^{-1}(Qi)$ contains a barrier C, and $f \restriction C$ would be a bad function on Q_i.

Say that $f : B \to Q$ is *perfect* if for all $s, t \in B$, if $s \lhd t$ then $f(s) \leqslant f(t)$.

LEMMA 1.7. *If B is a barrier and $f:B \to Q$ then there is a barrier $C \subseteq B$ with $f \restriction C$ bad or perfect.*

Proof. Let $B(2) = D_0 \cup D_1$, where $D_0 = \{b_1 \cup b_2 : b_1, b_2 \in B, b_1 \lhd b_2, f(b_1) \not\leqslant f(b_2)\}$, $D_1 = \{b_1 \cup b_2 : b_1, b_2 \in B, b_1 \lhd b_2, f(b_1) \leqslant f(b_2)\}$. Since the expression of $s \in B(2)$ as $b_1 \cup b_2, b_1, b_2 \in B, b_1 \lhd b_2$, is unique, $D_0 \cap D_1 = \varnothing$. By Lemma 1.5 there is a barrier D and an i with $D \subseteq D_i$. Take $C = B \restriction \bigcup D$. If $b_1, b_2 \in C$ and $b_1 \lhd b_2$ then $b_1 \cup b_2 \in D_i$. Thus if $i = 0$, $f \restriction C$ is bad, and if $i = 1$, $f \restriction C$ is perfect.

For quasi-ordered sets Q_i $(i \in I)$, let $\otimes_{i \in I} Q_i$ be the cartesian product of the Q_i's with the usual order. If $x \in \otimes_{i \in I} Q_i$ let $(x)_i$ be the projection of x onto the ith coordinate.

LEMMA 1.8. *If $n < \omega$ and Q_i, $i \leqslant n$, are quasi-ordered sets and $f:B \to \otimes_{i \leqslant n} Q_i$ is bad, then there is a barrier $C \subseteq B$ and an i such that, letting $f_i(b) = (f(b))_i$, $f_i \restriction C$ is bad. Hence if each Q_i is* bqo *then Q is* bqo.

Proof. Apply Lemma 1.7 successively to the f_i's, getting barriers $B \supseteq C_0 \supseteq C_1 \supseteq \cdots \supseteq C_n$. For some i, $f_i \restriction C_i$ must be bad; otherwise each $f_i \restriction C_i$ would be perfect and hence $f \restriction C_n$ would be perfect.

In [12], a proof is given that Q bqo implies $(Q)^{<\omega}$ bqo (which strengthens Lemma 1.8). This result is then used to show that if Q is bqo then $\mathscr{P}(Q)$ is bqo under $\leqslant_1$ (which strengthens Lemma 1.3). See [12] for Nash-Williams' original proofs of these results. Here, we will go directly to his forerunning method and derive by those means his result in [13] that Q bqo implies $\bigcup_{\alpha \in \mathrm{On}} (Q)^\alpha$ bqo; this will have the above results as corollaries.

DEFINITIONS. If B and C are barriers, then B *foreruns* C (written $B \sqsubseteq C$) if

(a) $\bigcup C \subseteq \bigcup B$, and
(b) for each $c \in C$ there is a $b \in B$ with $b \leqslant c$.

B strictly foreruns C (written $B \sqsubset C$) if $B \sqsubseteq C$ and there are $b \in B$, $c \in C$ with $b < c$ (and so $b \notin C$).

Suppose Q is quasi-ordered by $\leqslant$. A *partial ranking* on Q is a well-founded (irreflexive) partial order $<'$ on Q such that if $q, r \in Q$ and $q <' r$ then $q < r$.

If B, C are barriers, $f:B \to Q$, and $g:C \to Q$, and $<'$ is a partial ranking on Q, then f *foreruns* g (*strictly foreruns* g) with respect to $<'$ if $B \sqsubseteq C$ ($B \sqsubset C$) and

(a) $g(a) = f(a)$ $(a \in B \cap C)$,
(b) $g(c) <' f(b)$ $(b \in B, c \in C, b < c)$.

We use the notation $f \sqsubseteq g$ ($f \sqsubset g$) to denote that f (strictly) foreruns g (the partial ranking $<'$ will always be specified by context).

Some examples: if $C \subseteq B$ then $B \sqsubseteq C$; if $C \subseteq B \cup B(2)$ and $C \nsubseteq B$ then $B \sqsubset C$. Note that if $B \sqsubseteq C \sqsubseteq D$ ($f \sqsubseteq g \sqsubseteq h$) then $B \sqsubseteq D$ ($f \sqsubseteq h$), and if also $C \sqsubset D$ ($g \sqsubset h$) then $B \sqsubset D$ ($f \sqsubset h$). We check the $\sqsubset$ cases. Suppose $B \sqsubseteq C \sqsubset D$. Then $\bigcup B \supseteq \bigcup D$, and if $C \sqsubset D$ via the pair c, d, then $B \sqsubset D$ via the pair b, d, where b is the member of B that is $\leqslant c$. Suppose $f:B \to Q$, $g:C \to Q$, $h:D \to Q$ satisfy $f \sqsubseteq g \sqsubset h$, and suppose $b \in B$, $d \in D$, and $b \leqslant d$. Let c be the member of C which is $\leqslant d$. Since B is a barrier and $B \sqsubseteq C$ we must have $b \leqslant c$. Thus if $b = d$ then $f(b) = g(b) = h(b)$. If $b < d$ then either $b < c = d$, $b = c < d$, or $b < c < d$, giving respectively $f(b) <' g(c) = h(d)$, $f(b) = g(c) <' h(d)$, $f(b) <' g(c) <' h(d)$.

Suppose $<'$ is a partial ranking on Q. For C a barrier, $g:C \to Q$ is *minimal bad* if g is bad and there is no bad h with $g \sqsubset h$.

THEOREM 1.9. *Let Q be quasi-ordered by $\leqslant$, $<'$ a partial ranking on Q. Then for any bad f on Q there is a minimal bad g such that $f \sqsubseteq g$.*

A consequence is that to prove a quasi-order Q is bqo it suffices to find a partial ranking $<'$ on Q and show that for each alleged bad g on Q there is a bad h with $g \sqsubset h$.

Proof. For barriers B, C, say that B *warily foreruns* C ($B \sqsubset_W C$) if

(a) $B \sqsubset C$,
(b) there is a $b \in B - C$ such that if $n \in \bigcup B$ and $n \leqslant \max b$ then $n \in \bigcup C$.

If $f:B \to Q$, $g:B \to Q$ then f *warily foreruns* g ($f \sqsubset_W g$) if $B \sqsubset_W C$ and $f \sqsubset g$.

Note that, letting b be as in (b), there is a $c \in C$ with $b < c$, because $b \subseteq \bigcup C$, no $b' \leqslant b$ belongs to C, and C is a barrier.

Given a bad f on Q as in the theorem, we shall construct a transfinite sequence $f \sqsubset_W f_1 \sqsubset_W f_2 \sqsubset \cdots$, of bad functions, which ends at some (countable) ordinal γ; f_γ will be the desired g. We prove two lemmas that allow the construction of this sequence.

LEMMA 1. *If $f:B \to Q$, $g:C \to Q$ are bad and $f \sqsubset g$, then there is a bad h with $f \sqsubset_W h$.*

Proof (well-foundedness of $<'$ not involved). Since $B \sqsubset C$, there is a $b_0 \in B$ such that there exists a $c \in C$ with $b_0 < c$; pick such a b_0 with $\max b_0$ as small as possible. Let $e = \{n \in \bigcup B : n \leqslant \max b_0\}$, let $X = e \cup \bigcup C$, and let $D = C \cup \{b \in B : b \subseteq X, b \nsubseteq \bigcup C\}$. Since $B \sqsubset C$, $f(a) = g(a)$ for all $a \in B \cap C$;

therefore we can define $h: D \to Q$ by

$$h(d) = f(d) \qquad (d \in B),$$
$$h(d) = g(d) \qquad (d \in C).$$

We claim that $B \sqsubset_W D$ via b_0, $f \sqsubset_W h$, and h is bad.

(1) D is a barrier on X. Given $Y \in [X]^{\aleph_0}$, we want to find a $d \in D$ with $d < Y$. Since $C \subset D$ we may assume for each $c \in C$ that $c \not< Y$. Since C is a barrier, there is thus a least $m \in Y$ with $m \notin \bigcup C$. Since $Y \subseteq \bigcup B$ there is a $b \in B$ with $b < Y$; we will be done by showing $b \in D$, i.e., we need to show $b \subseteq X$ and $b \nsubseteq \bigcup C$. Since $b < Y$, $b \subseteq X$. Suppose we had $b \subseteq \bigcup C$. Then $m \notin b$. Then since $b < Y$ and $m \in Y$, $\max b < m$. Since $b < Y$, no $b' \leqslant b$ belongs to C. Since C is a barrier and $b \subseteq \bigcup C$, there is a $c \in C$ with $b \leqslant c$ or $c \leqslant b$, whence we must have $b < c$. By the minimality of b_0, then, $\max b \geqslant \max b_0$. We have $m \in X$ and $m \notin \bigcup C$ so, by definition of X, $m \leqslant \max b_0$. Thus $\max b < m \leqslant \max b_0 \leqslant \max b$, a contradiction.

Given $s, t \in D$ we show that $s \not\subset t$. If $s, t \in B$ or $s, t \in C$ then we are done. Suppose $s \in C$, $t \in B$, and $s \subseteq t$. Since $B \sqsubset C$, there is an $s' \in B$, $s' \leqslant s$, so $s = t$ since we cannot have $s' \subset t$. Finally, suppose $s \in B - C$ and $t \in C$. Then, since $s \in D$, $s \nsubseteq \bigcup C$, so $s \nsubseteq t$.

(2) $f \sqsubset_W h$. We check that $B \sqsubset D$. Since $D \subseteq B \cup C$ and $B \sqsubseteq C$, we have $B \sqsubseteq D$. Letting $B \sqsubset C$ via the pair (b, c), we have that $B \sqsubset D$ via (b, c) since $C \subseteq D$. We check that $B \sqsubset_W D$ via b_0. Since there is a $c \in C$ with $b_0 < c$, $b_0 \notin C$ and $b_0 \subseteq \bigcup C$. Thus $b_0 \notin D$. By the definition of X $(= \bigcup D)$, if $n \in \bigcup B$ and $n \leqslant \max b_0$ then $n \in X$. Finally, $f \sqsubset h$, since if $a \in B \cap D$ then $f(a) = h(a)$, and if $b \in B$, $d \in D$, and $b < d$ then, since $d \in C - B$, $h(d) = g(d) <' f(b)$.

(3) h is bad. Suppose $s, t \in D$, $s \lhd t$. We show $h(s) \not\leqslant h(t)$.

If $s, t \in B$, then $h(s) = f(s) \not\leqslant f(t) = h(t)$, and if $s, t \in C$ then $h(s) = g(s) \not\leqslant g(t) = h(t)$.

Suppose $s \in B$, $t \in C - B$. Since $B \sqsubseteq C$ there is a $t' \in B$ with $t' < t$, and, since $s \lhd t$, we must have $s \lhd t'$ lest $t' \subset s$. Then $h(s) = f(s) \not\leqslant f(t')$, and $h(t) = g(t) <' f(t')$, so $h(t) < f(t')$, from which we conclude $h(s) \not\leqslant h(t)$.

Finally, we claim that if $s \in C - B$, $t \in B - C$ $(s, t \in D)$, then $s \lhd t$ is impossible. Since $t \notin C$, by definition of D there is an $m \in t$, $m \notin \bigcup C$, whence, by definition of X, $m \leqslant \max b_0$. Suppose $s \lhd t$. Then, since $s \in C$, $m > \max s$. Since $B \sqsubset C$ there is an $s' \in B$ with $s' < s$. Thus $\max s' < \max s < m \leqslant \max b_0$. This contradicts the minimality of $\max b_0$, i.e., s' has smaller maximum and is a member of B with a proper extension in C.

LEMMA 2. (i) *If* $B \sqsubset_W C$ *and* $C \sqsubset_W D$ *then* $B \sqsubset_W D$. *If* $f \sqsubset_W g$ *and* $g \sqsubset_W h$ *then* $f \sqsubset_W h$.

(ii) *Let* λ *be a limit ordinal, and let* $f_\alpha: B_\alpha \to Q$ $(\alpha < \lambda)$ *satisfy* $f_\alpha \sqsubset_W f_\beta$ $(\alpha < \beta < \lambda)$. *Let* $B = \bigcup_\alpha \bigcap_{\beta \geqslant \alpha} B_\beta$, *and define* $f: B \to Q$ *by, for* $b \in B_\alpha$, $f(b) = f_\beta(b)$ *for some/all* $\beta \geqslant \alpha$. *Then* $f_\alpha \sqsubset_W f$ *for each* α, *and if each* f_α *is bad,* f *is bad.*

Proof. (i) Let $B \sqsubset_W C$ via $b \in B - C$, $C \sqsubset_W D$ by $c \in C - D$; we want to find a $b' \in B - D$ with $B \sqsubset_W D$ via b'. If $\max b \leqslant \max c$ then we may take $b' = b$, i.e., if $n \in \bigcup B$ and $n \leqslant \max b$ then $n \in \bigcup C$, and, since $n \leqslant \max c$, $n \in \bigcup D$. Suppose $\max b > \max c$. Take b' to be the member of B that is $\leqslant c$. Since $c \subseteq \bigcup D$ there is a $d \in D$ with $c \leqslant d$ or $d \leqslant c$, whence, since $c \notin D$ and $C \sqsubset D$, $c < d$. Thus $b' \notin D$. If $n \in \bigcup B$ and $n \leqslant \max b'$, then $n < \max b$, so $n \in \bigcup C$, and, since $n \leqslant \max c$, $n \in \bigcup D$. For the second part of (i) we have $f \sqsubset h$ by transitivity of $\sqsubset$, and $\operatorname{dom} f \sqsubset_W \operatorname{dom} h$ by the first part of (i).

(ii) The definition of f in (ii) makes sense by the definition of $\sqsubset_W$. Let $Y = \bigcap_\alpha (\bigcup B_\alpha)$. We show that B is a barrier on Y. For $\alpha < \beta$, let $b_{\alpha\beta} \in B_\alpha - B_\beta$ witness $B_\alpha \sqsubset_W B_\beta$. Note that, since $b_{\alpha\beta} \notin B_\beta$, $b_{\alpha\beta} \notin B_\gamma$ for every $\gamma > \beta$, for otherwise by $B_\beta \sqsubseteq B_\gamma$ there would be a $b \in B_\beta$ with $b < b_{\alpha\beta}$, contradicting $B_\alpha \sqsubseteq B_\beta$.

We first show that Y is infinite. Letting $i < \omega$, we find an $m \in Y$, $m > i$. By the finiteness of $\mathscr{P}(\{0, 1, \ldots, i\})$ there is an α_0 such that for all β with $\alpha_0 < \beta$ $\max b_{\alpha_0\beta} > i$. Among all β with $\alpha_0 < \beta$, pick β_0 such that $\max b_{\alpha_0\beta_0}$ is minimal. For every $n \in \bigcup B_{\alpha_0}$ with $n \leqslant \max b_{\alpha_0\beta_0}$, $n \in \bigcup B_\beta$ for all $\beta \geqslant \beta_0$ by definition of $\sqsubset_W$ and minimality of $\max b_{\alpha_0\beta_0}$. Hence $n \in \bigcup B_\delta$ for all δ since the $\bigcup B_\delta$'s are decreasing. Thus $\max b_{\alpha_0\beta_0} \in \bigcap_\delta (\bigcup B_\delta) = Y$.

Next we claim that for all $Z \in [Y]^{\aleph_0}$ there is a $b \in B$ with $b < Z$. For each α, since $\bigcup B_\alpha \supseteq Z$ there is a $b_\alpha \in B_\alpha$, $b_\alpha < Z$. If $\alpha < \beta$ then $b_\alpha \leqslant b_\beta$, and $f_\beta(b_\beta) \leqslant f_\alpha(b_\alpha)$, and if $b_\alpha < b_\beta$ then $f_\beta(b_\beta) <' f_\alpha(b_\alpha)$. Since $<'$ is well founded, the b_α's must thus eventually be a constant b, whence $b \in B$.

If $c, d \in B$ then $c, d \in B_\alpha$ for some α, and so $c \not\subset d$.

We check that for each α, $B_\alpha \sqsubset_W B$ via the $b_{\alpha\beta}$ with $\max b_{\alpha\beta}$ minimal. If $b \in B$ then $b \in B_\gamma$ for eventually all γ. Thus, since $B_\alpha \sqsubseteq B_\gamma$, there is a $b' \in B_\alpha$ with $b' \leqslant b$. There is a $b \in B_\beta$ with $b_{\alpha\beta} < b$; thus $b_{\alpha\beta} \notin B$. By the minimality of $\max b_{\alpha\beta}$, $\{n \in \bigcup B_\alpha : n \leqslant \max b_{\alpha\beta}\} \subseteq Y$, as seen above. Thus $B_\alpha \sqsubset_W B$. It follows that $f_\alpha \sqsubset_W f$.

Finally, suppose each f_α is bad. Then if $c, d \in B$ there is an α with $c, d \in B_\alpha$, $f(c) = f_\alpha(c)$, $f(d) = f_\alpha(d)$. So if $c \lhd d$ then $f(c) \not\leqslant f(d)$.

To prove the theorem, let the given bad function f be f_0. Choose a sequence $f_0, f_1, \ldots, f_\alpha, \ldots$ of bad functions such that $\alpha < \beta$ implies $f_\alpha \sqsubset_W f_\beta$, terminating the sequence when it cannot be further extended. Since $\operatorname{dom} f_\alpha \neq \operatorname{dom} f_\beta$ $(\alpha \neq \beta)$ the sequence must end, and by Lemma 2 the sequence has a last term f_γ (it may be seen in fact that γ is countable). To show $f_\gamma = g$ works for the theorem, suppose there were a bad h with $f_\gamma \sqsubset h$. Then by Lemma 1 there would be a bad $f_{\gamma+1}$ with $f_\gamma \sqsubset_W f_{\gamma+1}$, a contradiction. This completes the proof.

For Q a quasi-ordering, $J, K \in \bigcup_{\alpha \in \mathrm{On}} (Q)^\alpha$, let $J <' K$ mean that J is isomorphic to a subsequence of K and $\operatorname{dom} J < \operatorname{dom} K$. Then $<'$ is a partial ranking.

The second part of the following theorem is the main result of Nash-Williams [13].

THEOREM 1.10. *If Q is quasi-ordered, and $f: B \to \bigcup_{\alpha \in \mathrm{On}} (Q)^\alpha$ is bad, then there is a barrier E and a bad j with $f \sqsubseteq j$ and range $j \subseteq Q$ (we identify Q* with $(Q)^1$). *Hence if Q is* bqo *then* $\bigcup_{\alpha \in \mathrm{On}} (Q)^\alpha$ is bqo.

Proof. By Theorem 1.9, there is a barrier C and a minimal bad $g: C \to \bigcup_{\alpha \in \mathrm{On}} (Q)^\alpha$ with $f \sqsubseteq g$. We will find the desired j with $g \sqsubseteq j$.

By Corollary 1.5, there is a barrier $D \subseteq C$ such that either $\operatorname{dom} g(d)$ is a successor ordinal for all $d \in D$ or $\operatorname{dom} g(d)$ is a limit ordinal for all $d \in D$.

Suppose the successor case holds. For $d \in D$, write $g(d)$ as $g_1(d)^\frown g_2(d)$ where $g_2(d) \in Q$ is the last term of $g(d)$. By two applications of Lemma 1.7, there is a barrier $E \subseteq D$ such that $g_1 \restriction E$ is bad or perfect, and $g_2(E)$ is bad or perfect. Since g is bad, one of the $g_i \restriction E$'s must be bad. If $g_2 \restriction E$ is bad then $j = g_2 \restriction E$ is as desired. We are done then by showing $g_1 \restriction E$ cannot be bad. Suppose it were. Note that $g_1(e) <' g(e)$. We define an h on $E(2)$ as follows: for $s \in E(2)$ ($s = e_1 \cup e_2$, $e_1, e_2 \in E$, $e_1 \lhd e_2$) let $h(s) = g_1(e_1)$. Then h is bad (recall that for $s, t \in E_2$, $s \lhd t$ if and only if there are $e_1, e_2, e_3 \in E$ with $e_1 \lhd e_2 \lhd e_3$, $s = e_1 \cup e_2$, $t = e_2 \cup e_3$). But $C \sqsubseteq D \sqsubseteq E \sqsubset E(2)$, and for $s = e_1 \cup e_2$ in $E(2)$, $h(s) = g_1(e_1) <' g(e_1)$. Thus $g \sqsubset h$, a contradiction.

We claim the limit case cannot hold. Suppose it did. Note that if $J, K \in \bigcup_{\alpha \in \mathrm{On}} (Q)^\alpha$, $\operatorname{dom} J$ is a limit ordinal, and $J \nleqslant K$, then for some proper initial segment J' of J, $J' \nleqslant K$ (otherwise $J \leqslant K$ via the function $H(\beta) =$ the least δ with $\delta > H(\beta')$ ($\beta' < \beta$) and $J(\beta) \leqslant K(\delta)$). Define h on $D(2)$ as follows: for $d_1, d_2 \in D(2)$, $d_1 \lhd d_2$, let $h(d_1 \cup d_2)$ be a proper initial segment of $g(d_1)$ that is $\nleqslant g(d_2)$. Then h is bad. Since $h(d_1 \cup d_2) <' g(d_1)$, and since $C \sqsubseteq D \sqsubset D(2)$, we have $g \sqsubset h$, a contradiction.

We use Theorem 1.10 to prove a similar result for the power set operation under $\leqslant_1$. If $X \in \bigcup_{\alpha \in \mathrm{On}} \mathscr{P}^\alpha(Q)$, let rank X be the least α with $X \in \mathscr{P}^\alpha(Q)$. Then an induction on ranks shows that if $X \leqslant_1 Y$ then rank $X \leqslant$ rank Y. Order $\bigcup_{\alpha \in \mathrm{On}} \mathscr{P}^\alpha(Q)$ by: $X <' Y$ if and only if $X \in Y \cup \bigcup Y \cup \bigcup\bigcup Y \cup \cdots$ (Q considered as a set of atoms). Then $<'$ is a partial ranking.

THEOREM 1.11. *Suppose $f: B \to \bigcup_{\alpha \in \mathrm{On}} \mathscr{P}^\alpha(Q)$ is bad under $\leqslant_1$. Then there is a bad j with $f \sqsubseteq j$ and range $j \subseteq Q$. Hence if Q is bqo,* $\bigcup_{\alpha \in \mathrm{On}} \mathscr{P}^\alpha(Q)$ is bqo.

Proof. Let g be a minimal bad function with $f \sqsubseteq g$, given by Theorem 1.9. By Corollary 1.5, there is a barrier $C \subseteq \operatorname{dom} g$ such that either for all $c \in C$, $g(c) \in Q$, or for all $c \in C$, $g(c) \notin Q$. In the former case we are done by taking $j = g \upharpoonright C$. We show the second case cannot hold. Suppose it did. For each $c \in C$ let J_c be a 1–1 function with domain an ordinal and range $g(c)$. If $J_c \leqslant J_d$ (as members of $\bigcup_{\alpha \in \mathrm{On}} (R)^\alpha$, where $R = \bigcup_{c \in C} g(c)$) then $g(c) \leqslant_1 g(d)$. Thus the function $g'(c) = J_c$ is bad. Applying Theorem 1.10, there is a bad j with $g' \sqsubseteq j$ and range $j \subseteq R$. But then $g \sqsubset j$, a contradiction.

Remarks. It may be seen that Nash-Williams' proof [13] that if Q is bqo then $\bigcup_{\alpha \in \mathrm{On}} (Q)^\alpha$ is bqo also gives the effective version in Theorem 1.10. That Q bqo implies $\bigcup_{\alpha \in \mathrm{On}} (Q)^\alpha$ bqo is also a corollary of the main theorem of [8], that Q bqo implies the class of Q-labeled "$\mathscr{M}$-types" is bqo. The results in [8] were derived there from a Q-labeled version of the Nash-Williams infinite tree theorem (the trees needed in the application are infinite with no infinite paths). It is possible to derive the part of [8] that used trees [8, Corollary 4.4] from Theorem 1.9 above instead, thus avoiding explicit mention of trees in the proof of the main result of [8]. Briefly, to show that if Q is bqo then the class $\mathscr{H}(Q)$ of "hereditarily regular unbounded" Q-labeled $\mathscr{M}$-types (which is shown to coincide with the class of additively indecomposable Q-labeled $\mathscr{M}$-types; see also [9, Section 2] for a summary of the scattered case) is bqo, one assumes the existence of a bad $g: B \to \mathscr{H}(Q)$, then uses the ranked construction of $\mathscr{H}(Q)$ to get a bad h with $g \sqsubset h$ (in a manner analogous to the proof of Theorem 1.10), and concludes the theorem from Theorem 1.9. This is essentially the way the scattered case of [8, Theorem 4.4] is presented in Fraïssé's paper [2].

2. A Class of Trees

A *tree* is a set T on which a partial order $<_T$ is defined such that for every $t \in T$, $\{s \in T : s <_T t\}$ is well ordered under $<_T$. Thus $T = \bigcup_{\alpha \in \mathrm{On}} T_\alpha$, where T_α, the αth *level* of T, is the set of all $t \in T$ such that $\{s : s <_T t\}$ has type α. The *height* of T is the least α with $T_\alpha = \varnothing$. $T_{<\alpha}$, $T_{\leqslant \alpha}$ are $\bigcup_{\beta < \alpha} T_\beta$, $\bigcup_{\beta \leqslant \alpha} T_\beta$ respectively. A subset X of T is *downward closed* if $y <_T x \in X$ implies $y \in X$. A *path* in T is a linearly ordered downward closed subset of T. If $x \in T$ (P a path in T) let $S(x)$ ($S(P)$) be the set of immediate successors of x (P). A path is *maximal* in T if $S(P) = \varnothing$. Let $br_T(x)$ ($= br(x)$ if unambiguous) be $\{y \in T : x \leqslant_T y\}$, the *branch* above x.

We make the assumption (for notational convenience only) on our trees T that T have a *root* (minimum element) and that if P is a path in T with no largest element then $\operatorname{Card} S(P) \leqslant 1$. Let $\mathscr{T}$ be the class of all such trees.

If $T \in \mathscr{T}$, $s, t \in T$, there is a greatest lower bound of s and t in T; denote it by $s \wedge t$. $\mathscr{T}$ is quasi-ordered by the rule: $T_1 \leqslant T_2$ (T_1 is *embeddable* in T_2) if and only if there is an $H: T_1 \to T_2$ such that $H(s \wedge t) = H(s) \wedge H(t)$ for all $s \in T_1$, i.e., H is order preserving, and if s and t are distinct immediate successors of u in T_1 then $h(s)$ and $h(t)$ occur above distinct immediate successors of $h(u)$ in T_2 (this is really the "$\leqslant_1$" embedding; we drop the subscript since the "$\leqslant_{\mathrm{m}}$" embedding (see [9]) will not be used below).

A *Q-tree* is a pair (T, l) where $T \in \mathscr{T}$ and $l: T \to Q$. Order the class of Q-trees by: $(T_1, l_1) \leqslant (T_2, l_2)$ if $T_1 \leqslant T_2$ via an H such that $l_1(t) \leqslant l_2(H(t))$ for all $t \in T_1$.

If $\mathscr{U} \subseteq \mathscr{T}$, let $\mathscr{U}_Q$ be the collection of Q-trees (T, l) with $T \in \mathscr{U}$. Let $\mathscr{T}_{\leqslant \omega} = \{T \in \mathscr{T}: \text{height } T \leqslant \omega\}$. Then the main theorem of [12] is that $\mathscr{T}_{\leqslant \omega}$ is bqo, and the generalization that Q bqo implies $(\mathscr{T}_{\leqslant \omega})_Q$ bqo was verified in [8]. Let η be $(2)^{<\omega}$, the complete binary tree of height ω. Call $T \in \mathscr{T}$ *scattered* if $\eta \nleqslant T$, and let $\mathscr{S}$ be the class of scattered trees. Let $\mathscr{M}$ be the class of trees T such that there are T_n ($n < \omega$) with $T = \bigcup_{n<\omega} T_n$, each T_n downward closed in T, each $T_n \in \mathscr{S}$.

We state some facts about $\mathscr{S}$ and $\mathscr{M}$. $\mathscr{S}$ is closed under *grafting*, that is, if $T \in \mathscr{S}$ and to each path P in T, such that either P has a maximum element or P is maximal in T, is corresponded a set $\mathscr{U}(P) \subseteq \mathscr{S}$, then $T' \in \mathscr{S}$, where T' is obtained from T by adding, for each P, copies of the members of $\mathscr{U}(P)$, making their roots new immediate successors of P. Namely, if there were an embedding of η into T', then either it would have its range $\subseteq T$ or there would be a copy of η in one of the grafted trees. $\mathscr{S}$ is closed under *finite downward closed unions*, i.e., if $T \in \mathscr{T}$, $T = \bigcup_{i<n} T_n$, and each T_n is downward closed in T, then $T \in \mathscr{S}$. This can be seen, e.g., by starting with T_0 and grafting n times to get T. $\mathscr{M}$ is closed under countable (downward closed) unions.

Suppose $T \in \mathscr{T}$, P a path in T, $z \in P$. Then let $\tilde{P}(z) = \{br(y): y \in S(z) \text{ and } y \notin P\}$ If (T, l) is a Q-tree, let $\tilde{P}^l(z) = \{(T', l): T' \in \tilde{P}(z)\}$. The version, for trees, of Hausdorff's characterization of scattered order types is given by the following lemma.

LEMMA 2.1. *Let $\mathscr{S}_0 = \{$the empty tree, the one point tree$\}$, and for $\alpha \in \mathrm{On}$, let $\mathscr{S}_{\alpha+1} = \{T$: there is a maximal path P in T such that $\tilde{P}(z) \subseteq \mathscr{S}_\alpha$ for all $z \in P\}$, and for λ a limit ordinal let $\mathscr{S}_\lambda = \bigcup_{\alpha<\lambda} \mathscr{S}_\alpha$. Then $\mathscr{S} = \bigcup_{\alpha \in \mathrm{On}} \mathscr{S}_\alpha$.*

Proof. Let $\bar{\mathscr{S}} = \bigcup_{\alpha \in \mathrm{On}} \mathscr{S}_\alpha$. We claim $\bar{\mathscr{S}} \subseteq \mathscr{S}$. Namely, if P is a path in T and $\tilde{P}(z) \subseteq \mathscr{S}$ for all $z \in P$, then $T \in \mathscr{S}$, so the claim holds by induction on α.

We claim that if $T \in \mathscr{T}$, $x \in T$, and $br(x) \notin \bar{\mathscr{S}}$ then there are y, z with $x <_T y$, $x <_T z$, y incomparable with z, with $br(y) \notin \bar{\mathscr{S}}$, $br(z) \notin \bar{\mathscr{S}}$.

Suppose this failed for some x. Choose a path P_0 in $br(x)$ such that for all $z \in P_0$, $br(z) \notin \bar{\mathscr{S}}$, P_0 as large as possible in the sense that $\{br(z): z \in S(P_0)\} \subseteq \bar{\mathscr{S}}$. For each $z \in P_0$, $\tilde{P}_0(z) \subseteq \bar{\mathscr{S}}$ by the assumption on x. We have $br(x) = P_0 \cup \bigcup\{br(z): z \in S(P_0)\} \cup \bigcup_{z \in P_0} \cup \tilde{P}_0(z)$. Pick a $z \in S(P_0)$ (if $S(P_0) \neq \varnothing$) and let $P = P_0 \cup P_1$, P_1 a path in $br(z)$ witnessing $br(z) \in \bar{\mathscr{S}}$. Then P witnesses $br(x) \in \bar{\mathscr{S}}$, a contradiction.

Suppose now that there is a $T \in \mathscr{S} - \bar{\mathscr{S}}$. Using the claim, choose for each $j \in (2)^{<\omega}$ an $x_j \in T$, with $x_\varnothing$ the root of T, $br(x_j) \notin \bar{\mathscr{S}}$, $x_j <_T x_{j^\frown 0}$, $x_j <_T x_{j^\frown 1}$, and $x_{j^\frown 0}$ incomparable with $x_{j^\frown 1}$. Then $T \notin \mathscr{S}$, namely, the function $H(j) = x_{j^\frown 0} \wedge x_{j^\frown 1}$ witnesses $\eta \leqslant T$. This contradiction proves the lemma.

For $T \in \mathscr{S}$, let rank T be the least α with $T \in \mathscr{S}_\alpha$. By induction on ranks we have that if $T_1 \leqslant T_2$ then rank $T_1 \leqslant$ rank T_2. Namely, suppose T_2 has rank α via a maximal path P in T_2, and $T_1 \leqslant T_2$ via H. Then $H^{-1}(P)$ will be a path, not necessarily maximal, in T_1. Suppose that either $y \in S(H^{-1}(P))$ or for some $z \in H^{-1}(P)$, $y \in H^{-1}(P)(z)$. Then the induction hypothesis and H give that $\text{rank}(br(y)) < \alpha$. It follows that $H^{-1}(P)$ can be extended to a maximal path in T_1 that witnesses rank $T_1 \leqslant \alpha$.

Define a relation $<'$ on $\mathscr{S}_Q$ by: $(T_1, l_1) <' (T_2, l_2)$ if (T_1, l_1) is either $(\{x\}, l_2)$ or $(br(x), l_2)$ for some $x \in T_2$, and rank $T_1 <$ rank T_2. Then $<'$ is a partial ranking on $\mathscr{S}_Q$.

THEOREM 2.2. *If $f: B \to \mathscr{S}_Q$ is bad then there is a bad j with range $j \subseteq Q$ (identifying $q \in Q$ with the one point tree labeled by q) such that $f \sqsubseteq j$. Hence if Q bqo then $\mathscr{S}_Q$ is bqo.*

Proof. By Theorem 1.9, there is a minimal bad $g: B' \to \mathscr{S}_Q$ with $f \sqsubseteq g$. For $b \in B'$, let $g(b) = (T_b, l_b)$. By Corollary 1.5 there is a barrier $C \subseteq B'$ such that either for every $c \in C$, Card $T_c = 1$ or for every $c \in C$, Card $T_c > 1$. If the former case holds we are done by taking $j = g \restriction C$.

We claim the latter case cannot hold. Suppose it did. For $c \in C$, let P_c be a maximal path in T_c such that for each $x \in P_c$ and each $T' \in \tilde{P}_c(x)$, rank $T' <$ rank T_c. Let $J_c: P_c \to (Q \otimes \mathscr{P}(\mathscr{S}_Q))$ be defined by

$$J_c(x) = \langle l_c(x), \tilde{P}_c^{l_c}(x) \rangle.$$

Identifying P_c with its order type, J_c is a $(Q \otimes \mathscr{P}(\mathscr{S}_Q))$-sequence, where we take the $\leqslant_1$ quasi-order on $\mathscr{S}_Q$. If $J_c \leqslant J_d$ then $(T_c, l_c) \leqslant (T_d, l_d)$, namely, let $H: P_c \to P_d$ witness $J_c \leqslant J_d$, and for each $x \in P_c$ extend H to an embedding of each member of $\tilde{P}_c^{l_c}(x)$ into a member of $\tilde{P}_d^{l_d}(H(x))$, in a 1–1 fashion. Hence the function $\bar{g}(c) = J_c$ is bad. By Theorem 1.10 there is a barrier D with $C \sqsubseteq D$, and a bad $h: D \to (Q \otimes \mathscr{P}(\mathscr{S}_Q))$ such that if $c \in C$, $d \in D$ and $c \leqslant d$

then for some $x \in P_c$, $h(d) = \langle l_c(x), \tilde{P}_c^{l_c}(x)\rangle$. By Lemma 1.8 there is a barrier $E \subseteq D$ such that either $h_0 : E \to Q$ or $h_1 : E \to \mathscr{P}(\mathscr{S}_Q)$ is bad, where $h_i(e) = h(e)_i$.

Suppose h_0 is bad. Let $h_0' : E(2) \to Q$ be defined by $h_0'(e_1 \cup e_2) = h_0(e_1)$ $(e_1, e_2 \in E, e_1 \lhd e_2)$. Then h_0' is bad. We have that $\bigcup B' \supseteq \bigcup E(2)$, and that if $b \in B'$, $e_1, e_2 \in E$, $e_1 \lhd e_2$, and $b \leqslant e_1 \cup e_2$, then $b \leqslant e_1$, and $h_0'(e_1 \lhd e_2) = h_0(e_1) = l_b(x)$ for some $x \in T_b$. Hence $B' \sqsubset E(2)$, and $g \sqsubset h_0'$, a contradiction.

Suppose h_1 is bad. By Theorem 1.11 there is a barrier F with $E \sqsubset F$ and a bad $h_1' : F \to \mathscr{S}(Q)$ such that if $e \in E$, $s \in F$ and $e \leqslant s$ then $h_1'(s) \in h_1(e)$. Define h_1'' on $F(2)$ by $h_1''(s \cup t) = h_1'(s)$ $(s, t \in F, s \lhd t)$. Suppose $b \in B'$, $s, t \in F$, $s \lhd t$. Then $b \leqslant s$, and $h_1''(s \cup t)$ is a Q-tree whose underlying tree has rank less than rank T_b. Hence $g \sqsubset h_1''$, a contradiction.

This proves the theorem.

We give another characterization of $\mathscr{M}$ (which will not be used below). Let η' be the $T \in \mathscr{T}$, with T countable and of height $\omega + 1$, such that $T_{<\omega} = \eta$ and for every $x \in T_{<\omega}$ there is a $y \in T_\omega$ with $x <_T y$. (By Cantor's diagonal construction, all such trees are isomorphic.) Let $\mathscr{S}' = \{T \in \mathscr{T} : \eta' \nleqslant T\}$. Let $\mathscr{M}'$ be the class of $T \in \mathscr{T}$ such that T can be written as $\bigcup_{n<\omega} T_n$, each T_n downward closed in T, each $T_n \in \mathscr{S}'$.

THEOREM 2.3. $\mathscr{M} = \mathscr{M}'$.

Proof. We have $\mathscr{M} \subseteq \mathscr{M}'$. For the other direction, since $\mathscr{M}$ is closed under countable downwards closed unions, it suffices to show that if $T \in \mathscr{S}'$ then $T \in \mathscr{M}$. Suppose $T \in \mathscr{S}' - \mathscr{M}$.

We claim that if $x \in \mathscr{T}$ and $br(x) \notin \mathscr{M}$ then there is a $y \in T$ such that

$$\{u : x <_T u <_T y \text{ and for some } v \in S(u),\ v \nleqslant_T y \text{ and } br(v) \notin \mathscr{M}\}$$

is infinite. Suppose this failed for x. Then $br(x) = \bigcup_{n<\omega} Y_n$, where

$$Y_n = \{y : x \leqslant_T y \text{ and there are at most } n\ u\text{'s such that } x <_T u <_T y \text{ and for some } v \in S(u),\ v \nleqslant_T y \text{ and } br(v) \notin \mathscr{M}\}.$$

Each Y_n is closed downward in $br(x)$. Each $Y_n \in \mathscr{M}$ (if $V \notin \mathscr{M}$ then clearly there are $v_i (i \leqslant n + 1)$ in V with $v_i <_V v_{i+1}$ such that for each $i < n + 1$ there is a $w_i \in S(v_i)$ with $br_v(w_i) \notin \mathscr{M}$ and $w_i \nleqslant_v v_{i+1}$). Hence $br(x) \in \mathscr{M}$, a contradiction, proving the claim.

But now the claim may be used repeatedly in a diagonal construction of an embedding H of η' into T (range $H \restriction \eta \subseteq \{x \in T : br(x) \notin \mathscr{M}\}$). This contradiction yields the theorem

THEOREM 2.4. *If Q is* bqo *then $\mathscr{M}_Q$ is* bqo.

Proof. For each $T \in \mathscr{M}$, there is a decomposition $T = \bigcup_{n<\omega} T^n$, each T^n downward closed in T, each $T^n \in \mathscr{S}$, such that

(a) $T^0 \subseteq T^1 \subseteq \cdots \subseteq T^n \subseteq \cdots$,

(b) for each n with $T^n \neq \varnothing$, $T = T^n \cup \bigcup\{br_T(x): x$ a maximal node of $T^n\}$.

Namely, let $T = \bigcup_{n<\omega} S_n$, S_n downward closed, $S_n \in \mathscr{S}$. If $U \subseteq T$ is downward closed with $U \in \mathscr{S}$, let $U^* = \{x \in T\colon$ for all $y <_T x,\ y \in U\}$. Then $U^* \in \mathscr{S}$ by the grafting property. Then the sequence T^n, $n < \omega$, defined by $T^0 = (S_0)^*$, $T^{n+1} = (T^n \cup S_{n+1})^*$, is as desired.

Q is a fixed better-quasi-ordering throughout. Let $\mathscr{F}_Q$ be the class of all $(T, l) \in \mathscr{M}_Q$ such that there is no infinite sequence $x_0 <_T x_1 <_T \cdots <_T x_n <_T \cdots$ with

$$(br(x_0), l) > (br(x_1), l) > \cdots > (br(x_n), l) > \cdots.$$

The trees in $\mathscr{T}_{\leqslant\omega}$ with this property are called *descensionally finite* in [12]; as in [12], we first show that this class is bqo. The next definitions have the aim of reducing the problem of embedding (T, l) into (U, m) ($(T, l), (U, m) \in \mathscr{F}_Q$) to another embedding problem involving branches of (T, l) and (U, m) of lower rank.

Suppose $(T, l) \in \mathscr{F}_Q$, $x \in T$. Define

$$\begin{aligned} a_{(t,l)}(x) = \langle\{(br(y), l)\colon & y \in S(x) \text{ and } (br(y), l) < (T, l)\}, \\ & \operatorname{Card}\{y \in S(x)\colon (br(y), l) \equiv (T, l)\}\rangle. \end{aligned}$$

We have that $a_{(T,l)}(x) \in \mathscr{P}(\mathscr{F}_Q) \otimes$ Cardinals. This space is quasi-ordered in the canonical way, using the $\leqslant_1$ relation for the power set operation.

Define, for each $n < \omega$,

$$A_{(T,l)}(n) = (T^n, \bar{l}),$$

where $\bar{l}(z) = l(z)$ for all z that are nonmaximal in T^n, and for z maximal in T^n, $\bar{l}(z) = \langle l(z), a_{(T,l)}(z)\rangle$. We have $A_{(T,l)}(n) \in \mathscr{M}_R$, where $R = Q \cup (Q \otimes (\mathscr{P}(\mathscr{F}_Q) \otimes \text{Cardinals}))$. Quasi-order R in the canonical way, making each $q \in Q$ incomparable with each $v \in (Q \otimes (\mathscr{P}(\mathscr{F}_Q) \otimes \text{Cardinals}))$.

Finally, let

$$A((T, l)) = \{A_{(T,l)}(n)\colon n < \omega\}.$$

We have $A((T, l)) \in \mathscr{P}(\mathscr{M}_R)$, whose quasi-order comes from the order on $\mathscr{M}_R$ and the $\leqslant_m$ order.

LEMMA 1. *$A((T, l)) \leqslant A((U, m))$ implies $(T, l) \leqslant (U, m)$.*

Proof. We will construct an embedding H of (T, l) into (U, m) in ω steps. The induction hypothesis at stage n is that H has been defined on a downward closed $Y_n \supseteq T^n$, such that

(a) if $t \in T - Y_n$ then there is a $y <_T t$ with y maximal in Y_n,

(b) if y is maximal in Y_n, then there is a 1-1 function J_y $(=J): S(y) \to S(H(y))$ such that if $z \in S(y)$ and $(br(z), l) < (T, l)$ then $(br(z), l) \leqslant (br(J(z)), m)$, and if $(br(z), l) \equiv (T, l)$ then $(br(J(z)), m) \equiv (U, m)$.

Suppose the nth stage has been defined, y is maximal in Y_n, and $z \in S(y)$. We want to extend H above z.

If $(br(z), l) < (T, l)$, then extend H to an embedding of $(br(z), l)$ into $(\mathrm{br}(J(z)), m)$.

Suppose then that $(br(z), l) \equiv (T, l)$, whence $(br(J(z)), m) \equiv (U, m)$. Since $A((T, l)) \leqslant A((U, m))$ there are i, K with $A_{(T,l)}(n + 1) \leqslant A_{(U,m)}(i)$ via K. Let L embed (U, m) into $(br(J(z)), m)$. Then extend H to $br(z) \cap T^{n+1}$ by

$$H \restriction br(z) \cap T^{n+1} = LK \restriction br(z) \cap T^{n+1}.$$

Since K in particular embeds $(br(z) \cap T^{n+1}, l)$ into (U, m), LK is an embedding. Suppose now that y' is maximal in $T^{n+1} \cap br(z)$. Then $a_{(T,l)}(y') \leqslant a_{(U,m)}(K(y'))$. This relation and the embedding L induce a 1–1 $J_{y'}: S(y') \to S(LK(y'))$ such that, for all $z' \in S(y')$, if $(br(z'), l) < (T, l)$ then $(br(z'), l) \leqslant (br(J_{y'}(z')), m)$, and if $(br(z'), l) \equiv (T, l)$, then $(br(J_{y'}(z'), m) \equiv (U, m)$. The induction hypothesis at stage $n + 1$ is thus preserved.

LEMMA 2. *$\mathscr{F}_Q$ is* bqo.

Proof. Define an ordering on $\mathscr{F}_Q$ by: $(U, m) <' (T, l)$ if and only if for some $x \in T$, $(U, m) = (br(x), l) < (T, l)$. Then $<'$ is a partial ranking on $\mathscr{F}_Q$. We will be done by Theorem 1.9 if, on the assumption that $g: B \to \mathscr{F}_Q$ is bad, a bad j with $g \sqsubset j$ is found.

For $b \in B$, let $g(b) = (T_b, l_b)$. Let $\bar{g}(b) = A((T_b, l_b))$. By Lemma 1, $\bar{g}$ is bad. By Lemma 1.3, there is a barrier $C \subseteq B(2)$ and an h defined on C such that for $c \in C$ $(c = b_1 \cup b_2,\ b_1, b_2 \in B,\ b_1 \lhd b_2)$, $h(c) \in \bar{g}(b_1)$, with h bad. Thus (with b_1, b_2, c as above) $h(c)$ is an $\mathscr{S}$ tree, labeled by elements of $Q \cup (\mathscr{P}(Z_{b_1}) \otimes \text{Cardinals})$, where for $b \in B$,

$$Z_b = \{(br(x), l_b): x \in T_b \text{ and } (br(x), l_b) < (T_b, l_b)\}.$$

By Corollary 1.5 and the fact that Q is bqo, find a barrier $C' \subseteq C$ such that for $b \in B$, $c' \in C'$ with $b \leqslant c'$ (whence $b < c'$), we have $h(c') \in \mathscr{P}(Z_b) \otimes$ Cardinals. By Lemma 1.2 and Lemma 1.8 there is a barrier $C'' \subseteq C'$ such that the function $h_1(c'') = (h(c''))_1$ is bad. By Theorem 1.11 there is a D with $C'' \sqsubset D$ and a bad function $j: D \to \mathscr{F}_Q$ such that, if $c'' \in C'$, $d \in D$ with $c'' \leqslant d$, then

$c'' < d$ and $j(d) \in h_1(c'')$. We have that $B \sqsubset D$, and if $b \in B$, $d \in D$ and $b \leqslant d$ then $b < d$ and $j(d) = (br(x), l_b) < (T_b, l_b)$ for some $x \in T_b$. Thus $g \sqsubset j$, giving the lemma.

Our aim is to prove that $\mathscr{F}_Q = \mathscr{M}_Q$, which will give the theorem.

For $(T, l) \in \mathscr{M}_Q$, we define objects $b(x)$, $B_{(T,l)}(n)$, $B((T, l))$ analogous to $a(x)$, $A_{(T,l)}(n)$, $A((T, l))$ and to which we associate the same quasi-orders. For $x \in T$, let

$$b_{(T,l)}(x) = b(x) = \langle \{(br(y), l): y \in S(x), (br(y), l) \in \mathscr{F}_Q\}, \\ \text{Card}\{y \in S(x): (br(y), l) \notin \mathscr{F}_Q\}\rangle.$$

For $n < \omega$, define

$$B_{(T,l)}(n) = (T^n, \tilde{l}),$$

where $\tilde{l}(z) = l(z)$ for all z that are nonmaximal in T^n, and for z maximal in T^n, $\tilde{l}(z) = \langle l(z), b_{(T,l)}(z)\rangle$. Lastly, define as before

$$B((T, l)) = \{B_{(T,l)}(i): i < \omega\}.$$

LEMMA 3. *Suppose that* $(T, l), (U, m) \in \mathscr{M}_Q - \mathscr{F}_Q$, *and that* $B((T, l)) \leqslant B((br(u), m))$ *for every* $u \in U$ *with* $(br(u), m) \notin \mathscr{F}_Q$. *Then* $(T, l) \leqslant (U, m)$.

Proof. We construct an embedding I of (T, l) into (U, m) by means similar to the construction of H in Lemma 1. The induction hypothesis at stage n is that I has been defined on a downward closed $Y_n \supseteq T^n$ such that

(a) if $t \in T - Y_n$ then there is a $y <_T t$ with y maximal in Y_n,

(b) if y is maximal in Y_n then there is a 1–1 function $J_y (=J): S(y) \to S(I(y))$ such that if $z \in S(y)$ and $(br(z), l) \in \mathscr{F}_Q$ then $(br(z), l) \leqslant (br(J(z)), m)$, and if $(br(z), l) \notin \mathscr{F}_Q$ then $(br(J(z)), m) \notin \mathscr{F}_Q$.

Suppose the nth stage has been defined, y is maximal in Y_n, and $z \in S(y)$. We want to extend I above z.

If $(br(z), l) \in \mathscr{F}_Q$, then extend I to an embedding of $(br(z), l)$ into $(br(J(z)), m)$.

Suppose then that $(br(z), l) \notin \mathscr{F}_Q$, whence $(br(J(z)), m) \notin \mathscr{F}_Q$ and $B((T, l)) \leqslant B((br(J(z)), m))$. Choose i, K such that $B_{(T,l)}(n+1) \leqslant B_{(br(J(z)),m)}(i)$ via K. Then extend I to $br(z) \cap T^{n+1}$ by

$$I \upharpoonright br(z) \cap T^{n+1} = K \upharpoonright br(z) \cap T^{n+1}.$$

Now if y' is maximal in $br(z) \cap T^{n+1}$, the relation $b_{(T,l)}(y') \leqslant b_{(U,m)}(I(y'))$ gives the desired $J_{y'}$ on $S(y')$, and the induction hypothesis is preserved.

LEMMA 4. $\mathscr{F}_Q = \mathscr{M}_Q$.

Proof. Suppose $(T, l) \in \mathscr{M}_Q - \mathscr{F}_Q$. Without loss of generality we may assume that for each $x \in T$ and $n < \omega$, $(br(x))^n = br(x) \cap T^n$. It follows that if $t, u \in T$ and $t \leqslant u$ then $B((br(u), l)) \leqslant B((br(t), l))$, namely, each $B_{(br(u),l)}(n) \leqslant B_{(br(t),l)}(n)$ via the identity embedding. From Lemma 2, various parts of Section 1, and Theorem 2.2, it follows that $\{B((U, m)) : (U, m) \in \mathscr{M}_Q\}$ is bqo, in particular, it is well founded. Thus there is a $t \in T$ such that $(br(t), l) \notin \mathscr{F}_Q$ and for every u with $t <_T u$ and $(br(u), l) \notin \mathscr{F}_Q$, we have $B((br(u), l)) \nless B((br(t), l))$ (whence $B((br(u), l)) \equiv B((br(t), l))$). Fix such a t.

Since $(br(t), l) \notin \mathscr{F}_Q$ there exists v with $t <_T v$ such that $(br(v), l) < (br(t), l)$ and $(br(v), l) \notin \mathscr{F}_Q$. Now apply Lemma 3 to $(br(t), l)$, $(br(v), l)$ to conclude that $(br(t), l) \leqslant (br(v), l)$, a contradiction.

References

1. E. Ellentuck, A new proof that analytic sets are Ramsey, *J. Symbolic Logic* **39** (1974), 163–165.
2. R. Fraïssé, Abritement entre relations et specialement entre chaînes, *Ist. Naz. Alta Mat. (Bologna), Symp. Math.* **5** (1970), 203–251.
3. F. Galvin and K. Prikry, Borel sets and Ramsey's theorem, *J. Symbolic Logic* **38** (1973), 193–198.
4. G. Higman, Ordering by divisibility in abstract algebras, *Proc. London Math. Soc.* **2** (1952), 326–336.
5. J. Kruskal, The theory of well-partially-ordered sets. Ph.D. Thesis, Princeton Univ., 1954.
6. J. Kruskal, Well-quasi-ordering, the tree theorem, and Vazsonyi's conjecture, *Trans. Amer. Math. Soc.* **95** (1960), 210–225.
7. J. Kruskal, The theory of well-quasi-ordering: a frequently discovered concept, *J. Combinatorial Theory Ser. A* **13** (1972), 297–305.
8. R. Laver, On Fraïssé's order type conjecture, *Ann. of Math.* **93** (1971), 89–111.
9. R. Laver, An order type decomposition theorem, *Ann. of Math.* **98** (1973), 96–119.
10. R. Laver, Well-quasi-orderings and sets of finite sequences, *Proc. Camb. Phil. Soc.* **79** (1976), 1–10.
11. C. St. J. A. Nash-Williams, On well-quasi-ordering transfinite sequences, *Proc. Camb. Phil. Soc.* **61** (1965), 33–39.
12. C. St. J. A. Nash-Williams, On well-quasi-ordering infinite trees, *Proc. Camb. Phil. Soc.* **61** (1965), 697–720.
13. C. St. J. A. Nash-Williams, On better-quasi-ordering transfinite sequences, *Proc. Camb. Phil. Soc.* **64** (1968), 273–290.
14. R. Rado, Partial well ordering of sets of vectors, *Mathematika* **1** (1954), 89–95.
15. J. Silver, Every analytic set is Ramsey, *J. Symbolic Logic* **35** (1970), 60–64.

AMS (MOS) 1970 subject classifications: 02KO4, 02K06.

STUDIES IN FOUNDATIONS AND COMBINATORICS
ADVANCES IN MATHEMATICS SUPPLEMENTARY STUDIES, VOL. 1

Three Cryptoisomorphism Theorems

G. A. EDGAR

Department of Mathematics
Ohio State University
Columbus, Ohio

1. INTRODUCTION

The notion of topological space is not general enough to include some kinds of convergence that arise naturally in mathematics. The convergences for partially ordered sets in [1, Chapter 10] are examples. Real-valued measurable functions on the unit interval with "convergence almost everywhere" is another. In a complete Boolean algebra (or a σ-complete vector lattice), the Boolean operations (resp., vector lattice operations) are continuous with respect to order-star convergence, but need not be continuous with respect to any "reasonable" topology [1, Exercise 12, p. 250, and p. 362]. Other examples can be found, for example, in [2, 3, 14, 18]. In order to include such examples, several generalizations of the notion of a topological space have been proposed in the last 25 years. Many of these generalizations are equivalent (or "cryptoisomorphic" in the terminology of Birkhoff [1, p. 154]); it is hoped that the present paper will clarify this situation, as well as serve as an introduction to these generalizations.

Three similar theorems are stated below. Theorem 1, which concerns topological spaces, can be considered classical. Eight ways of defining the category of topological spaces are stated. Theorem 2 deals with pretopological spaces (in the terminology of [3]), and Theorem 3 with pseudotopological spaces ($= L^*$ spaces). Most (but not all) of the assertions of these theorems can be found in the literature. We believe that collecting them all in one place and giving a unified proof will serve a useful purpose.

Of the three theorems, only Theorem 2 will be proved here, since the proofs are all similar. As we shall see, the proof is a straightforward but lengthy verification. The proof given here is in terms of nonstandard analysis. It seems to be the most unified of the many proofs possible.

For further reading, we recommend the following papers. For filter-convergence spaces, see [3, 4, 6, 9]. For net-convergence spaces, see [14, 17, 18, 19]. Closure spaces are the primary topic of the gigantic book [2].

This work constitutes a portion of the author's Ph.D. thesis [5], written under the direction of G. Birkhoff.

ISBN 0-12-599101-0

2. Preliminaries

Terminology from set theory is as in [7]. Note also: ran(f) denotes the range of the function f. Also $\mathfrak{P}(X)$ denotes the power set of the set X; $\mathfrak{F}(X)$ denotes the set of all filters on the set X; and $\mathfrak{U}(X)$ denotes the set of all ultrafilters on the set X. We will write $\langle a, b\rangle$ for an ordered pair, and similarly $\langle a_i\rangle_{i\in I}$ or $\langle a_i : i\in I\rangle$. If $\langle A_i : i\in I\rangle$ is a family of sets, then $\prod_{i\in I} A_i$ denotes the cartesian product $\{\langle a_i\rangle_{i\in I} : a_i\in A_i \text{ for all } i\in I\}$, and $\coprod_{i\in I} A_i$ denotes the disjoint union $\{\langle i, a\rangle : i\in I, a\in A_i\}$.

Terminology from category theory is as in [13]. Note also: $\mathbf{S}$ denotes the category of sets. A *concrete category* is a category $\mathbf{C}$ together with a faithful functor $\mathscr{S}:\mathbf{C}\to\mathbf{S}$. Two concrete categories $\langle\mathbf{C}, \mathscr{S}\rangle$, $\langle\mathbf{D}, \mathscr{T}\rangle$ are said to be *isomorphic over* $\mathbf{S}$ iff there is an isomorphism of categories $\mathscr{F}:\mathbf{C}\to\mathbf{D}$ such that $\mathscr{S} = \mathscr{T}\mathscr{F}$. This is (one interpretation of) the *cryptoisomorphism* of Birkhoff [1, p. 154].

Terminology from topology is as in [8]. Note also: A *filtered set* is a set S together with a filter $\mathscr{S}$ on S. If $\langle S, \mathscr{S}\rangle$ is a filtered set and $f:S\to X$ is a function, we write $f(\mathscr{S})$ for the filter on X generated by the filterbase $\{f(A):A\in\mathscr{S}\}$. A function $\varphi:S\to T$ between filtered sets $\langle S, \mathscr{S}\rangle$ and $\langle T, \mathscr{T}\rangle$ is said to be a *filter map* iff $\varphi^{-1}(A)\in\mathscr{S}$ for every $A\in\mathscr{T}$. An *ultrafiltered set* is a filtered set $\langle S, \mathscr{S}\rangle$ such that $\mathscr{S}$ is an ultrafilter. A function $\mathbf{x}:S\to X$ defined on an ultrafiltered set is called an *ultranet*. An ultranet $\mathbf{y}$ is said to be a *subultranet* of an ultranet $\mathbf{x}$ iff there exists a filter map φ such that $\mathbf{y} = \mathbf{x}\circ\varphi$.

Terminology from nonstandard analysis is as in [15] or [12]. We fix at the beginning an enlargement of a portion of mathematics containing everything we are interested in. An object is assumed to be standard unless the contrary is indicated in some way. If P is an object in the original portion of mathematics, then $^*P$ denotes the object in the enlargement "with the same name;" the star is usually omitted for objects not being considered sets. The object P (or the corresponding object $^*P$) is called *standard* to distinguish it from those objects of the enlargement that do not arise in this manner, i.e., do not have "names."

If $\langle D, \leqslant\rangle$ is a directed set, then an *element $m\in {}^*D$ is said to be *infinite* iff $m\geqslant n$ for all $n\in D$ (i.e., for all standard n). The order relation is concurrent, so infinite *elements exist. If $\mathbf{x}:D\to X$ is a net, then the *tail* of $\mathbf{x}$ is defined to be

$$\tau(\mathbf{x}) = \{\mathbf{x}(m):m\in {}^*D \text{ is infinite}\}.$$

If $\mathscr{S}$ is a family of subsets of X, its *nucleus* [12] is

$$\mathrm{Nuc}(\mathscr{S}) = \bigcap\{{}^*A:A\in\mathscr{S}\}.$$

A *set $A\subseteq {}^*X$ is called *nuclear* iff it is the nucleus of some family of sets, or equivalently iff, for every $a\in {}^*X\backslash A$, there is a (standard) set $B\subseteq X$ such that $A \subseteq {}^*B$ and $a \notin {}^*B$. If $\langle S, \mathscr{S}\rangle$ is an ultrafiltered set and $\mathbf{x}: S \to X$ is an ultranet, then the *tail* of $\mathbf{x}$ is

$$\tau(\mathbf{x}) = \mathbf{x}(\mathrm{Nuc}(\mathscr{S})).$$

If $A \subseteq {}^*X$, then the *filter* of A is

$$\mathrm{Fil}(A) = \{B \subseteq X : {}^*B \supseteq A\}.$$

Now [12] $A \subseteq {}^*X$ is nuclear if and only if $A = \mathrm{Nuc}(\mathrm{Fil}(A))$; and $\mathscr{S} \subseteq \mathfrak{P}(X)$ is a filter if and only if $\mathscr{S} \neq \mathfrak{P}(X)$ and $\mathscr{S} = \mathrm{Fil}(\mathrm{Nuc}(\mathscr{S}))$. The smallest nuclear *set containing a *point $a \in {}^*X$ is $\hat{a} = \mathrm{Nuc}(\mathrm{Fil}(\{a\}))$. A *set $A \subseteq {}^*X$ is said to be *saturated* iff $\hat{a} \subseteq A$ whenever $a \in A$. (The nonstandard theory of filters is discussed further in [11 and 12], and the theory of nets in [5, Section 9, and 15, p. 97].)

3. Topology

For completeness, we begin with the definitions relevant to Theorem 1.

(a) A *topological space* is a set X together with a subset $\mathscr{T} \subseteq \mathfrak{P}(X)$ such that

(a1) $\varnothing, X \in \mathscr{T}$;
(a2) if $A, B \in \mathscr{T}$, then $A \cap B \in \mathscr{T}$;
(a3) if $\mathscr{S} \subseteq \mathscr{T}$, then $\bigcup \mathscr{S} \in \mathscr{T}$.

The elements of $\mathscr{T}$ are called *open sets*. A function $f: X \to Y$ between topological spaces is said to be *continuous* iff $f^{-1}(A)$ is open whenever $A \subseteq Y$ is open. Let $\mathbf{T}_{\mathrm{a}}$ denote the category of topological spaces and continuous maps.

(b) A *topological closure space* is a set X together with a function $\eta: \mathfrak{P}(X) \to \mathfrak{P}(X)$ such that

(b1) $A \subseteq \eta(A)$;
(b2) $\eta(\varnothing) = \varnothing$ and $\eta(A \cup B) = \eta(A) \cup \eta(B)$;
(b3) $\eta(\eta(A)) = \eta(A)$.

$\eta(A)$ is called the *closure* of the set A. A function $f: X \to Y$ between closure spaces is said to be *continuous* iff $f(\eta(A)) \subseteq \eta(f(A))$ for all $A \subseteq X$. Let $\mathbf{T}_{\mathrm{b}}$ denote the category of topological closure spaces and continuous maps.

(c) A *topological neighborhood space* is a set X together with a function $v: X \to \mathfrak{P}(\mathfrak{P}(X))$ such that

(c1) $a \in A$ for all $A \in v(a)$;
(c2) $v(a) \in \mathfrak{F}(X)$;

(c3) if $A \in \nu(a)$, then there exists $B \in \nu(a)$ such that $B \subseteq A$ and $B \in \nu(b)$ for all $b \in B$.

The elements of $\nu(a)$ are called *neighborhoods* of the point a. A function $f: X \to Y$ between neighborhood spaces is said to be *continuous* iff $f^{-1}(A)$ is a neighborhood of a whenever A is a neighborhood of $f(a)$. Let $\mathbf{T}_c$ denote the category of topological neighborhood spaces and continuous maps.

(d) A *topological net-convergence space* (cf. "convergence class" of Kelley [8, p. 73]) is a set X together with a relation $\to$ between nets in X and points of X such that

(d1) if $\mathbf{x}$ is a constant net with value a, then $\mathbf{x} \to a$;

(d2) $\mathbf{x} \to a$ if and only if every subnet $\mathbf{y}$ of $\mathbf{x}$ has a subnet $\mathbf{z}$ such that $\mathbf{z} \to a$;

(d3) if $\mathbf{x}: D \to X$ is a net such that $\mathbf{x} \to a$, and for each $n \in D$, if $\mathbf{y}_n: E_n \to X$ is a net such that $\mathbf{y}_n \to \mathbf{x}(n)$, then $\mathbf{z} \to a$, where $\mathbf{z}: D \times \prod_{n \in D} E_n \to X$ is the "iterated net" defined by $\mathbf{z}(n, f) = \mathbf{y}_n(f(n))$.

If $\mathbf{x} \to a$, then the net $\mathbf{x}$ is said to *converge* to the point a. A function $f: X \to Y$ between net-convergence spaces is said to be *continuous* iff $\mathbf{x} \to a$ implies $f \circ \mathbf{x} \to f(a)$. Let $\mathbf{T}_d$ denote the category of topological net-convergence spaces and continuous maps.

(e) A *topological filter-convergence space* (see [4]) is a set X together with a relation $\to$ between $\mathfrak{F}(X)$ and X such that

(e1) $\{A \in \mathfrak{P}(X): a \in A\} \to a$;

(e2) $\mathcal{U} \to a$ if and only if, for every filter $\mathcal{V} \supseteq \mathcal{U}$, there is a filter $\mathcal{W} \supseteq \mathcal{V}$ such that $\mathcal{W} \to a$;

(e3) if I is a set, $\psi: I \to X$, $\mathcal{U} \in \mathfrak{F}(I)$, $\psi(\mathcal{U}) \to a$, and for each $i \in I$, $\mathcal{V}_i \in \mathfrak{F}(X)$ and $\mathcal{V}_i \to \psi(i)$, then $\mathcal{W} \to a$, where $\mathcal{W}$ is the "compression"

$$\mathcal{W} = \left\{ \bigcup_{i \in J} A_i : J \in \mathcal{U},\ A_i \in \mathcal{V}_i \text{ for } i \in J \right\}.$$

If $\mathcal{U} \to a$, then the filter $\mathcal{U}$ is said to *converge* to the point a. A function $f: X \to Y$ between filter-convergence spaces is said to be *continuous* iff $\mathcal{U} \to a$ implies $f(\mathcal{U}) \to f(a)$. Let $\mathbf{T}_e$ denote the category of topological filtercon-gence spaces and continuous maps.

(f) A *topological ultranet-convergence space* [5] is a set X together with a relation $\to$ between ultranets in X and points of X such that

(f1) if $\mathbf{x}$ is a constant ultranet with value a, then $\mathbf{x} \to a$;

(f2) if $\mathbf{x}$ is a subultranet of $\mathbf{y}$, then $\mathbf{x} \to a$ if and only if $\mathbf{y} \to a$;

(f3) if $\mathbf{x} \to a$, where $\mathbf{x}$ is an ultranet on the ultrafiltered set $\langle S, \mathcal{S} \rangle$, and, for each $s \in S$, $\mathbf{y}_s \to \mathbf{x}(s)$, where $\mathbf{y}_s$ is an ultranet on $\langle T_s, \mathcal{T}_s \rangle$, then $\mathbf{z} \to a$, where $\mathbf{z}$ is the "iterated ultranet" defined by $\mathbf{z}(s, t) = \mathbf{y}_s(t)$ on the ultrafiltered set

$$\left\langle \coprod_{s \in S} T_s, \left\{ \coprod_{s \in A} B_s : A \in \mathscr{S}, B_s \in \mathscr{T}_s \text{ for } s \in A \right\} \right\rangle.$$

If $\mathbf{x} \to a$, then the ultranet $\mathbf{x}$ is said to *converge* to the point a. A function $f: X \to Y$ between ultranet-convergence spaces is said to be *continuous* iff $\mathbf{x} \to a$ implies $f \circ \mathbf{x} \to f(a)$. Let $\mathbf{T}_f$ denote the category of topological ultranet-convergence spaces and continuous maps.

(g) A *topological ultrafilter-convergence space* [3] is a set X together with a relation $\to$ between $\mathfrak{U}(X)$ and X such that

(g1) $\{A \in \mathfrak{P}(X): a \in A\} \to a$;

(g2) if I is a set, $\psi: I \to X$, $\mathscr{U} \in \mathfrak{U}(I)$, $\psi(\mathscr{U}) \to a$, and for each $i \in I$, $\mathscr{V}_i \in \mathfrak{U}(X)$ and $\mathscr{V}_i \to \psi(i)$, then $\mathscr{W} \to a$, where $\mathscr{W}$ is the "compression" ultrafilter

$$\mathscr{W} = \left\{ \bigcup_{i \in J} A_i : J \in \mathscr{U}, A_i \in \mathscr{V}_i \text{ for } i \in J \right\}.$$

If $\mathscr{U} \to a$, then the ultrafilter $\mathscr{U}$ is said to *converge* to the point a. A function $f: X \to Y$ between ultrafilter-convergence spaces is said to be *continuous* iff $\mathscr{U} \to a$ implies $f(\mathscr{U}) \to f(a)$. Let $\mathbf{T}_g$ denote the category of topological ultrafilter-convergence spaces and continuous maps.

(h) A *nonstandard topological space* (cf. [15, Chapter IV; 12, p. 27]) is a set X together with a function $\mu: X \to \mathfrak{P}(^*X)$ such that

(h1) $a \in \mu(a)$;

(h2) $\mu(a)$ is nuclear;

(h3) if $A \subseteq X$ and $^*A \supseteq \mu(a)$, then there exists $B \subseteq A$ such that $^*B \supseteq \mu(a)$ and $^*B \supseteq \mu(b)$ for all $b \in B$.

The *set $\mu(a)$ is called the *monad* of a. A function $f: X \to Y$ between nonstandard topological spaces is said to be *continuous* iff $f(\mu(a)) \subseteq \mu(f(a))$ for all $a \in X$. Let $\mathbf{T}_h$ denote the category of nonstandard topological spaces and continuous maps.

THEOREM 1. *The eight notions* (a)–(h) *are cryptoisomorphic in the sense that the concrete categories* $\mathbf{T}_a, \ldots, \mathbf{T}_h$ *are isomorphic over* $\mathbf{S}$.

For those parts of the proof that are not found in [8], imitate the proof of Theorem 2, below, or see the references quoted in parts (d) through (h), above.

4. Pretopology

The theorem on pretopological spaces is similar to Theorem 1. Certain definitions (such as the definitions of continuity) are not repeated.

(b) A *closure space* (Čech [2]; called "pré-adhérence" in [3]; see also [18] and "liaison space" in [16]) is a set X together with a function $\eta : \mathfrak{P}(X) \to \mathfrak{P}(X)$ such that

(b1) $A \subseteq \eta(A)$;

(b2) $\eta(\varnothing) = \varnothing$ and $\eta(A \cup B) = \eta(A) \cup \eta(B)$.

Let $\mathbf{P}_b$ denote the category of closure spaces and continuous maps.

(c) A *neighborhood space* (called "mehrstufige Topologie" in [10]; see also [6, 18]) is a set X together with a function $v : X \to \mathfrak{P}(\mathfrak{P}(X))$ such that

(c1) $a \in A$ for all $A \in v(a)$;

(c2) $v(a)$ is a filter on X.

Let $\mathbf{P}_c$ denote the category of neighborhood spaces and continuous maps.

(d) A *pretopological net-convergence space* (cf. [18, 17]) is a set X together with a relation $\to$ between nets in X and points of X such that

(d1) if $\mathbf{x}$ is a constant net with value a, then $\mathbf{x} \to a$;

(d2) $\mathbf{x} \to a$ if and only if, for every subnet $\mathbf{y}$ of $\mathbf{x}$, there is a net $\mathbf{z}$ in $\operatorname{ran}(\mathbf{y})$ such that $\mathbf{z} \to a$.

Let $\mathbf{P}_d$ denote the category of pretopological net-convergence spaces and continuous maps.

(e) A *principal-ideal filter-convergence space* [6] is a set X together with a relation $\to$ between $\mathfrak{F}(X)$ and X such that

(e1) $\{A \in \mathfrak{P}(X) : a \in A\} \to a$;

(e2) $\{\mathcal{U} \in \mathfrak{F}(X) : \mathcal{U} \to a\}$ is a principal dual ideal in $\mathfrak{F}(X)$.

Let $\mathbf{P}_e$ be the category of principal-ideal filter-convergence spaces and continuous maps.

(f) A *pretopological ultranet-convergence space* is a set X together with a relation $\to$ between ultranets in X and points of X such that

(f1) if $\mathbf{x}$ is a constant ultranet with value a, then $\mathbf{x} \to a$;

(f2) $\mathbf{x} \to a$ if and only if, for every subultranet $\mathbf{y}$ of $\mathbf{x}$, there is an ultranet $\mathbf{z}$ in $\operatorname{ran}(\mathbf{y})$ such that $\mathbf{z} \to a$.

Let $\mathbf{P}_f$ denote the category of pretopological ultranet-convergence spaces and continuous maps.

(g) A *pretopological space* [3] is a set X together with a relation $\rightarrow$ between $\mathfrak{U}(X)$ and X such that

(g1) $\{A \in \mathfrak{P}(X): a \in A\} \rightarrow a$;

(g2) $\mathscr{U} \rightarrow a$ if and only if, for every $A \in \mathscr{U}$, there exists $\mathscr{V} \in \mathfrak{U}(X)$ such that $A \in \mathscr{V} \rightarrow a$.

Let $\mathbf{P}_g$ denote the category of pretopological spaces and continuous maps.

(h) A *nonstandard pretopological space* is a set X together with a function $\mu: X \rightarrow \mathfrak{P}(^*X)$ such that

(h1) $a \in \mu(a)$;

(h2) $\mu(a)$ is nuclear.

Let $\mathbf{P}_h$ denote the category of nonstandard pretopological spaces and continuous maps.

THEOREM 2. *The seven notions* (b)–(h) *are cryptoisomorphic in the sense that the concrete categories* $\mathbf{P}_b, \ldots, \mathbf{P}_h$ *are isomorphic over* $\mathbf{S}$.

Certain parts of this theorem are proved in the literature. The equivalence of (essentially) (b), (c), and (d) is proved in [18], the equivalence of (b) and (c) in [2], the equivalence of (e) and (g) in [6], the equivalence of (b) and (g) in [3]. We give here a unified proof of the whole theorem.

Proof. We proceed by showing each of the categories isomorphic to the category $\mathbf{P}_h$ of nonstandard pretopological spaces.

(b) Let $\langle X, \mu \rangle \in \mathbf{P}_h$. For $A \subseteq X$, define

$$\eta(A) = \{a \in X: \mu(a) \cap {}^*A \neq \varnothing\}.$$

If $a \in A$, then $a \in \mu(a) \cap {}^*A$, so $a \in \eta(A)$. Thus $A \subseteq \eta(A)$. Also, $\eta(\varnothing) = \varnothing$ since ${}^*\varnothing = \varnothing$. Finally,

$$\begin{aligned}\eta(A \cup B) &= \{a \in X: \mu(a) \cap {}^*(A \cup B) \neq \varnothing\}\\ &= \{a \in X: (\mu(a) \cap {}^*A) \cup (\mu(a) \cap {}^*B) \neq \varnothing\}\\ &= \{a \in X: \mu(a) \cap {}^*A \neq \varnothing\} \cup \{a \in X: \mu(a) \cap {}^*B \neq \varnothing\}\\ &= \eta(A) \cup \eta(B).\end{aligned}$$

Thus $\langle X, \eta \rangle \in \mathbf{P}_b$; call it $\mathscr{F}(\langle X, \mu \rangle)$. Suppose $f: X \rightarrow Y$ is a continuous function of nonstandard pretopological spaces. Let $A \subseteq X$, and suppose $a \in \eta(A)$. Then there exists $b \in \mu(a) \cap {}^*A$. Now $f(b) \in f(\mu(a) \cap {}^*A) \subseteq f(\mu(a)) \cap f({}^*A) \subseteq \mu(f(a)) \cap {}^*(f(A))$, so that $f(a) \in \eta(f(A))$. Thus $f(\eta(A)) \subseteq \eta(f(A))$, so that f is a continuous map $\mathscr{F}(X) \rightarrow \mathscr{F}(Y)$. Therefore $\mathscr{F}$ is a functor $\mathbf{P}_h \rightarrow \mathbf{P}_b$.

Now let $\langle X, \eta\rangle \in \mathbf{P}_b$. For $a \in X$, define

$$\mu(a) = \bigcap\{{}^*A : A \subseteq X \text{ and } a \notin \eta(X \backslash A)\}.$$

If $a \notin A$, then $a \in X \backslash A \subseteq \eta(X \backslash A)$. Thus $a \in \mu(a)$. Also, $\mu(a)$ is nuclear since it is an intersection of standard sets. Thus $\langle X, \mu\rangle \in \mathbf{P}_h$; call it $\mathscr{G}(\langle X, \eta\rangle)$. Suppose $f: X \to Y$ is a continuous map of closure spaces. Let $a \in X$ and suppose $b \in \mu(a)$. Suppose $B \subseteq Y$ and $f(a) \notin \eta(Y \backslash B)$. Then, since $f(\eta(X \backslash f^{-1}(B))) \subseteq \eta(f(X \backslash f^{-1}(B))) \subseteq \eta(Y \backslash B)$, we have $f(b) \in B$. Therefore $f(\mu(a)) \subseteq \mu(f(a))$, and f is a continuous map $\mathscr{G}(X) \to \mathscr{G}(Y)$. Thus $\mathscr{G}$ is a functor $\mathbf{P}_b \to \mathbf{P}_h$.

Let $\langle X, \eta\rangle \in \mathbf{P}_b$, and let $\langle X, \eta'\rangle = \mathscr{F}\mathscr{G}(\langle X, \eta\rangle)$. Let $A \subseteq X$. If $a \in \eta'(A)$, then $\mu(a) \cap {}^*A \neq \varnothing$, so $\mu(a) \subseteq {}^*(X \backslash A)$, and thus $a \in \eta(A)$. Therefore $\eta'(A) \subseteq \eta(A)$. If $a \notin \eta'(A)$, then $\mu(a) \cap {}^*A = \varnothing$, so $\mu(a) \subseteq {}^*(X \backslash A)$, and thus $a \notin \eta(A)$. Therefore $\eta(A) \subseteq \eta'(A)$. This shows $\eta = \eta'$, so that $\mathscr{F}\mathscr{G}$ is an identity functor.

Let $\langle X, \mu\rangle \in \mathbf{P}_h$, and let $\langle X, \mu'\rangle = \mathscr{G}\mathscr{F}(\langle X, \mu\rangle)$. For $a \in X$,

$$\begin{aligned} \mu'(a) &= \bigcap\{{}^*A : A \subseteq X \text{ and } a \notin \eta(X \backslash A)\} \\ &= \bigcap\{{}^*A : A \subseteq X \text{ and } \mu(a) \cap {}^*(X \backslash A) = \varnothing\} \\ &= \bigcap\{{}^*A : A \subseteq X \text{ and } \mu(a) \subseteq {}^*A\} \\ &= \mu(a) \end{aligned}$$

since $\mu(a)$ is nuclear. Thus $\mathscr{G}\mathscr{F}$ is an identity functor.

(c) Let $\langle X, \mu\rangle \in \mathbf{P}_h$. For $a \in X$, define $v(a) = \mathrm{Fil}(\mu(a))$. Now $a \in \mu(a)$, so $a \in A$ for all $A \in v(a)$. Thus $\langle X, v\rangle \in \mathbf{P}_c$; call it $\mathscr{F}(\langle X, \mu\rangle)$. Suppose $f: X \to Y$ is a continuous map of nonstandard pretopological spaces. If $a \in X$, then for every $B \in v(f(a))$, we have ${}^*B \supseteq \mu(f(a)) \supseteq f(\mu(a))$, so ${}^*(f^{-1}(B)) = f^{-1}({}^*B) \supseteq \mu(a)$, and $f^{-1}(B) \in v(a)$. Therefore f is a continuous map $\mathscr{F}(X) \to \mathscr{F}(Y)$. Thus $\mathscr{F}$ is a functor $\mathbf{P}_h \to \mathbf{P}_c$.

Now let $\langle X, v\rangle \in \mathbf{P}_c$. For $a \in X$, define $\mu(a) = \mathrm{Nuc}(v(a))$. Clearly $a \in \mu(a)$, and $\mu(a)$ is nuclear, so $\mathscr{G}(\langle X, v\rangle) = \langle X, \mu\rangle \in \mathbf{P}_h$. Suppose $f: X \to Y$ is a continuous map of neighborhood spaces. If $a \in X$ and $b \in \mu(a)$, then $b \in {}^*A$ for every $A \in v(a)$. But then $b \in f^{-1}({}^*B)$ for every $B \in v(f(a))$, so $f(b) \in {}^*B$ for every $B \in v(f(a))$, i.e., $f(b) \in \mu(f(a))$. Hence $f(\mu(a)) \subseteq \mu(f(a))$, so f is a continuous map $\mathscr{G}(X) \to \mathscr{G}(Y)$. Thus $\mathscr{G}$ is a functor $\mathbf{P}_c \to \mathbf{P}_h$.

Since $\mathrm{Nuc}(\mathrm{Fil}(A)) = A$ for every nuclear *set A and $\mathrm{Fil}(\mathrm{Nuc}(\mathscr{S})) = \mathscr{S}$ for every filter $\mathscr{S}$ [12], we have $\mathscr{F}\mathscr{G}$ and $\mathscr{G}\mathscr{F}$ are identity functors.

(d) Let $\langle X, \mu\rangle \in \mathbf{P}_h$. If $\mathbf{x}$ is a net in X and $a \in X$, define $\mathbf{x} \to a$ iff $\tau(x) \subseteq \mu(a)$. If $\mathbf{x}$ is a constant net with value a, then $\tau(\mathbf{x}) = \{a\} \subseteq \mu(a)$, so $\mathbf{x} \to a$. If $\mathbf{x} \to a$ and $\mathbf{y}$ is a subnet of $\mathbf{x}$, then $\tau(\mathbf{y}) \subseteq \tau(\mathbf{x}) \subseteq \mu(a)$, so $\mathbf{y} \to a$. If $\mathbf{x} \not\to a$, then there exists A such that ${}^*A \supseteq \mu(a)$ and $\mathbf{x}$ is frequently in $X \backslash A$. Then $\mathbf{x}$ has a subnet $\mathbf{y}$ in $X \backslash A$; thus if $\mathbf{z}$ is a net in $\mathrm{ran}(\mathbf{y}) \subseteq X \backslash A$, $\mathbf{z} \not\to a$. Thus $\langle X, \to\rangle \in \mathbf{P}_d$; call it $\mathscr{F}(\langle X, \mu \rangle)$. If $f: X \to Y$ is a continuous map of nonstandard pretopological spaces, we have: if $\mathbf{x} \to a$, then $\tau(f \circ \mathbf{x}) = f(\tau(\mathbf{x})) \subseteq f(\mu(a)) \subseteq \mu(f(a))$, and therefore $f \circ \mathbf{x} \to f(a)$. Thus f is a continuous map $\mathscr{F}(X) \to \mathscr{F}(Y)$. Hence $\mathscr{F}$ is a functor $\mathbf{P}_{\mathrm{h}} \to \mathbf{P}_{\mathrm{d}}$.

Now let $\langle X, \to \rangle \in \mathbf{P}_{\mathrm{d}}$. For $a \in X$, define

$$\mu(a) = \bigcup \{\tau(x) : \mathbf{x} \to a\}.$$

Then $a \in \mu(a)$ since a constant net converges to its value. To show $\mu(a)$ is nuclear, suppose $b \notin \mu(a)$. Let $\mathbf{x}: D \to X$ be a net with $\tau(\mathbf{x}) = \hat{b}$ (e.g., $\mathbf{x}(A, a) = a$ on the directed set $E = \coprod_{A \in \mathrm{Fil}(\{b\})} A$, ordered by $\langle A_1, a_1 \rangle \geqslant \langle A_2, a_2 \rangle$ iff $A_1 \subseteq A_2$). Now $\tau(\mathbf{x}) \cap \mu(a) = \varnothing$ since $\mu(a)$ is saturated. There is a subnet $\mathbf{y}$ of $\mathbf{x}$ such that $\mathbf{z} \to a$ for no net $\mathbf{z}$ in ran($\mathbf{y}$), i.e., ${}^*(\mathrm{ran}(y)) \cap \mu(a) = \varnothing$. But $b \in \hat{b} = \tau(\mathbf{x}) = \tau(\mathbf{y}) \subseteq {}^*(\mathrm{ran}(\mathbf{y}))$. Thus $b \notin \bigcap \{{}^*A : A \subseteq X,\ \mu(a) \subseteq {}^*A\}$. This shows $\bigcap \{{}^*A : A \subseteq X,\ \mu(a) \subseteq {}^*A\} = \mu(a)$, so that $\mu(a)$ is nuclear. Write $\mathscr{G}(\langle X, \to \rangle) = \langle X, \mu \rangle \in \mathbf{P}_{\mathrm{h}}$. Now if $f: X \to Y$ is a continuous map of pretopological net-convergence spaces, we have:

$$\begin{aligned} f(\mu(a)) &\subseteq \bigcup \{\tau(f \circ \mathbf{x}) : \mathbf{x} \to a\} \\ &\subseteq \bigcup \{\tau(\mathbf{y}) : \mathbf{y} \to f(a)\} \\ &= \mu(f(a)). \end{aligned}$$

Thus $\mathscr{G}$ is a functor $\mathbf{P}_{\mathrm{d}} \to \mathbf{P}_{\mathrm{h}}$.

Let $\langle X, \mu \rangle \in \mathbf{P}_{\mathrm{h}}$, and $\langle X, \mu' \rangle = \mathscr{G}\mathscr{F}(\langle X, \mu \rangle)$. Now

$$\begin{aligned} \mu'(a) &= \bigcup \{\tau(\mathbf{x}) : \mathbf{x} \to a\} \\ &= \bigcup \{\tau(\mathbf{x}) : \mathbf{x} \text{ is a net with } \tau(\mathbf{x}) \subseteq \mu(a)\} \\ &\subseteq \mu(a). \end{aligned}$$

Conversely, if $b \in \mu(a)$, then there is a net $\mathbf{x}$ with $\tau(\mathbf{x}) = \hat{b}$, so $b \in \mu'(a)$. Thus $\mathscr{G}\mathscr{F}$ is an identity functor.

Finally, let $\langle X, \to \rangle \in \mathbf{P}_{\mathrm{d}}$, and $\langle X, \Rightarrow \rangle = \mathscr{F}\mathscr{G}(\langle X, \to \rangle)$. Now if $\mathbf{x} \Rightarrow a$, then $\tau(\mathbf{x}) \subseteq \mu(a) = \bigcup \{\tau(\mathbf{y}) : \mathbf{y} \to a\}$. Let $\mathbf{x}'$ be a subnet of $\mathbf{x}$; we have $\tau(\mathbf{x}') \subseteq \mu(a)$, so there is $\mathbf{y} \to a$ with $\tau(\mathbf{x}') \cap \tau(\mathbf{y}) \neq \varnothing$. But then $\mathbf{x}'$ and $\mathbf{y}$ have a common subnet [5], call it $\mathbf{x}''$; and $\mathbf{x}'' \to a$ since it is a subnet of $\mathbf{y}$. Now every subnet $\mathbf{x}'$ of $\mathbf{x}$ has a subnet $\mathbf{x}''$ with $\mathbf{x}'' \to a$, so $\mathbf{x} \to a$. Conversely, if $\mathbf{x} \to a$, then $\tau(\mathbf{x}) \subseteq \mu(a)$, so $\mathbf{x} \Rightarrow a$. Thus $\mathscr{F}\mathscr{G}$ is an identity functor.

(e) Let $\langle X, \mu \rangle \in \mathbf{P}_{\mathrm{h}}$. If $\mathscr{U} \in \mathfrak{F}(X)$ and $a \in X$, define $\mathscr{U} \to a$ iff $\mathrm{Nuc}(\mathscr{U}) \subseteq \mu(a)$. Now $\mathrm{Nuc}(\{A \in \mathfrak{P}(X) : a \in A\}) = \{a\} \subseteq \mu(a)$, so $\{A \in \mathfrak{P}(X) : a \in A\} \to a$. Since $\mu(a)$ is nuclear, $\mathrm{Nuc}(\mathscr{U}) \subseteq \mu(a)$ if and only if $\mathrm{Fil}(\mu(a)) \subseteq \mathscr{U}$, so $\{\mathscr{U} \in \mathfrak{F}(X) : \mathscr{U} \to a\}$ is the principal dual ideal in $\mathfrak{F}(X)$ generated by the filter $\mathrm{Fil}(\mu(a))$. Thus $\langle X, \to \rangle \in \mathbf{P}_{\mathrm{e}}$; call it $\mathscr{F}(\langle X, \mu \rangle)$. If $f: X \to Y$ is a continuous map of nonstandard pretopological spaces, and if $\mathscr{U} \to a$ in X, then

$\mathrm{Nuc}(f(\mathscr{U})) = f(\mathrm{Nuc}(\mathscr{U})) \subseteq f(\mu(a)) \subseteq \mu(f(a))$, so that $f(\mathscr{U}) \to f(a)$. Thus f is a continuous map $\mathscr{F}(X) \to \mathscr{F}(Y)$. Hence $\mathscr{F}$ is a functor $\mathbf{P}_{\mathrm{h}} \to \mathbf{P}_{\mathrm{e}}$.

Now let $\langle X, \to\rangle \in \mathbf{P}_{\mathrm{e}}$. For $a \in X$, define

$$\mu(a) = \bigcup\{\mathrm{Nuc}(\mathscr{U}) : \mathscr{U} \to a\}.$$

Now $\{A \in \mathfrak{P}(X) : a \in A\} \to a$, so $a \in \mu(a)$. Also $\{\mathscr{U} : \mathscr{U} \to a\}$ is a principal dual ideal in $\mathfrak{F}(X)$, say generated by $\mathscr{V}$. Thus $\mu(a) = \mathrm{Nuc}(\mathscr{V})$, so that $\mu(a)$ is nuclear. Thus $\langle X, \mu\rangle \in \mathbf{P}_{\mathrm{h}}$; write $\mathscr{G}(\langle X, \to\rangle) = \langle X, \mu\rangle$. If $f : X \to Y$ is a continuous map of principal-ideal filter-convergence spaces, then $f(\mu(a)) \subseteq \bigcup\{f(\mathrm{Nuc}(\mathscr{U})) : \mathscr{U} \to a\} = \bigcup\{\mathrm{Nuc}(f(\mathscr{U})) : \mathscr{U} \to a\} \subseteq \bigcup\{\mathrm{Nuc}(\mathscr{V}) : \mathscr{V} \to f(a)\} = \mu(f(a))$. Thus $\mathscr{G}$ is a functor $\mathbf{P}_{\mathrm{e}} \to \mathbf{P}_{\mathrm{h}}$.

Let $\langle X, \mu\rangle \in \mathbf{P}_{\mathrm{h}}$, and $\langle X, \mu'\rangle = \mathscr{G}\mathscr{F}(\langle X, \mu\rangle)$. Now $\mu'(a) = \bigcup\{\mathrm{Nuc}(\mathscr{U}) : \mathscr{U} \to a\} = \bigcup\{\mathrm{Nuc}(\mathscr{U}) : \mathrm{Nuc}(\mathscr{U}) \subseteq \mu(a)\} \subseteq \mu(a)$. Also $\{A \in \mathfrak{P}(X) : {}^*A \supseteq \mu(a)\} \to a$, so $\mu'(a) \supseteq \mu(a)$. Thus $\mu = \mu'$, so $\mathscr{G}\mathscr{F}$ is an identity functor.

Finally, let $\langle X, \to\rangle \in \mathbf{P}_{\mathrm{e}}$, and $\langle X, \Rightarrow\rangle = \mathscr{F}\mathscr{G}(\langle X, \to\rangle)$. Now if $\mathscr{U} \to a$, then $\mu(a) \supseteq \mathrm{Nuc}(\mathscr{U})$, so $\mathscr{U} \Rightarrow a$. Conversely, if $\mathscr{U} \Rightarrow a$, then $\mathrm{Nuc}(\mathscr{U}) \subseteq \mu(a)$. But $\{\mathscr{V} \in \mathfrak{F}(X) : \mathscr{V} \to a\}$ is the principal dual ideal in $\mathfrak{F}(X)$ generated by $\{A \in \mathfrak{P}(A) : {}^*A \supseteq \mu(a)\}$, so $\mathscr{U} \to a$. Thus $\mathscr{F}\mathscr{G}$ is an identity functor.

(f) The definitions are: $\mathbf{x} \to a$ iff $\tau(\mathbf{x}) \subseteq \mu(a)$, and $\mu(\mathrm{a}) = \bigcup\{\tau(\mathbf{x}) : \mathbf{x} \to a\}$; the proof is like that of (d) and (e), above.

(g) The definitions are: $\mathscr{U} \to a$ iff $\mathrm{Nuc}(\mathscr{U}) \subseteq \mu(a)$, and $\mu(a) = \bigcup\{\mathrm{Nuc}(\mathscr{U}) : \mathscr{U} \to a\}$; and the proof is like that of (d) and (e), above.

5. Pseudotopology

Finally, we discuss pseudotopological spaces ($= L^*$ spaces).

(d) An *L^* space* (Ordman [14]; called limit* space in Taylor [18]) is a set X together with a relation $\to$ between nets in X and points of X such that

(d1) if $\mathbf{x}$ is a constant net with value a, then $\mathbf{x} \to a$;

(d2) $\mathbf{x} \to a$ if and only if every subnet $\mathbf{y}$ of $\mathbf{x}$ has a subnet $\mathbf{z}$ such that $\mathbf{z} \to a$.

Let $\mathbf{M}_{\mathrm{d}}$ denote the category of L^* spaces and continuous maps.

(e) A *pseudotopological filter-convergence space* (see [4]) is a set X together with a relation $\to$ between $\mathfrak{F}(X)$ and X such that

(e1) $\{A \in \mathfrak{P}(X) : a \in A\} \to a$;

(e2) $\mathscr{U} \to a$ if and only if, for every filter $\mathscr{V} \supseteq \mathscr{U}$, there is a filter $\mathscr{W} \supseteq \mathscr{V}$ such that $\mathscr{W} \to a$.

Let $\mathbf{M}_{\mathrm{e}}$ denote the category of pseudotopological filter-convergence spaces and continuous maps.

(f) A *pseudotopological ultranet-convergence space* is a set X together with a relation $\rightarrow$ between ultranets in X and points of X satisfying

(f1) if $\mathbf{x}$ is a constant ultranet with value a, then $\mathbf{x} \rightarrow a$;

(f2) if $\mathbf{x}$ is a subultranet of $\mathbf{y}$, then $\mathbf{x} \rightarrow a$ if and only if $\mathbf{y} \rightarrow a$.

Let $\mathbf{M}_{\mathrm{f}}$ denote the category of pseudotopological ultranet-convergence spaces and continuous maps.

(g) A *pseudotopological space* (Choquet [3]) is a set X together with a relation $\rightarrow$ between $\mathfrak{U}(X)$ and X such that

(g1) $\{A \in \mathfrak{P}(X): a \in A\} \rightarrow a$.

Let $\mathbf{M}_{\mathrm{g}}$ denote the category of pseudotopological spaces and continuous maps.

(h) A *nonstandard pseudotopological space* is a set X together with a function $\mu: X \rightarrow \mathfrak{P}(^*X)$ such that

(h1) $a \in \mu(a)$;

(h2) $\mu(a)$ is saturated.

Let $\mathbf{P}_{\mathrm{h}}$ be the category of nonstandard pseudotopological spaces and continuous maps.

THEOREM 3. *The five notions* (d)–(h) *are cryptoisomorphic in the sense that the concrete categories* $\mathbf{M}_{\mathrm{d}}, \ldots, \mathbf{M}_{\mathrm{h}}$ *are isomorphic over* $\mathbf{S}$.

REFERENCES

1. G. BIRKHOFF, "Lattice Theory," 3rd ed., *Amer. Math. Soc.*, Providence, Rhode Island, 1967.
2. E. ČECH, "Topological Spaces," Czechoslovak Academy of Sciences, Prague, 1966.
3. G. CHOQUET, Convergences, *Ann. Univ. Grenoble. Sér. Sci.* **23** (1948), 57–112.
4. C. H. COOK AND H. R. FISCHER, Regular convergence spaces, *Math. Ann.* **174** (1967). 1–7.
5. G. EDGAR, Convergence from an algebraic point of view. Ph.D. Thesis, Harvard Univ., 1973.
6. H. R. FISCHER, Limesräume, *Math. Ann.* **137** (1959), 269–303.
7. P. R. HALMOS, "Naive Set Theory," Van Nostrand–Reinhold, Princeton, New Jersey, 1965.
8. J. L. KELLEY, "General Topology," Van Nostrand–Reinhold, Princeton, New Jersey, 1955.
9. D. C. KENT, Convergence functions and their related topologies, *Fund. Math.* **54** (1964), 125–133.
10. H.-J. KOWALSKY, Beiträge zur topologischen Algebra, *Math. Nachr.* **11** (1954), 143–186.
11. W. A. J. LUXEMBURG, A general theory of monads, *in* "Applications of Model Theory to Algebra, Analysis, and Probability" (W. A. J. Luxemburg, ed.), Holt, New York, 1969.
12. M. MACHOVER AND J. HIRSCHFELD, "Lectures on Nonstandard Analysis," Lecture Notes in Mathematics No. 94, Springer–Verlag, Berlin and New York, 1969.
13. S. MACLANE, "Category Theory for the Working Mathematician," Springer–Verlag, Berlin and New York, 1971.

14. E. T. Ordman, Convergence and abstract spaces in functional analysis, *J. Undergraduate Math.* **1** (1969) 79–96.
15. A. Robinson, "Non-standard Analysis," North-Holland Publ., Amsterdam, 1966.
16. C. Silva Rehrmann, Espaces de liaison, *Ciencias Ser. 1 Mat.* (1970), 13 pp.
17. W. Taylor, Convergence in relational structures, *Math. Ann.* **186** (1970) 215–227.
18. J. W. Tukey, "Convergence and Uniformity in Topology," Annals of Mathematics Study No. 2, Princeton Univ. Press, Princeton, New Jersey, 1940.
19. K. Wichterle, On *W*-convergence spaces, *Czechoslovak Math. J.* **18** (1968), 569–588.

AMS (MOS) 1970 subject classifications: 54A05, 54A20.

STUDIES IN FOUNDATIONS AND COMBINATORICS
ADVANCES IN MATHEMATICS SUPPLEMENTARY STUDIES, VOL. 1

Topological Duality for Prevarieties of Universal Algebras

BRIAN A. DAVEY

Department of Mathematics
La Trobe University
Bundoora, Victoria Australia

Many duality theorems for categories of universal algebras arise in the following manner: The category **A** is the prevariety generated by an algebra A which has a compact topology compatible with its operations (for example, if A is the circle group, the two-element Boolean algebra, or the two-element meet-semilattice with unit, then **A** is, respectively, the category of abelian groups, the category of Boolean algebras, or the category of meet-semilattices with unit): the category **X**, which is dual to **A**, is a category of compact topological spaces with some additional structure (for the examples given above **X** is, respectively, the category of compact topological abelian groups, the category of Boolean spaces, or the category of totally disconnected compact topological meet-semilattices with unit); in some sense the object A lies both in the category **A** and the category **X**, and the duality between **A** and **X** is provided by the Hom-set functors

$$\mathbf{A}(-, A):\mathbf{A} \to \mathbf{X}^{\mathrm{op}} \qquad \text{and} \qquad \mathbf{X}(-, A):\mathbf{X}^{\mathrm{op}} \to \mathbf{A}.$$

The study of dualities of this type is the central theme of this work. Of course, duality theories have been studied from a purely categorical point of view (see [21, 27, 28, 32, 34]), but the gap between the general functorial setting and concrete adjunctions or duality theories is frequently vast; as a rule it is a nontrivial matter to bridge this gap. Our aim is to tread a path between the abstract functorial theories and the concrete individual duality theories known in the literature. The path is heavily biased in favour of algebra, for indeed this work was originally motivated by a desire to use topological methods in order to answer algebraic questions.

The preliminaries are presented in Section 1. In Section 2 the general theory is presented: We assume that **X** is a category with a faithful functor into the category of compact topological spaces; under certain natural restrictions on **X**, necessary and sufficient conditions are found for **X** to be dual to **A**. The results of Section 2 are applied in Section 3 to the case where

ISBN 0-12-599101-0

$\mathbf{X}$ is a category of compact topological partial algebras derived in a natural way from the category $\mathbf{A}$. The approach developed in Section 3 is utilized in Section 4 to provide proofs of Pontryagin's duality for abelian groups, Stone's duality for Boolean algebras, and the duality for meet-semilattices with unit (and zero). Proofs are also obtained for two more recent duality theorems: the duality for an equational class generated by a primal algebra due to Hu, and the duality for the category of distributive lattices with unit (and zero) due to Priestley. The section closes with the presentation of a new duality theorem for the category of Stone algebras.

1. Preliminaries

Our references for category theory, universal algebra, and lattice theory will be MacLane [35], Grätzer [15], and Grätzer [17] respectively; for our general topological requirements we refer to Dugundji [9], and for a discussion of Boolean spaces we refer to Halmos [20].

If $\mathbf{A}$ is a class of similar algebras, then by the category $\mathbf{A}$ we mean the category whose objects are the algebras in $\mathbf{A}$ and whose morphisms are all of the homomorphisms between algebras in $\mathbf{A}$, that is $\mathbf{A}(A, B) = \mathrm{Hom}(A, B)$ for all $A, B \in \mathbf{A}$. We shall make no notational distinction between an algebra (or a partial algebra) and its underlying set.

If f is an m-ary operation on a partial algebra A, then

$$\mathscr{D}(f, A) = \{\langle a_0, \ldots, a_{m-1}\rangle \in A^m : f(a_0, \ldots, a_{m-1}) \text{ is defined}\}$$

is called the *domain of f in A*.

Let A be a partial algebra whose base set is endowed with a Hausdorff topology. Then A is a *topological partial algebra* if each operation on A is continuous on its domain; that is, if f is an m-ary operation on A, $\langle a_0, \ldots, a_{m-1}\rangle \in \mathscr{D}(f, A)$, and U is an open neighborhood of $f(a_0, \ldots, a_{m-1})$, then there exist open neighborhoods $U_0, \ldots, U_{m-1}$ of $a_0, \ldots, a_{m-1}$ respectively such that

$$f(b_0, \ldots, b_{m-1}) \in U \quad \text{for all} \quad \langle b_0, \ldots, b_{m-1}\rangle \in (U_0 \times \cdots \times U_{m-1}) \cap \mathscr{D}(f, A).$$

Let $\mathbf{I}$, $\mathbf{H}$, $\mathbf{S}$, and $\mathbf{P}$ denote the usual class operators corresponding respectively to the formation of all isomorphic copies, homomorphic images, subalgebras, and direct products. We shall call a class $\mathbf{A}$ of similar algebras a *prevariety* if $\mathbf{A} = \mathbf{ISP}(\mathbf{A})$. Since we allow empty products it follows that a prevariety $\mathbf{A}$ contains the trivial (i.e., one-element) algebras. Clearly the prevariety generated by an algebra A is $\mathbf{A} = \mathbf{ISP}(\{A\})$. If $\mathbf{A}$ is a prevariety then, by [15, Corollary 1, p. 167], all *free algebras* exist in $\mathbf{A}$; for any ordinal κ, let $\mathscr{F}_{\mathbf{A}}(\kappa)$ denote the free algebra in $\mathbf{A}$ with free generators $\{x_\gamma : \gamma < \kappa\}$.

Let **A** and **B** be categories. If there is a faithful functor $|\text{-}|:\mathbf{A} \to \mathbf{B}$, then **A** *is grounded in* **B** and $|\text{-}|$ is called a *grounding*. The category **A** is naturally grounded if there exists $A_1 \in \mathbf{A}$ such that $\mathbf{A}(A_1, \text{-}):\mathbf{A} \to \mathbf{Set}$ is a grounding. If $F:\mathbf{B} \to \mathbf{A}$ is left adjoint to the grounding $|\text{-}|:\mathbf{A} \to \mathbf{B}$, then F is called a **B**-*free functor* for **A**; if $\mathbf{B} = \mathbf{Set}$, then F is simply called a *free functor*. Forgetful functors are the most accessible examples of groundings, and the formation of free algebras in a prevariety is a typical example of a free functor. If $F:\mathbf{Set} \to \mathbf{A}$ is a free functor for a category **A**, then for all $A \in \mathbf{A}$,

$$\mathbf{A}(F(\{\varnothing\}), A) \simeq \mathbf{Set}(\{\varnothing\}, |A|) \simeq |A|,$$

and hence $A_1 = F(\{\varnothing\})$ provides a natural grounding naturally isomorphic to $|\text{-}|$; in particular, if **A** is a prevariety, then $A_1 = \mathscr{F}_{\mathbf{A}}(1)$ provides a natural grounding naturally isomorphic to the forgetful functor.

If $|\text{-}|:\mathbf{A} \to \mathbf{Set}$ is a grounding, then a morphism $g \in \mathbf{A}$ is an **A**-*injection* (**A**-*surjection*) or simply an *injection* (*surjection*) provided $|g|$ is one-to-one (onto). Clearly, injections are monic and surjections are epic, and if **A** is naturally grounded, then g is an injection if and only if it is monic.

Suppose that **A** is grounded in **Set**; let $\mathscr{I}$ be a class of **A**-injections and let $\mathscr{S}$ be a class of **A**-surjections. An object $A \in \mathbf{A}$ is said to be $\mathscr{I}$-injective in **A** if for all $g:B \to C$ in $\mathscr{I}$ and each $h \in \mathbf{A}(B, A)$ there exists $h' \in \mathbf{A}(C, A)$ such that $gh' = h$, and A is said to be $\mathscr{S}$-*projective* in **A** if for all $g:C \to B$ in $\mathscr{S}$ and each $h \in \mathbf{A}(A, B)$ there exists $h' \in \mathbf{A}(A, C)$ such that $h'g = h$. If A is $\mathscr{S}$-projective in **A** where $\mathscr{S}$ is the class of all **A**-surjections, then we say that A is *sur-projective* in **A**. If **A** is a category of universal algebras and A is $\mathscr{I}$-injective where $\mathscr{I}$ is the class of all **A**-injections (i.e., embeddings), then we drop the prefix and say that A is *injective* in **A**.

A topological space will always be denoted by its underlying set. The category of compact spaces and continuous maps is denoted by **Comp**, and the full subcategory of Boolean spaces, that is, zero-dimensional compact spaces, or equivalently totally disconnected compact spaces, is denoted by **ZComp**. The Hom-sets $\mathbf{Comp}(X, Y)$ and $\mathbf{ZComp}(X, Y)$ are both abbreviated to $\mathscr{C}(X, Y)$. As usual, the Stone–Čech compactification functor is denoted by $\beta:\mathbf{Set} \to \mathbf{ZComp}(\mathbf{Comp})$.

Since we often work with product spaces we introduce the following notation: If $U \subseteq X_\lambda$, then

$$\langle \lambda; U \rangle = \{a \in \textstyle\prod(X_\gamma | \gamma < \kappa): a_\lambda \in U\};$$

similarly, if $U \subseteq A$ and $b \in B$, then

$$\langle b; U \rangle = \{g \in A^B: bg \in U\}.$$

Hence, if U is open in X_λ, then $\langle \lambda; U \rangle$ is a subbasic open set in $\prod(X_\gamma : \gamma < \kappa)$, and if U is open in A, then $\langle b; U \rangle$ is a subbasic open set in A^B.

Finally, a continuous map $\phi \in \mathscr{C}(X^{\kappa}, X)$ has a *finite support* if it factors through a finite power of X; i.e., there exist $\gamma_0, \ldots, \gamma_{n-1} < \kappa$ such that for all $a, b \in X^{\kappa}$, $a_{\gamma_j} = b_{\gamma_j}$ $(j < n)$ implies that $a\phi = b\phi$.

2. Dualities via Structured Compact Spaces

The concept of structure referred to here is an intuitive one; formally, a category of structured compact spaces will be one that is grounded in **Comp**.

Since we are primarily concerned with representing algebras as algebras of continuous functions our emphasis is on dualities, rather than full dualities, in the following sense.

2.1 Definition. Let **A** and **X** be categories and assume that $D: \mathbf{A} \to \mathbf{X}^{\mathrm{op}}$ is left adjoint to $E: \mathbf{X}^{\mathrm{op}} \to \mathbf{A}$. Then $\langle D, E\rangle$ is a *duality* (*between* **A** *and* **X**) if the unit $\eta: \mathrm{id}_{\mathbf{A}} \to ED$ of the adjunction is a natural isomorphism, and is a *full duality* if the counit $\epsilon: \mathrm{id}_{\mathbf{X}} \to DE$ is also a natural isomorphism.

The following result, which is due to Hofmann and Keimel [23], says, in essense, that when considering a pair $\langle D, E\rangle$ of adjoint contravariant functors between "concrete" categories **A** and **X**, one necessarily has an object A that lies in both categories, D and E are given (up to a natural isomorphism) by Hom-functors in the contravariant argument, and $D(\text{-}) = \mathbf{A}(\text{-}, A)$ and $E(\text{-}) = \mathbf{X}(\text{-}, A)$ inherit their structure from an appropriate power of A.

2.2 Proposition. *Let* **A** *and* **X** *be naturally grounded categories and suppose that* $D: \mathbf{A} \to \mathbf{X}^{\mathrm{op}}$ *is left adjoint to* $E: \mathbf{X}^{\mathrm{op}} \to \mathbf{A}$. *Let*

$$|\text{-}|_{\mathbf{A}} = \mathbf{A}(A_1, \text{-}) \qquad \text{and} \qquad |\text{-}|_{\mathbf{X}} = \mathbf{X}(X_1, \text{-})$$

be the grounding functors and set $A = EX_1$ *and* $A^0 = DA_1$. *Then*

(i) $|A|_{\mathbf{A}} \simeq |A^0|_{\mathbf{X}}$,
(ii) $|D(\text{-})|_{\mathbf{X}} \simeq \mathbf{A}(\text{-}, A)$ *and* $|E(\text{-})|_{\mathbf{A}} \simeq \mathbf{X}(\text{-}, A^0)$, *and*
(iii) *if* **X** *has products, then there exists a monic natural transformation*

$$D \to (A^0)^{|\text{-}|_{\mathbf{A}}} : \mathbf{A} \to \mathbf{X}^{\mathrm{op}}.$$

The development in this, and the next, section should be viewed in the light of this result.

The proof of the next lemma is straightforward and so is omitted.

2.3 Lemma. *Let A be a topological algebra.*

(i) *For every topological space* X, $\mathscr{C}(X, A)$ *is a subalgebra of* A^X.
(ii) *For every algebra B of the same type as A,* $\mathrm{Hom}(B, A)$ *is a closed subspace of* A^B.

For the remainder of this section we assume that A is a nontrivial compact topological algebra and that $\mathbf{A} = \mathbf{ISP}(\{A\})$ is the prevariety generated by A. In view of Lemma 2.3, for all $B \in \mathbf{A}$, $\mathbf{A}(B, A)$ is a compact space, and if A is finite, then $\mathbf{A}(B, A)$ is a Boolean space. Thus it is easily seen that $\mathbf{A}(\text{-}, A)\colon \mathbf{A} \to \mathbf{Comp}^{\mathrm{op}}$ is a well-defined functor, where $g\mathbf{A}(h, A) = hg$, and $\mathscr{C}(\text{-}, A)\colon \mathbf{Comp}^{\mathrm{op}} \to \mathbf{A}$ is a well-defined functor, where $\phi\mathscr{C}(\psi, A) = \psi\phi$.

For each $B \in \mathbf{A}$ define $\eta_B\colon B \to \mathscr{C}(\mathbf{A}(B, A), A)$ by $b\eta_B = \Gamma_b$, where $g\Gamma_b = bg$ for all $g \in \mathbf{A}(B, A)$; for each $X \in \mathbf{Comp}$ define $\epsilon_X\colon X \to \mathbf{A}(\mathscr{C}(X, A), A)$ by $x\epsilon_X = \Gamma_x$, where $\phi\Gamma_x = x\phi$ for all $\phi \in \mathscr{C}(X, A)$.

2.4 PROPOSITION. *$\langle \mathbf{A}(\text{-}, A), \mathscr{C}(\text{-}, A); \eta, \epsilon\rangle$ is an adjunction from $\mathbf{A}$ to $\mathbf{Comp}^{\mathrm{op}}$.*

Proof. By [35, Theorem 2, p. 81] it is sufficient to prove that η is a well-defined natural transformation and that, for all $B \in \mathbf{A}$, $\eta_B\colon B \to \mathscr{C}(\mathbf{A}(B, A), A)$ is universal to $\mathscr{C}(\text{-}, A)$ from B.

If U is open in A, then $U(b\eta_B)^{-1} = \langle b; U\rangle \cap \mathbf{A}(B, A)$ and hence $b\eta_B$ is continuous for all $b \in B$. Furthermore, for each (m-ary) operation f,

$$\begin{aligned} g(f(b_0, \ldots, b_{m-1})\eta_B) &= f(b_0, \ldots, b_{m-1})g \\ &= f(b_0 g, \ldots, b_{m-1} g) \\ &= f(g(b_0\eta_B), \ldots, g(b_{m-1}\eta_B)) \\ &= gf(b_0\eta_B, \ldots, b_{m-1}\eta_B), \end{aligned}$$

and so η_B is a homomorphism. To see that η is a natural transformation we must show that if $h \in \mathbf{A}(B, C)$, then $\eta_B\mathscr{C}(\mathbf{A}(h, A), A) = h\eta_C$. But if $b \in B$ and $g \in \mathbf{A}(C, A)$, then

$$g(b\eta_B\mathscr{C}(\mathbf{A}(h, A), A)) = g(\mathbf{A}(h, A)b\eta_B) = (hg)b\eta_B = b(hg) = (bh)g = g(bh\eta_C),$$

as required.

Now let $B \in \mathbf{A}$, let $X \in \mathbf{Comp}$, and let $\alpha\colon B \to \mathscr{C}(X, A)$ be a homomorphism. The unique fill-in map $\beta\colon X \to \mathbf{A}(B, A)$ such that $\eta_B\mathscr{C}(\beta, A) = \alpha$ is defined by $b(x\beta) = x(b\alpha)$. It is easily checked that for all $x \in X$, $x\beta\colon B \to A$ is a homomorphism and thus β is well defined. It remains to prove that β is continuous; but for any open subset U of A, $\langle b; U\rangle\beta^{-1} = U(b\alpha)^{-1}$, which is open in X since $b\alpha$ is continuous. ∎

If A is finite we may restrict the codomain of $\mathbf{A}(\text{-}, A)$ and the domain of $\mathscr{C}(\text{-}, A)$ to the category $\mathbf{ZComp}$ of Boolean spaces.

2.5 COROLLARY. *If A is finite, then $\langle \mathbf{A}(\text{-}, A), \mathscr{C}(\text{-}, A); \eta, \epsilon\rangle$ is an adjunction from $\mathbf{A}$ to $\mathbf{ZComp}^{\mathrm{op}}$.*

2.6 PROPOSITION. (i) *If* $h \in \mathbf{A}(B, C)$ *is a surjection, then* $\mathbf{A}(h, A): \mathbf{A}(C, A) \to \mathbf{A}(B, A)$ *is a homeomorphism onto a closed subspace.*

(ii) *For all* $\kappa > 0$ *the map* $\rho_\kappa: \mathbf{A}(\mathscr{F}_\mathbf{A}(\kappa), A) \to A^\kappa$, *defined by* $g\rho_\kappa = \langle x_\gamma g \rangle_{\gamma < \kappa}$, *is a homeomorphism.*

(iii) *Assume that* $B \in \mathbf{A}$ *is generated by* $\{b_\gamma : \gamma < \kappa\}$ *and let* $h: \mathscr{F}_\mathbf{A}(\kappa) \to B$ *be the homomorphism determined by* $x_\gamma h = b_\gamma$ $(\gamma < \kappa)$. *Then* $\mathbf{A}(h, A)\rho_\kappa: \mathbf{A}(B, A) \to A^\kappa$ *is a homeomorphism onto a closed subspace.*

Proof. (i) Clearly $\mathbf{A}(h, A)$ is one-to-one and since $\mathbf{A}(C, A)$ is compact and $\mathbf{A}(B, A)$ is Hausdorff we need only prove that $\mathbf{A}(h, A)$ is continuous. But if U is open in A and $b \in B$, then $\langle b; U \rangle \mathbf{A}(h, A)^{-1} = \langle bh; U \rangle$, which is open in $\mathbf{A}(C, A)$.

(ii) By the definition of a free algebra it follows that ρ_κ is a bijection. Again it is sufficient to prove that ρ_κ is continuous; but since $\langle \gamma; U \rangle \rho_\kappa^{-1} = \langle x_\gamma; U \rangle$ for every open set U in A, this is immediate.

(iii) This follows directly from (i) and (ii). ∎

2.7 *Remark.* In the sequel we shall abbreviate $\rho_1: \mathbf{A}(\mathscr{F}_\mathbf{A}(1), A) \to A$ to ρ.

When one sets out to find a concrete dual category $\mathbf{X}$ for a prevariety $\mathbf{A}$, often the first step is to find a proto-dual in the following sense.

2.8 DEFINITION. Let $D: \mathbf{A} \to \mathbf{X}^{\mathrm{op}}$ be a functor into a category $\mathbf{X}$ that has a grounding $|-|: \mathbf{X} \to \mathbf{Comp}$ into the category of compact spaces. Then $\langle D, \mathbf{X}, |-| \rangle$ *is a proto-dual of* $\mathbf{A}$ if

(i) $|D(\text{-})| = \mathbf{A}(\text{-}, A)$, and

(ii) for all $X \in \mathbf{X}$, $E(X) = \{|\phi|\rho : \phi \in \mathbf{X}(X, D\mathscr{F}_\mathbf{A}(1))\}$ is a subalgebra of $A^{|X|}$ (and therefore is a subalgebra of $\mathscr{C}(|X|, A)$).

2.9 *Remark.* By analogy with Proposition 2.2 we denote $D\mathscr{F}_\mathbf{A}(1)$ by A^0. Note that, by Proposition 2.6, $|A^0| = \mathbf{A}(\mathscr{F}_\mathbf{A}(1), A)$ is homeomorphic to A.

Suppose that $\langle D, \mathbf{X}, |-| \rangle$ is a proto-dual of $\mathbf{A}$. If f is an m-ary operation and $\phi_0, \ldots, \phi_{m-1} \in \mathbf{X}(X, A^0)$, then, by 2.8(ii), there exists $\phi \in \mathbf{X}(X, A^0)$ such that $|\phi|\rho = f(|\phi_0|\rho, \ldots, |\phi_{m-1}|\rho)$. Since $|-|$ is faithful, ϕ is unique and will be denoted by $f(\phi_0, \ldots, \phi_{m-1})$. Clearly, this defines an algebraic structure on the set $\mathbf{X}(X, A^0)$ in such a way that the map $\phi \to |\phi|\rho$ is an isomorphism between $\mathbf{X}(X, A^0)$ and $E(X)$. If $\psi \in \mathbf{X}(X, Y)$, then

$$\mathbf{X}(\psi, A^0): \mathbf{X}(Y, A^0) \to \mathbf{X}(X, A^0)$$

is defined by $\phi\mathbf{X}(\psi, A^0) = \psi\phi$ for all $\phi \in \mathbf{X}(Y, A^0)$, and $E(\psi): E(Y) \to E(X)$ is defined by $\phi E(\psi) = |\psi|\phi$ for all $\phi \in E(Y)$.

The following lemma is obvious.

2.10 LEMMA. *If* $\langle D, \mathbf{X}, |\text{-}| \rangle$ *is a proto-dual of* $\mathbf{A}$, *then* $\mathbf{X}(\text{-}, A^0): \mathbf{X}^{\text{op}} \to \mathbf{A}$ *and* $E: \mathbf{X}^{\text{op}} \to \mathbf{A}$ *are naturally isomorphic functors.*

We shall work with the functor E rather than with $\mathbf{X}(\text{-}, A^0)$ as this makes some of the proofs a little less technical. Observe that, like the functor D, E maps surjections to injections. For the remainder of this section, unless otherwise stated, $\langle D, \mathbf{X}, |\text{-}| \rangle$ is a fixed proto-dual of $\mathbf{A}$.

Our aim at the moment is to prove an analog of Proposition 2.4 for the categories $\mathbf{A}$ and $\mathbf{X}$; the next result shows that we may define the natural transformation $\eta: \text{id}_{\mathbf{A}} \to ED$ exactly as before.

2.11 LEMMA. *For all* $B \in \mathbf{A}$ *and each* $b \in B$ *the map* $\Gamma_b: \mathbf{A}(B, A) \to A$, *defined by* $g\Gamma_b = bg$ *for all* $g \in \mathbf{A}(B, A)$, *is an element of* $ED(B)$. *In fact,* $\Gamma_b = |D(h_b)|\rho$, *where* $h_b: \mathscr{F}_{\mathbf{A}}(1) \to B$ *is the homomorphism determined by* $x_0 h_b = b$.

Proof. $g|D(h_b)|\rho = g\mathbf{A}(h_b, A)\rho = (h_b g)\rho = bg = g\Gamma_b$. ∎

We say that a continuous map $\psi: |X| \to |Y|$ *lifts to* $\mathbf{X}$ if there is a morphism $\psi' \in \mathbf{X}(X, Y)$ such that $|\psi'| = \psi$.

2.12 PROPOSITION. *For all* $B \in \mathbf{A}$ *define* $\eta_B: B \to ED(B)$ *by* $b\eta_B = \Gamma_b$; *then* $\eta: \text{id}_{\mathbf{A}} \to ED$ *is a natural transformation. Furthermore,* D *is left adjoint to* E *with* η *as the unit of the adjunction if and only if for all* $B \in \mathbf{X}$, *all* $X \in \mathbf{X}$, *and every homomorphism* $\alpha: B \to E(X)$ *the continuous map* $\beta: |X| \to |D(B)| = \mathbf{A}(B, A)$, *defined by* $b(x\beta) = x(b\alpha)$, *lifts to* $\mathbf{X}$.

Proof. That η is a natural transformation follows as in the proof of Proposition 2.4. Let $\alpha: B \to E(X)$ be a homomorphism, let $\beta': X \to D(B)$ be a morphism of $\mathbf{X}$, and let $\beta = |\beta'|: |X| \to |D(E)| = \mathbf{A}(B, A)$. Clearly we have $\eta_B E(\beta') = \alpha$ if and only if

$$x(b\alpha) = x(b\eta_B E(\beta')) = x(|\beta'|\Gamma_b) = x(\beta\Gamma_b) = b(x\beta),$$

for all $b \in B$ and all $x \in |X|$. The result follows at once. ∎

2.13 *Remark.* If D is left adjoint to E with η as the unit of the adjunction, then the counit $\epsilon: \text{id}_{\mathbf{X}} \to DE$ of the adjunction satisfies $x|\epsilon_X| = \Gamma_x$, where $\Gamma_x \in |DE(X)| = \mathbf{A}(E(X), A)$ is defined by $\phi\Gamma_x = x\phi$ for all $\phi \in E(X)$.

The main theorem of this section can now be stated; its proof will consist of a series of lemmas.

2.14 THEOREM. *$\langle D, E\rangle$ is a duality between* **A** *and* **X** *if and only if*

(D_0) *for all $B \in \mathbf{A}$, all $X \in \mathbf{X}$, and every $\alpha \in \mathbf{A}(B, E(X))$, the map $\beta \in \mathscr{C}(|X|, \mathbf{A}(B, A))$, defined by $b(x\beta) = x(b\alpha)$, lifts to* **X**,

(D_1) *there is a class $\mathscr{I}$ of* **X**-*injections, containing the image under D of the class of all* **A**-*surjections, such that A^0 is $\mathscr{I}$-injective in* **X**,

(D_2) *if $1 \leqslant n < \omega$ and $\phi \in \mathbf{X}(D\mathscr{F}_{\mathbf{A}}(n), A^0)$, then the map $\rho_n^{-1}|\phi|\rho: A^n \to A$ is a polynomial function, and*

(D_3) *for all $\kappa \geqslant \omega$ and each $\phi \in \mathbf{X}(D\mathscr{F}_{\mathbf{A}}(\kappa), A^0)$, the map $\rho_\kappa^{-1}|\phi|\rho: A^\kappa \to A$ has a finite support.*

Furthermore, if A is finite, then $\langle D, E\rangle$ is a duality if and only if (D_0), (D_1), *and* (D_2) *hold.*

The necessity of (D_0) follows from Proposition 2.12; (D_2) is clearly necessary for η_B to be an isomorphism whenever B is a finitely generated free algebra; and (D_3) is necessary for η_B to be an isomorphism whenever B is an infinitely generated free algebra since every κ-ary polynomial function has a finite support. Finally, the necessity of (D_1) follows from the next proposition and its corollary. Observe that condition (D_1) is meaningful since the image under D of an **A**-surjection is an **X**-injection by Proposition 2.6.

2.15 LEMMA. *If $\langle D, E\rangle$ is a duality, then D is full and faithful.*

Proof. For any $h \in \mathbf{A}(B, C)$ we have $\eta_B ED(h) = h\eta_C$. Thus if $h, k \in \mathbf{A}(B, C)$ satisfy $D(h) = D(k)$, then $h\eta_C = \eta_B ED(h) = \eta_B ED(k) = k\eta_C$, and hence $h = k$ since η_C is monic.

Let $\phi \in \mathbf{X}(D(C), D(B))$. We shall find $h \in \mathbf{A}(B, C)$ satisfying $|D(h)| = \mathbf{A}(h, A) = |\phi|$, from which it follows that $D(h) = \phi$ since $|\text{-}|$ is faithful. Let $h = \eta_B E(\phi)\eta_C^{-1}$ and note that for all $b \in B$, $bh = (|\phi|\Gamma_b)\eta_C^{-1}$. Now for each $\psi \in ED(C)$ we have, for all $g \in \mathbf{A}(C, A)$,

$$g\psi = g(\psi\eta_C^{-1}\eta_C) = g\Gamma_{\psi\eta_C^{-1}} = (\psi\eta_C^{-1})g.$$

Hence, setting $\psi = |\phi|\Gamma_b$, we have

$$b(hg) = (bh)g = ((|\phi|\Gamma_b)\eta_C^{-1})g = g(|\phi|\Gamma_b) = (g|\phi|)\Gamma_b = b(g|\phi|).$$

This gives $g|\phi| = hg = g\mathbf{A}(h, A)$ for all $g \in \mathbf{A}(C, A)$, whence $|\phi| = \mathbf{A}(h, A) = |D(h)|$. ∎

2.16 COROLLARY. *If $\langle D, E\rangle$ is a duality, then A^0 is $\mathscr{I}$-injective, where $\mathscr{I} = \{D(h): h$ is an* **A**-*surjection*$\}$.

Proof. As we have already noted, $\mathscr{I}$-injectivity is well defined. The $\mathscr{I}$-*injectivity* of $A^0 = D\mathscr{F}_{\mathbf{A}}(1)$ follows immediately from the sur-projectivity of $\mathscr{F}_{\mathbf{A}}(1)$ and the fullness of D. ∎

We now establish the sufficiency of the conditions (D_0), (D_1), (D_2), and (D_3). For all $B \in \mathbf{A}$, $\mathbf{A}(B, A)$ separates the points of B since $\mathbf{A} = \mathbf{ISP}(\{A\})$, and hence η_B is an embedding. Thus it remains to show that η_B is a surjection for all $B \in \mathbf{A}$.

2.17 Lemma. *If* (D_2) *and* (D_3) *hold, then* η_B *is an isomorphism whenever* B *is a free algebra in* $\mathbf{A}$.

Proof. Clearly, (D_2) implies that η_B is a surjection for every finitely generated free algebra $B = \mathscr{F}_{\mathbf{A}}(n)$. Let $\kappa \geqslant \omega$ and let $\phi \in \mathbf{X}(D\mathscr{F}_{\mathbf{A}}(\kappa), A^0)$. Then, by (D_3), $\sigma = \rho_\kappa^{-1}|\phi|\rho : A^\kappa \to A$ has a finite support, say $\{\gamma_0, \ldots, \gamma_{n-1}\}$. Define $h : \mathscr{F}_{\mathbf{A}}(\kappa) \to \mathscr{F}_{\mathbf{A}}(n)$ by $x_{\gamma_j} h = x_j$ $(j < n)$ and h arbitrary on all other generators. Then $D(h)\phi \in \mathbf{X}(D\mathscr{F}_{\mathbf{A}}(n), A^0)$ and hence $\lambda = \rho_n^{-1}|D(h)\phi|\rho : A^n \to A$ is a polynomial function, by (D_1). Let $p(x_0, \ldots, x_{n-1})$ by an n-ary polynomial that induces the function λ. It is easily seen that the κ-ary polynomial $q(x) = p(x_{\gamma_0}, \ldots, x_{\gamma_{n-1}})$ induces the function σ, and hence η_B is a surjection for $B = \mathscr{F}_{\mathbf{A}}(\kappa)$. ∎

Condition (D_1) now allows us to extend to arbitrary algebras in $\mathbf{A}$ by taking homomorphic images.

2.18 Lemma. *If* (D_1) *holds and* η_B *is an isomorphism for* $B = \mathscr{F}_{\mathbf{A}}(\kappa)$, *then* η_B *is an isomorphism for every* κ-*generated algebra* $B \in \mathbf{A}$.

Proof. Suppose that B is generated by $\{b_\gamma | \gamma < \kappa\}$ and let $h : \mathscr{F}_{\mathbf{A}}(\kappa) \to B$ be the homomorphism determined by $x_\gamma h = b_\gamma$ $(\gamma < \kappa)$. Since h is a surjection, for every $\phi \in \mathbf{X}(D(B), A^0)$ there exists $\psi \in \mathbf{X}(D\mathscr{F}_{\mathbf{A}}(\kappa), A^0)$ with $\phi = D(h)\psi$, by (D_1). There is a κ-ary polynomial, say q, with $q = \rho_\kappa^{-1}|\psi|\rho$. For all $g \in \mathbf{A}(B, A)$ we have

$$\begin{aligned} g(|\phi|\rho) &= g(|D(h)|\, |\psi|\rho) = g(\mathbf{A}(h, A)|\psi|\rho) = hg(|\psi|\rho) \\ &= (hg)\rho_\kappa \rho_\kappa^{-1}|\psi|\rho = q(\langle b_\gamma g\rangle_{\gamma<\kappa}) = q(\langle b_\gamma\rangle_{\gamma<\kappa})g = g\Gamma_b, \end{aligned}$$

where $b = q(\langle b_\gamma\rangle_{\gamma<\kappa})$. Hence $|\phi|\rho = \Gamma_b = b\eta_B$, and thus η_B is a surjection. ∎

The sufficiency of the conditions now follows from Proposition 2.12, Lemma 2.17, and Lemma 2.18. That (D_3) is superfluous when A is finite follows from the following observation.

2.19 LEMMA. *Let A be a finite discrete topological space. Then every continuous map $\phi: A^\kappa \to A$ has a finite support.*

Proof. It is not difficult to see that a nonempty set U is clopen in A^κ if and only if there exist $\gamma_{ij} < \kappa$ and $a_{ij} \in A$ such that

$$U = \bigcup(\bigcap(\langle\gamma_{ij}; \{a_{ij}\}\rangle : j < n_i) : i < n);$$

we say that $\{\gamma_{ij} : j < n_i; i < n\}$ *fixes* U.

Now $\{a\phi^{-1} : a \in \mathrm{Im}(\phi)\}$ is a finite partition of A^κ into clopen sets. For each $a \in \mathrm{Im}(\phi)$, let Λ_a be a finite set of indices that fixes the clopen set $a\phi^{-1}$. Clearly $\bigcup(\Lambda_a \mid a \in \mathrm{Im}(\phi))$ is a finite support for ϕ. ∎

If $\langle D, E\rangle$ is a duality, then since $\mathscr{I}$ contains the image under D of the class of all **A**-surjections it follows, by Proposition 2.6, that for all $B \in \mathbf{A}$ there exists κ and an injection $\tau: D(B) \to D\mathscr{F}_{\mathbf{A}}(\kappa)$ contained in $\mathscr{I}$. If for every $X \in \mathbf{X}$ there exists κ and an injection $\tau: X \to D\mathscr{F}_{\mathbf{A}}(\kappa)$ contained in $\mathscr{I}$ (which must happen if $\langle D, E\rangle$ is a full duality), then we can slightly weaken the assumption that $\langle D, \mathbf{X}, |\text{-}|\rangle$ is a proto-dual of **A**; namely, the assumption that $E(X)$ is a subalgebra of $A^{|X|}$ can be dropped.

2.20 THEOREM. *Let $D: \mathbf{A} \to \mathbf{X}^{\mathrm{op}}$ be a functor into a category* **X** *that has a grounding $|\text{-}|: \mathbf{X} \to \mathbf{Comp}$, and assume that $|D(\text{-})| = \mathbf{A}(\text{-}, A)$. Suppose that* ($\mathrm{D}_0$), ($\mathrm{D}_1$), ($\mathrm{D}_2$), *and* ($\mathrm{D}_3$) *hold, and that for every $X \in \mathbf{X}$ there exists κ and an injection $\tau: X \to D\mathscr{F}_{\mathbf{A}}(\kappa)$ contained in $\mathscr{I}$. Then $\langle D, \mathbf{X}, |\text{-}|\rangle$ is a proto-dual of* **A**, *and $\langle D, E\rangle$ is a duality.*

Proof. We must show that for all $X \in \mathbf{A}$, $E(X)$ is a subalgebra of $A^{|X|}$. Let $\tau: X \to D\mathscr{F}_{\mathbf{A}}(\kappa)$ be an injection contained in $\mathscr{I}$. Let $\phi_0, \ldots, \phi_{m-1} \in \mathbf{X}(X, A^0)$ and let f be an m-ary operation. Since A^0 is $\mathscr{I}$-injective there exist $\psi_0, \ldots, \psi_{m-1} \in \mathbf{X}(D\mathscr{F}_{\mathbf{A}}(\kappa), A^0)$ with $\phi_j = \tau\psi_j$ for all $j < m$. Let $Y = \mathbf{A}(\mathscr{F}_{\mathbf{A}}(\kappa), A)$; now ($\mathrm{D}_2$) and ($\mathrm{D}_3$) imply that $ED\mathscr{F}_{\mathbf{A}}(\kappa)$ is a subalgebra of A^Y isomorphic to $\mathscr{F}_{\mathbf{A}}(\kappa)$, as in the proof of Lemma 2.17. Thus $f(|\psi_0|\rho, \ldots, |\psi_{m-1}|\rho) \in ED\mathscr{F}_{\mathbf{A}}(\kappa)$ and hence there exists $\psi \in \mathbf{X}(D\mathscr{F}_{\mathbf{A}}(\kappa), A^0)$ with $|\psi|\rho = f(|\psi_0|\rho, \ldots, |\psi_{m-1}|\rho)$. Hence

$$\begin{aligned} f(|\phi_0|\rho, \ldots, |\phi_{m-1}|\rho) &= f(|\tau|\,|\psi_0|\rho, \ldots, |\tau|\,|\psi_{m-1}|\rho) \\ &= |\tau| f(|\psi_0|\rho, \ldots, |\psi_{m-1}|\rho) = |\tau|\,|\psi|\rho = |\tau\psi|\rho \in E(X), \end{aligned}$$

and thus $E(X)$ is a subalgebra of $A^{|X|}$. ∎

We now give a brief discussion of conditions under which a duality $\langle D, E\rangle$ between **A** and **X** will be a full duality.

2.21 LEMMA. *Suppose that* $\langle D, E\rangle$ *is a duality. Then* $\langle D, E\rangle$ *is a full duality if and only if for each* $X \in \mathbf{X}$ *the map* $|\epsilon_X|$ *is a homeomorphism and there exists* $\psi \in \mathbf{X}(DE(X), X)$ *such that* $|\psi| = |\epsilon_X|^{-1}$*, that is,* $|\epsilon_X|^{-1}$ *lifts to* $\mathbf{X}$.

Proof. Only the sufficiency requires proof. Clearly we must show that $\psi = \epsilon_X^{-1}$. But

$$|\mathrm{id}_X| = \mathrm{id}_{|X|} = |\epsilon_X|\,|\epsilon_X|^{-1} = |\epsilon_X|\,|\psi| = |\epsilon_X\psi|,$$

and hence $\mathrm{id}_X = \epsilon_X\psi$ since $|\text{-}|$ is faithful. Similarly, $\mathrm{id}_{DE(X)} = \psi\epsilon_X$. ∎

If $\langle D, E\rangle$ is a full duality, then for each $X \in \mathbf{X}$, the map $|\epsilon_X|$ is, in particular, a homeomorphism onto a closed subspace.

2.22 LEMMA. *Suppose that condition* (D_0) *holds. Then the following are equivalent for each* $X \in \mathbf{X}$:

(i) $|\epsilon_X|$ *is a homeomorphism onto a closed subspace;*
(ii) $E(X)$ *separates the points of* $|X|$;
(iii) *there exists* κ *and a morphism* $\tau \in \mathbf{X}(X, D\mathscr{F}_{\mathbf{A}}(\kappa))$ *such that* $|\tau|\rho_\kappa : |X| \to A^\kappa$ *is a homeomorphism onto a closed subspace.*

Proof. (i)⇔(ii). This follows immediately from Remark 2.13.

(i)⇒(iii). Assume that $E(X)$ is generated by $\{b_\gamma : \gamma < \kappa\}$ and let $h : \mathscr{F}_{\mathbf{A}}(\kappa) \to E(X)$ be the homomorphism determined by $x_\gamma h = b_\gamma$ ($\gamma < \kappa$). Then $D(h)$ is an injection by Proposition 2.6, and $\tau = \epsilon_X D(h)$ is the required injection.

(iii)⇒(ii). Let $x, y \in |X|$ be distinct. Since

$$\langle a_\gamma\rangle_{\gamma<\kappa} = x|\tau|\rho_\kappa \neq y|\tau|\rho_\kappa = \langle b_\gamma\rangle_{\gamma<\kappa},$$

there exists $\lambda < \kappa$ such that $a_\lambda \neq b_\lambda$. Let $h_\lambda : \mathscr{F}_{\mathbf{A}}(1) \to \mathscr{F}_{\mathbf{A}}(\kappa)$ be the homomorphism determined by $x_0 h_\lambda = x_\lambda$. Then $\phi = \tau D(h_\lambda) \in \mathbf{X}(X, A^0)$ and

$$x|\phi| = x|\tau|\mathbf{A}(h_\lambda, A) = h_\lambda(x|\tau|) = a_\lambda,$$

and similarly $y|\phi| = b_\lambda$. Hence $x|\phi|\rho \neq y|\phi|\rho$ since ρ is one-to-one; and so $E(X)$ separates the points of $|X|$. ∎

If $\langle D, E\rangle$ is a full duality between $\mathbf{A}$ and $\mathbf{X}$, then we can describe the left adjoint to the grounding $|\text{-}| : \mathbf{X} \to \mathbf{Comp}$, that is, the $\mathbf{Comp}$-free functor for $\mathbf{X}$.

2.23 PROPOSITION. *The following statements are related by* (i)⇒(ii)⇒(iii).

(i) $\langle D, E\rangle$ *is a full duality.*
(ii) $D\mathscr{C}(\text{-}, A) : \mathbf{Comp} \to \mathbf{X}$ *is left adjoint to* $|\text{-}| : \mathbf{X} \to \mathbf{Comp}$.
(iii) $D(A^{(\text{-})}) : \mathbf{Set} \to \mathbf{X}$ *is left adjoint to the forgetful functor* $|\text{-}| : \mathbf{X} \to \mathbf{Set}$.

Proof. (i) ⇒ (ii). $\mathscr{C}(\text{-}, A)\colon \mathbf{Comp} \to \mathbf{A}^{\mathrm{op}}$ is left adjoint to $\mathbf{A}(\text{-}, A)\colon \mathbf{A}^{\mathrm{op}} \to \mathbf{Comp}$ by Proposition 2.4, and if $\langle D, E\rangle$ is a full duality then $D\colon \mathbf{A}^{\mathrm{op}} \to \mathbf{X}$ is left adjoint to $E\colon \mathbf{X} \to \mathbf{A}^{\mathrm{op}}$. Since left adjoints compose, (ii) follows.

(ii) ⇒ (iii). Since left adjoints compose it follows that $D\mathscr{C}(\beta(\text{-}), A)\colon \mathbf{Set} \to \mathbf{X}$ is left adjoint to $|\text{-}|\colon \mathbf{X} \to \mathbf{Set}$, where $\beta\colon \mathbf{Set} \to \mathbf{Comp}$ is the Stone–Čech compactification functor. Since $\mathscr{C}(\beta S, A) \simeq \mathscr{C}(S, A) \simeq A^S$, (iii) follows. ∎

If A is finite and $\langle D, E\rangle$ is a full duality, then $|X|$ is a Boolean space for all $X \in \mathbf{X}$ and hence we may replace (ii) by

(ii)′ $D\mathscr{C}(\text{-}, A)\colon \mathbf{ZComp} \to \mathbf{X}$ *is left adjoint to* $|\text{-}|\colon \mathbf{X} \to \mathbf{ZComp}$.

Examples in which $D\mathscr{C}(\text{-}, A)\colon \mathbf{ZComp} \to \mathbf{X}$ is left adjoint to $|\text{-}|\colon \mathbf{X} \to \mathbf{ZComp}$ while $\langle D, E\rangle$ is a duality but not a full duality may be found in Davey [8]. If A is injective in $\mathbf{A}$, then we can obtain a partial converse of Proposition 2.23.

2.24 PROPOSITION. *Suppose that $\langle D, E\rangle$ is a duality. If A is injective in $\mathbf{A}$, $E(X)$ separates the points of X for each $X \in \mathbf{X}$, and $D\mathscr{C}(\text{-}, A)$ is left adjoint to $|\text{-}|$, then $|\epsilon_X|\colon |X| \to |DE(X)|$ is a homeomorphism for every $X \in \mathbf{X}$.*

Proof. Since $E(X)$ separates the points of $|X|$, by Lemma 2.22 it is sufficient to show that $|\epsilon_X|$ is a surjection for each $X \in \mathbf{X}$. Firstly we shall show that $|\epsilon_{D(B)}|$ is a surjection for each $B \in \mathbf{A}$; in fact, we shall show that $\epsilon_{D(B)}$ is an isomorphism in $\mathbf{X}$. Since D and E are adjoint functors it follows that $\epsilon_{D(B)}D(\eta_B) = \mathrm{id}_{D(B)}$ (see [35, Theorem 2, p. 81). But since η_B is invertible, $D(\eta_B)$ is invertible and consequently $\epsilon_{D(B)} = \mathrm{id}_{D(B)}D(\eta_B)^{-1} = D(\eta_B)^{-1}$ is invertible.

Denote the unit and counit of the adjoint pair $\langle |\text{-}|, D\mathscr{C}(\text{-}, A)\rangle$ by

$$\zeta\colon \mathrm{id}_{\mathbf{Comp}} \to |D\mathscr{C}(\text{-}, A)| \qquad \text{and} \qquad \xi\colon D\mathscr{C}(|\text{-}|, A) \to \mathrm{id}_{\mathbf{X}}.$$

Since $\zeta_{|X|}|\xi_X| = \mathrm{id}_{|X|}$ (see [35, Theorem 2, p. 81]) it follows that $|\xi_X|$ is a surjection and hence $E(\xi_X)\colon E(X) \to ED\mathscr{C}(|X|, A)$ is an $\mathbf{A}$-injection.

We now show that $|\epsilon_X|\colon |X| \to |DE(X)|$ is a surjection. Let $g \in \mathbf{A}(E(X), A)$. Since $E(\xi_X)$ is an injection and A is injective in $\mathbf{A}$, there exists $g'\colon ED\mathscr{C}(|X|, A) \to A$ with $g = E(\xi_X)g'$. Let $B = \mathscr{C}(|X|, A)$. Since $|\epsilon_{D(B)}|$ is a surjection it follows that there exists $\phi \in |D(B)| = \mathbf{A}(\mathscr{C}(|X|, A), A)$ with $\Gamma_\phi = \phi|\epsilon_{D(B)}| = g'$. Set $x = \phi|\xi_X|$ and let $\psi \in E(X) \subseteq \mathscr{C}(|X|, A)$. Then

$$\begin{aligned}\psi(x|\epsilon_X|) &= \psi\Gamma_x = x\psi = (\phi|\xi_X|)\psi = \phi(|\xi_X|\psi)\\ &= (|\xi_X|\psi)\Gamma_\phi = (|\xi_X|\psi)g' = (\psi E(\xi_X))g' = \psi(E(\xi_X)g') = \psi g,\end{aligned}$$

and hence $x|\epsilon_X| = g$, as required. ∎

By collecting together 2.21–2.24, observing that the proof of Proposition 2.24 requires only that A be injective with respect to the images under

E of the $\mathbf{X}$-surjections, and noting that if $\langle D, E\rangle$ is a full duality, then E is full (and faithful), we obtain the following characterization of full dualities.

2.25 THEOREM. *Suppose that $\langle D, E\rangle$ is a duality. Then $\langle D, E\rangle$ is a full duality if and only if*

(E_1) *$E(X)$ separates the points of $|X|$ for all $X \in \mathbf{X}$,*
(E_2) *there is a class $\mathcal{J}$ of $\mathbf{A}$-injections, containing the image under E of the class of $\mathbf{X}$-surjections, such that A is $\mathcal{J}$-injection in $\mathbf{A}$,*
(E_3) *$D\mathscr{C}(\text{-}, A)$ is left adjoint to $|\text{-}|$, and*
(E_4) *the homeomorphism $|\epsilon|^{-1}:|DE(X)| \to |X|$ lifts to $\mathbf{X}$ for each $X \in \mathbf{X}$.*

Proof. Note that (E_1), (E_2), and (E_3) guarantee that (E_4) is meaningful. Only the necessity of (E_2) has not already been established. From (E_3) it follows that $D(A^{(-)})$ is a free functor for $\mathbf{X}$ and hence $D(A)$ is sur-projective in $\mathbf{X}$ (alternatively, see Proposition 2.26 below). Let $\mathcal{J} = \{E(\psi):\psi$ is an $\mathbf{X}$-surjection$\}$; then the $\mathcal{J}$-injectivity of A follows from the sur-projectivity of $D(A)$ and the fullness of E. ∎

We close this section with some remarks on injectives and projectives. The following useful result is proved Hofmann *et al.* [24].

2.26 PROPOSITION. *If $\mathbf{X}$ is a category which is grounded in* **Comp** (**ZComp**) *and F:* **Comp** $\to$ **X** *(F:* **ZComp** $\to$ **X**) *is a* **Comp**-*free* (**ZComp**-*free*) *functor for* **X**, *then the following are equivalent*:

(i) *P is sur-projective in $\mathbf{X}$;*
(ii) *P is a retract of $F(\beta S)$ for some set S;*
(iii) *P is a retract of $F(X)$ for some compact extremally disconnected space X.*

2.27 PROPOSITION. *Suppose that $\langle D, E\rangle$ is a full duality. If A is injective in $\mathbf{A}$, then the following are equivalent*:

(i) *I is injective in $\mathbf{A}$;*
(ii) *$D(I)$ is sur-projective in $\mathbf{X}$;*
(iii) *I is isomorphic to $E(P)$ for some P that is sur-projective in $\mathbf{X}$;*
(iv) *I is a retract of $\mathscr{C}(X, A)$ for some compact extremally disconnected space X.*

Proof. (i)$\Rightarrow$(ii). Let $\phi \in \mathbf{X}(X, Y)$ be a surjection and let $\psi \in \mathbf{X}(D(I), Y)$. Then $E(\phi)$ is an injection and hence, since $ED(I) \simeq I$ is injective, there exists $g \in \mathbf{A}(E(X), ED(I))$ with $E(\phi)g = E(\psi)$. A simple calculation shows that $\psi' = \epsilon_{D(I)}D(g)\epsilon_X^{-1}$ satisfies $\psi'\phi = \psi$, as required.

(ii) ⇒ (iii). This is trivial.

(iii) ⇒ (iv). Since $\langle D, E\rangle$ is a full duality, by Proposition 2.23 $D\mathscr{C}(\text{-}, A)$ is a **Comp**-free functor for **X**. Thus combining (iii) and Proposition 2.26 we have that I is isomorphic to $E(P)$, where P is a retract of $D\mathscr{C}(X, A)$ for some compact extremally disconnected space X. Since functors preserve retractions it follows that I is a retract of $ED\mathscr{C}(X, A) \simeq \mathscr{C}(X, A)$, as required.

(iv) ⇒ (i). By (iv), $D(I)$ is a retract of $D\mathscr{C}(X, A)$ and hence $D(I)$ is sur-projective in **X** by Proposition 2.26, again using the fact that $D\mathscr{C}(\text{-}, A)$ is a **Comp**-free functor for **X**. Let $g \in \mathbf{A}(B, C)$ be an injection and let $h \in \mathbf{A}(B, I)$. Since A is injective in **A**, $D(g)$ is a surjection, and since $D(I)$ is sur-projective there exists $\phi \in \mathbf{X}(D(I), D(C))$ with $\phi D(g) = D(h)$. Thus $h' = \eta_C E(\phi)\eta_I^{-1}$ satisfies $gh' = h$, as required. ∎

3. Dualities via Compact Topological Partial Algebras

As in the previous section, we assume that A is a compact topological algebra and that $\mathbf{A} = \mathbf{ISP}(\{A\})$ is the prevariety generated by A.

For any $B \in \mathbf{A}$, and any n-ary polynomial p we may regard p, defined pointwise, as an operation on the set A^B. By relativization, p becomes a partial operation on the subset $\mathbf{A}(B, A)$. Hence, for all $g_0, \ldots, g_{n-1} \in \mathbf{A}(B, A)$, $\langle g_0, \ldots, g_{n-1}\rangle \in \mathscr{D}(p, \mathbf{A}(B, A))$ if and only if $p(g_0, \ldots, g_{n-1})$, defined pointwise, is a homomorphism.

3.1 *Remark.* (i) Clearly, $\langle g_0, \ldots, g_{n-1}\rangle \in \mathscr{D}(p, \mathbf{A}(B, A))$ if and only if for every m-ary operation f and all $b_0, \ldots, b_{m-1} \in B$,

$$\begin{aligned} &p(f(b_0 g_0, \ldots, b_{m-1} g_0), \ldots, f(b_0 g_{n-1}, \ldots, b_{m-1} g_{n-1})) \\ &\quad = f(p(b_0 g_0, \ldots, b_0 g_{n-1}), \ldots, p(b_{m-1} g_0, \ldots, b_{m-1} g_{n-1})). \end{aligned}$$

(ii) If $\mathscr{D}(p, \mathbf{A}(B, A))$ is nonempty for some $B \in \mathbf{A}$, then for every nullary operation 0, $p(0, 0, \ldots, 0) = 0$ is an equation in A and so holds identically in **A**.

(iii) For each nullary polynomial p, $\mathscr{D}(p, \mathbf{A}(B, A))$ is nonempty if and only if $\{p\}$ is a one-element subalgebra of A; hence if p is defined in $\mathbf{A}(B, A)$ for some $B \in \mathbf{A}$, then p is defined in $\mathbf{A}(B, A)$ for all $B \in \mathbf{A}$. In particular, if 0 and 1 are distinct nullary operations on A, then for all $B \in \mathbf{A}$, $\mathscr{D}(0, \mathbf{A}(B, A))$ and $\mathscr{D}(1, \mathbf{A}(B, A))$ are both empty.

Throughout the remainder of the chapter, $\mathscr{Q}$ will denote a fixed set of finitary polynomials (of the type of the algebra A) and whenever we refer to the partial algebra $\mathbf{A}(B, A)$ it will be implied that the operations on $\mathbf{A}(B, A)$ are exactly those that arise, as described above, from polynomials in $\mathscr{Q}$.

Every $p \in \mathcal{Q}$ induces a continuous (full) operation on A^B and hence the restriction of p to $\mathbf{A}(B, A)$ is also continuous. Thus, for every $B \in \mathbf{A}$, $\mathbf{A}(B, A)$ is a compact topological partial algebra (see Lemma 2.3 and the remark following it).

3.2 DEFINITION. For each $\kappa > 0$ define a partial algebra structure on the compact space A^κ as follows: If $p \in \mathcal{Q}$ is n-ary, then for all $a^0, \ldots, a^{n-1} \in A^\kappa$, $\langle a^0, \ldots, a^{n-1} \rangle \in \mathcal{D}(p, A^\kappa)$ if and only if, for every κ-ary polynomial q,

$$p(q(a^0), \ldots, q(a^{n-1})) = q(p(a^0, \ldots, a^{n-1}));$$

in particular, if $p \in \mathcal{Q}$ is a nullary polynomial then p is a nullary operation defined on the partial algebra A^κ if and only if $\{p\}$ is a one-element subalgebra of A (see Remark 3.1(iii)). If $\langle a^0, \ldots, a^{n-1} \rangle \in \mathcal{D}(p, A^\kappa)$, then $p(a^0, \ldots, a^{n-1})$ is defined pointwise and hence the partial algebra A^κ is a weak subalgebra of the full algebra $\langle A^\kappa; \mathcal{Q} \rangle$.

3.3 DEFINITION. A map $\psi: X \to Y$ between two similar topological partial algebras is an *iseomorphism* if it is both an isomorphism and a homeomorphism.

Define a category $\mathbf{X} = \mathbf{X}_{\mathcal{Q}}$ of compact partial algebras by declaring that X is an object of $\mathbf{X}$ if it is iseomorphic to a closed subalgebra of the compact partial algebra A^κ for some κ, and a map $\psi: X \to Y$ is a morphism of $\mathbf{X}$ if it is a continuous homomorphism.

We can now prove a stronger version of Proposition 2.6; in essence, this result says that the range of the functor $\mathbf{A}(\text{-}, A)$ can be lifted from **Comp** to $\mathbf{X}$. For notational simplicity we shall often abbreviate a κ-tuple $\langle x_\gamma \rangle_{\gamma < \kappa}$ to $\langle x_\gamma \rangle$.

3.4 PROPOSITION. (i) *For all $h \in \mathbf{A}(B, C)$, the map $\mathbf{A}(h, A)$ is a continuous homomorphism. If h is a surjection, then $\mathbf{A}(h, A)$ is an iseomorphism onto a closed subalgebra.*

(ii) *For all $\kappa > 0$, the map $\rho_\kappa: \mathbf{A}(\mathcal{F}_{\mathbf{A}}(\kappa), A) \to A^\kappa$ is an iseomorphism.*

(iii) *Assume that $B \in \mathbf{A}$ is generated by $\{b_\gamma: \gamma < \kappa\}$ and let $h: \mathcal{F}_{\mathbf{A}}(\kappa) \to B$ be the homomorphism determined by $x_\gamma h = b_\gamma$ $(\gamma < \kappa)$. Then $\mathbf{A}(h, A)\rho_\kappa$ is an iseomorphism onto a closed subalgebra.*

Proof. (i) It is easily verified that $\mathbf{A}(h, A)$ is a continuous homomorphism. Assume that h is a surjection; by Proposition 2.6 it remains to prove that for all $p \in \mathcal{Q}$, $p(g_0, \ldots, g_{n-1})$ is defined in $\mathbf{A}(C, A)$ if and only if $p(hg_0, \ldots, hg_{n-1})$ is defined in $\mathbf{A}(B, A)$. But if $p(g_0, \ldots, g_{n-1})$ is a homomorphism, then $hp(g_0, \ldots, g_{n-1}) = p(hg_0, \ldots, hg_{n-1})$ is a homomorphism, and conversely, if $p(hg_0, \ldots, hg_{n-1}) = hp(g_0, \ldots, g_{n-1})$ is a homomorphism then, since h is onto, $p(g_0, \ldots, g_{n-1})$ is also a homomorphism.

(ii) Let $p \in \mathcal{Q}$ be n-ary, let $g_0, \ldots, g_{n-1} \in \mathbf{A}(\mathscr{F}_{\mathbf{A}}(\kappa), A)$, and for all $j < n$ define $a^j \in A^\kappa$ by $a_\gamma{}^j = x_\gamma g_j$ for all $\gamma < \kappa$. Then $g_j \rho_\kappa = a^j$ for all $j < n$.

Now $p(g_0, \ldots, g_{n-1})$ is a homomorphism if and only if it is determined by its values on the generators, that is, if and only if for each κ-ary polynomial q,

$$q(\langle x_\gamma \rangle) p(g_0, \ldots, g_{n-1}) = q(\langle x_\gamma p(g_0, \ldots, g_{n-1}) \rangle),$$

i.e.,

$$p(q(\langle x_\gamma \rangle) g_0, \ldots, q(\langle x_\gamma \rangle) g_{n-1}) = q(\langle p(x_\gamma g_0, \ldots, x_\gamma g_{n-1} \rangle),$$

i.e.,

$$p(q(\langle x_\gamma g_0 \rangle), \ldots, q(\langle x_\gamma g_{n-1} \rangle)) = q(p(\langle x_\gamma g_0 \rangle), \ldots, p(\langle x_\gamma g_{n-1} \rangle),$$

i.e.,

$$p(q(a^0), \ldots, q(a^{n-1})) = q(p(a^0, \ldots, a^{n-1})).$$

Thus $p(g_0, \ldots, g_{n-1})$ is defined in $\mathbf{A}(\mathscr{F}_{\mathbf{A}}(\kappa), A)$ if and only if $p(g_0 \rho_\kappa, \ldots, g_{n-1} \rho_\kappa)$ is defined in A^κ. Since ρ_κ is a homeomorphism by Proposition 2.6, the result follows.

(iii) This follows directly from (i) and (ii). ∎

No notational distinction will be drawn between an object $X \in \mathbf{X}$ and its underlying compact space, nor between a morphism $\psi \in \mathbf{X}$ and its underlying continuous map; that is, we shall suppress the obvious grounding $|-|: \mathbf{X} \to$ **Comp**. Hence

$$D = \mathbf{A}(\text{-}, A): \mathbf{A} \to \mathbf{X}^{\mathrm{op}}$$

is a well-defined functor.

Clearly, when applying the results of the previous section we may replace $A^0 = D\mathscr{F}_{\mathbf{A}}(1)$ by its iseomorphic copy A (regarded as a compact partial algebra by putting $\kappa = 1$ in Definition 3.2). This abuse of notation should cause no problems since in any given situation it will be clear whether A is acting as an object of $\mathbf{A}$ or as an object of $\mathbf{X}$.

In general $\langle D, \mathbf{X}, |-| \rangle$ need not be a proto-dual of $\mathbf{A}$, for we do not know that $\mathbf{X}(X, A)$ is a subalgebra of A^X. If $\mathbf{X}(X, A)$ is a subalgebra of A^X for all $X \in \mathbf{X}$, then

$$E = \mathbf{X}(\text{-}, A): \mathbf{X}^{\mathrm{op}} \to \mathbf{A}$$

is a well-defined functor, by Lemma 2.10. As before, $\eta_B: B \to ED(B)$ is defined by $b\eta_B = \Gamma_b$, where $g\Gamma_b = bg$ for all $g \in D(B)$, and $\epsilon_X: X \to DE(X)$ is defined by $x\epsilon_X = \Gamma_x$, where $\phi\Gamma_x = x\phi$ for all $\phi \in E(X)$.

3.5 PROPOSITION. *Suppose that* $\mathbf{X}(X, A)$ *is a subalgebra of* A^X *for all* $X \in \mathbf{X}$. *Then* $\langle D, E; \eta, \epsilon\rangle$ *is an adjunction from* $\mathbf{A}$ *to* $\mathbf{X}^{\mathrm{op}}$.

Proof. By Proposition 2.12 it remains to be shown that for all $B \in \mathbf{A}$, all $X \in \mathbf{X}$, and every $\alpha \in \mathbf{A}(B, \mathbf{X}(X, A))$, the map $\beta: X \to \mathbf{A}(B, A)$, defined by $b(x\beta) = x(b\alpha)$, is a homomorphism.

As we noted in the proof of Proposition 2.4, for all $x \in X$, $x\beta: B \to A$ is a homomorphism; in particular, if $p \in \mathcal{Q}$ is n-ary, $x^0, \ldots, x^{n-1} \in X$, and $p(x^0, \ldots, x^{n-1})$ is defined in X, then $p(x^0, \ldots, x^{n-1})\beta$ is a homomorphism. But

$$\begin{aligned} b(p(x^0, \ldots, x^{n-1})\beta) &= p(x^0, \ldots, x^{n-1})(b\alpha) \\ &= p(x^0(b\alpha), \ldots, x^{n-1}(b\alpha)) \qquad (\text{since } b\alpha \in \mathbf{X}(X, A)) \\ &= p(b(x^0\beta), \ldots, b(x^{n-1}\beta)) = bp(x^0\beta, \ldots, x^{n-1}\beta), \end{aligned}$$

and hence $p(x^0\beta, \ldots, x^{n-1}\beta)$ is a homomorphism since it is pointwise equal to the homomorphism $p(x^0, \ldots, x^{n-1})\beta$. Thus $p(x^0\beta, \ldots, x^{n-1}\beta)$ is defined in $\mathbf{A}(B, A)$ and equals $p(x^0, \ldots, x^{n-1})\beta$. Hence β is a homomorphism. ∎

Theorem 2.14 may now be applied.

3.6 PROPOSITION. *Suppose that* $\mathbf{X}(X, A)$ *is a subalgebra of* A^X *for all* $X \in \mathbf{X}$. *Then* $\langle D, E\rangle$ *is a duality between* $\mathbf{A}$ *and* $\mathbf{X}$ *if and only if*

(D_1) *there is a class* $\mathcal{I}$ *of* $\mathbf{X}$*-injections, containing the image under* D *of the class of all* $\mathbf{A}$*-surjections, such that* A *is* $\mathcal{I}$*-injective in* $\mathbf{X}$,

(D_2) *for all* n, *with* $1 \leqslant n < \omega$, *every* $\phi \in \mathbf{X}(A^n, A)$ *is a polynomial function*,

(D_3) *for all* $\kappa \geqslant \omega$, *every* $\phi \in \mathbf{X}(A^\kappa, A)$ *has a finite support*.

Furthermore, if A *is finite, then* $\langle D, E\rangle$ *is duality if and only if* (D_1) *and* (D_2) *hold*.

If I is $\mathcal{I}$-injective in $\mathbf{X}$ where $\mathcal{I}$ is the class of all iseomorphisms onto closed subalgebras, then we shall simply say that I is *injective* in $\mathbf{X}$. If A is injective in $\mathbf{X}$, then Theorem 2.20 is applicable.

3.7 THEOREM. $E: \mathbf{X}^{\mathrm{op}} \to \mathbf{A}$ *is a well-defined functor and* $\langle D, E\rangle$ *is a duality between* $\mathbf{A}$ *and* $\mathbf{X}$ *whenever*

(D_1)′ A *is injective in* $\mathbf{X}$,

(D_2) *for all* n, *with* $1 \leqslant n < \omega$, *every* $\phi \in \mathbf{X}(A^n, A)$ *is a polynomial function*,

(D_3) *for all* $\kappa \geqslant \omega$ *every* $\phi \in \mathbf{X}(A^\kappa, A)$ *has a finite support*.

Furthermore, if A *is finite*, (D_1)′ *and* (D_2) *will suffice*.

In the present situation we have a somewhat more satisfactory characterization of full dualities.

3.8 THEOREM. *Assume that* $\mathbf{X}(X, A)$ *is a subalgebra of* A^X *for all* $X \in \mathbf{X}$, *and that* $\langle D, E\rangle$ *is a duality between* $\mathbf{A}$ *and* $\mathbf{X}$. *Then* $\langle D, E\rangle$ *is a full duality if and only if* $\epsilon_X: X \to DE(X)$ *is a surjection for all* $X \in \mathbf{X}$.

Proof. It is sufficient to prove that if $\langle D, E\rangle$ is a duality, then ϵ_X is an iseomorphism onto a closed subalgebra for all $X \in \mathbf{X}$. But ϵ_X is a homeomorphism onto a closed subspace by Lemma 2.22 and is a homomorphism since $\epsilon_X \in \mathbf{X}$. Thus we must show that for all $p \in \mathcal{Q}$ and all $x^0, \ldots, x^{n-1} \in X$, $p(x^0, \ldots, x^{n-1})$ is defined in X whenever $p(x^0\epsilon_X, \ldots, x^{n-1}\epsilon_X)$ is defined in $DE(X)$.

Without loss of generality we may assume that X is a closed subalgebra of A^κ for some κ. Note that the γth projection $\pi_\gamma: X \to A$ is a continuous homomorphism since it is the restriction of the continuous homomorphism $\rho_\kappa^{-1} D(h_\gamma)\rho: A^\kappa \to A$, where $h_\gamma: \mathscr{F}_{\mathbf{A}}(1) \to \mathscr{F}_{\mathbf{A}}(\kappa)$ is the homomorphism determined by $x_0 h_\gamma = x_\gamma$.

If $p(x^0\epsilon_X, \ldots, x^{n-1}\epsilon_X)$ is defined in $DE(X)$, then

$$p(x^0\epsilon_X, \ldots, x^{n-1}\epsilon_X): \mathbf{X}(X, A) \to A$$

is a homomorphism. Hence if q is a ν-ary polynomial and $\phi_\gamma \in \mathbf{X}(X, A)$ for all $\gamma < \nu$, then

$$q(\langle\phi_\gamma\rangle)p(x^0\epsilon_X, \ldots, x^{n-1}\epsilon_X) = q(\langle\phi_\gamma p(x^0\epsilon_X, \ldots, x^{n-1}\epsilon_X)\rangle).$$

In particular, let q be any κ-ary polynomial, and let $\phi_\gamma = \pi_\gamma$ for all $\gamma < \kappa$. Then

$$\begin{aligned}
p(q(x^0), \ldots, q(x^{n-1})) &= p(q(\langle x^0\pi_\gamma\rangle), \ldots, q(\langle x^{n-1}\pi_\gamma\rangle)) \\
&= p(x^0 q(\langle\pi_\gamma\rangle), \ldots, x^{n-1}q(\langle\pi_\gamma\rangle)) \\
&= p(q(\langle\pi_\gamma\rangle)x^0\epsilon_X, \ldots, q(\langle\pi_\gamma\rangle)x^{n-1}\epsilon_X) \\
&= q(\langle\pi_\gamma\rangle)p(x^0\epsilon_X, \ldots, x^{n-1}\epsilon_X) \\
&= q(\langle\pi_\gamma p(x^0\epsilon_X, \ldots, x^{n-1}\epsilon_X)\rangle) \\
&= q(\langle p(\pi_\gamma x^0\epsilon_X, \ldots, \pi_\gamma x^{n-1}\epsilon_X)\rangle) \\
&= q(\langle p(x^0\pi_\gamma, \ldots, x^{n-1}\pi_\gamma)\rangle) \\
&= q(p(x^0, \ldots, x^{n-1})).
\end{aligned}$$

It follows that $p(x^0, \ldots, x^{n-1})$ is defined in A^κ and hence, since X is a subalgebra of A^κ, $p(x^0, \ldots, x^{n-1})$ is defined in X. ∎

We may now combine this result with Proposition 2.24.

3.9 THEOREM. *Assume that* $\mathbf{X}(X, A)$ *is a subalgebra of* A^X *for all* $X \in \mathbf{X}$, *and that* $\langle D, E\rangle$ *is a duality between* $\mathbf{A}$ *and* $\mathbf{X}$. *If* $D\mathscr{C}(\text{-}, A)$: **Comp (ZComp)** $\to \mathbf{X}$

is left adjoint to the forgetful functor from **X** *into* **Comp** (**ZComp**) *and A is injective in* **A**, *then $\langle D, E\rangle$ is a full duality.*

3.10 *Remark.* Fatjlowicz [11, 12, 13] has also considered duality theory for prevarieties. Rather than consider compact topological partial algebras he assumes that both the category **A** and the category **X** are prevarieties; but a proper class of possibly infinitary operations is allowed. The results presented here intersect with his results in the case where **X** is a category of compact topological algebras of the same type as A; we now turn our attention to this case.

If each operation in the type of A gives rise to a full operation on $\mathbf{A}(B, A)$ for every $B \in \mathbf{A}$, then, by choosing $\mathcal{Q}$ to be the set of all operations on A, the category $\mathbf{X} = \mathbf{X}_{\mathcal{Q}}$ becomes a category of compact topological algebras of the same type as A.

3.11 PROPOSITION. (Evans [10]). *Let f_0 be an n-ary operation in the type of A. Then the following are equivalent*:

(i) *f_0 induces a full operation on $\mathbf{A}(B, A)$ for every $B \in \mathbf{A}$*;
(ii) *f_0 induces a full operation on $\mathbf{A}(\mathcal{F}_{\mathbf{A}}(\omega), A)$*;
(iii) *f_0 induces a full operation on $\mathbf{A}(\mathcal{F}_{\mathbf{A}}(m), A)$ for all $m < \omega$*;
(iv) *for all $m < \omega$, for every m-ary operation f_1 in the type of A, and for all $a^0, \ldots, a^{m-1} \in A^n$,*

$$f_0(f_1(a^0, \ldots, a^{m-1})) = f_1(f_0(a^0), \ldots, f_0(a^{m-1})).$$

Proof. (i) ⇒ (ii). This is trivial.

(ii) ⇒ (iii). $\mathcal{F}_{\mathbf{A}}(m)$ is (isomorphic to) a subalgebra of $\mathcal{F}_{\mathbf{A}}(\omega)$ and any homomorphism $g: \mathcal{F}_{\mathbf{A}}(m) \to A$ extends to a homomorphism $g': \mathcal{F}_{\mathbf{A}}(\omega) \to A$. If $g_0, \ldots, g_{n-1} \in \mathbf{A}(\mathcal{F}_{\mathbf{A}}(m), A)$, then $f_0(g_0, \ldots, g_{n-1})$ is the restriction of $f_0(g_0', \ldots, g_{n-1}')$ to $\mathcal{F}_{\mathbf{A}}(m)$. Since $f_0(g_0', \ldots, g_{n-1}')$ is a homomorphism, so is $f_0(g_0, \ldots, g_{n-1})$.

(iii) ⇒ (iv). This follows from Proposition 3.4(ii) with $\kappa = m$.

(iv) ⇒ (i). Let $g_0, \ldots, g_{n-1} \in \mathbf{A}(B, A)$. Since $f_0(g_0, \ldots, g_{n-1})$ is defined pointwise, the equations of (iv) imply that $f_0(g_0, \ldots, g_{n-1})$ is a homomorphism. ∎

3.12 *Remark.* The operation f_0 may be defined on Hom(B, C) for every algebra C of the same type as A. Since condition (iv) is a set of equations, (i)–(iv) are equivalent to

(v) *f_0 induces a full operation on* Hom(B, C) *for all B, C in the equational class generated by A.*

If $\mathcal{Q}$ is the set of all operations in the type of A and $\mathbf{A}(B, A)$ is a full algebra for all $B \in \mathbf{A}$, then $\mathbf{X}(X, A)$ is a subalgebra of A^X for all $X \in \mathbf{X}_{\mathcal{Q}}$, as the following slightly more general lemma implies.

3.13 LEMMA. *Assume that each operation in the type of A induces a full operation on $\mathbf{A}(B, A)$ for all $B \in \mathbf{A}$. Let $\mathbf{X}$ be any full subcategory of the category of all compact topological algebras of the same type as A, and assume that $A \in \mathbf{X}$ and that $\mathbf{X}(X, A)$ is nonempty for all $X \in \mathbf{X}$. Then $\mathbf{X}(X, A)$ is a subalgebra of A^X for all $X \in \mathbf{X}$, and hence $E = \mathbf{X}(\text{-}, A): \mathbf{X}^{\text{op}} \to \mathbf{A}$ is a well-defined functor.*

Proof. Let $G: \mathbf{X} \to \mathbf{A}$ be the forgetful functor. Let f be an m-ary operation and let $\phi_0, \ldots, \phi_{m-1} \in \mathbf{X}(X, A)$. Then $f(\phi_0, \ldots, \phi_{m-1})$ is a homomorphism since $\mathbf{A}(G(X), A)$ is a full algebra and $\mathbf{X}(X, A) \subseteq \mathbf{A}(G(X), A)$. Similarly, $f(\phi_0, \ldots, \phi_{m-1})$ is continuous since $\mathbf{X}(X, A) \subseteq \mathscr{C}(X, A)$. Hence $f(\phi_0, \ldots, \phi_{m-1}) \in \mathbf{X}(X, A)$ since $\mathbf{X}$ is a full subcategory of the category of all compact topologica algebras of the same type as A. ▮

We now describe an analog of the *Bohr compactification functor* from the theory of compact abelian groups (see [22]).

3.14 THEOREM. *Let $\mathcal{Q}$ be the set of operations in the type of A and assume that $\mathbf{A}(B, A)$ is a full algebra for all $B \in \mathbf{A}$. If $\langle D, E \rangle$ is a full duality between $\mathbf{A}$ and $\mathbf{X} = \mathbf{X}_{\mathcal{Q}}$, then $DGD: \mathbf{A} \to \mathbf{X}$ is left adjoint to the forgetful functor $G: \mathbf{X} \to \mathbf{A}$.*

Proof. (Sketch). The unit $\mu: \mathrm{id}_{\mathbf{A}} \to GDGD$ of the adjunction is given by $g(b\mu_B) = bg$ for all $b \in B$ and all $g \in GD(B) = G\mathbf{A}(B, A)$. Since every $X \in \mathbf{X}$ is iseomorphic to $D(C)$ for some $C \in \mathbf{A}$, it is sufficient to prove that for each homomorphism $\alpha: B \to GD(C)$, there is a unique morphism $\beta: DGD(B) \to D(C)$ in $\mathbf{X}$ with $\mu_B G(\beta) = \alpha$. Define $\delta: C \to G\mathbf{A}(B, A)$ by $b(c\delta) = c(b\alpha)$; then $\beta = D(\delta)$ is the required morphism. ▮

We close by noting an important restriction on the applicability of the theory expounded in this section.

3.15 PROPOSITION. *If $\langle D, E \rangle$ is a duality between $\mathbf{A}$ and $\mathbf{X} = \mathbf{X}_{\mathcal{Q}}$, then the algebra A has at most one one-element subalgebra.*

Proof. If $\langle D, E \rangle$ is a duality then $\eta_A: A \to \mathbf{X}(\mathbf{A}(A, A), A)$ is an isomorphism. If $\{b, c\}$ are one-element subalgebras of A then the constant endomorphisms $\bar{b}, \bar{c}: A \to A$ onto $\{b\}$ and $\{c\}$ respectively are elements of $\mathbf{A}(A, A)$.

If $p \in \mathscr{Q}$ is nullary, then $b = c$ since any one-element subalgebra of A must equal $\{p\}$. Thus we may assume that every $p \in \mathscr{Q}$ has arity greater than zero. For each $p \in \mathscr{Q}$, $p(b, \ldots, b)$ is defined in the partial algebra A since for any unary polynomial q,

$$p(q(b), \ldots, q(b)) = p(b, \ldots, b) = b = q(b) = q(p(b, \ldots, b)).$$

Thus the constant map $\hat{b}: \mathbf{A}(A, A) \to A$ onto $\{b\}$ is an element of $\mathbf{X}(\mathbf{A}(A, A), A)$ and hence there exists $a \in A$ with $\Gamma_a = \hat{b}$. Thus $b = \bar{c}$ and $\hat{b} = \bar{c}\Gamma_a = a\bar{c} = c$. ∎

4. Examples

4.1 *Pontryagin's Duality for Abelian Groups*

This famous duality theorem is due to Pontryagin [37, 38]. The main distinguishing feature of a proof based on Theorem 3.7 and Theorem 3.8 is that we need prove the duality only for $\mathscr{F}_{\mathbf{A}}(1)$ (i.e., $\mathbb{Z}$) rather than for all elementary groups, which is the usual technique of proof.

Let **Ab** be the category of abelian groups and let **K** be the category of compact abelian groups. Denote by T the compact abelian group $\mathbb{R}/\mathbb{Z}$.

4.1.1 Theorem. *$\langle$**Ab**$(\text{-}, T)$, **K**$(\text{-}, T)\rangle$ is a full duality between **Ab** and **K**.*

Proof. Since $\{0\}$ is a one-element subalgebra of T, $-(a + b) = (-a) + (-b)$, and $(a + b) + (c + d) = (a + c) + (b + d)$, $\mathbf{Ab}(B, T)$ is an abelian group for all $B \in \mathbf{Ab}$ (Proposition 3.11), and hence both $\mathbf{Ab}(\text{-}, T)$ and $\mathbf{K}(\text{-}, T)$ are well-defined functors (Lemma 3.13).

Since all subdirectly irreducible abelian groups, the cyclic groups of prime-power order and the Prüfer groups, are isomorphic to subgroups of T it follows that $\mathbf{Ab} = \mathbf{ISPS}(\{T\}) = \mathbf{ISP}(\{T\})$. That every compact abelian group X is iseomorphic to a closed subalgebra of a power of T follows from the fact that $\mathbf{K}(X, T)$ separates the points of X, unfortunately, a proof of this result requires some nontrivial representation theory for (locally) compact abelian groups (see [39, C, p. 241] or [22, Theorem (22.17), p. 345]). That T is injective in **K** is also nontrivial. By an application of the Stone–Weierstrass theorem, it is shown in [22, Theorem (23.20), p. 364] that if Y is a compact abelian group and B is a subgroup of $\mathbf{K}(Y, T)$ that separates the points of Y, then $B = \mathbf{K}(Y, T)$. Now let Y be a closed subgroup of a compact abelian group X and let $B = \{\phi | Y : \phi \in \mathbf{K}(X, T)\}$. Since B separates the points of Y it follows that $B = \mathbf{K}(Y, T)$ and hence every character $\phi \in \mathbf{K}(Y, T)$ is the restriction to Y of a character $\phi' \in \mathbf{K}(X, T)$; whence T is injective in **K**.

It is clear that $\mathbf{K}(T^n, T) \simeq \sum_{j<n} \mathbf{K}(T, T)$, for all $1 \leqslant n < \omega$. Hence to show that each $\phi \in \mathbf{K}(T^n, T)$ is a polynomial function it is sufficient to show that each $\phi \in \mathbf{K}(T, T)$ is a polynomial function; that is, since $\mathscr{F}_{\mathbf{A}}(1) \simeq \mathbb{Z}$ it is sufficient to show that $\eta_{\mathbb{Z}}: \mathbb{Z} \to \mathbf{K}(T, T)$, defined by $a(n\eta_{\mathbb{Z}}) = na$ for all $a \in T$, is an isomorphism. For this we refer to [39, C, p. 247].

We now show that for all $\kappa \geqslant \omega$, every $\phi \in \mathbf{K}(T^\kappa, T)$ has a finite support. Note that $\{\Lambda_n : 1 \leqslant n < \omega\}$ is a neighborhood basis of zero, where $\Lambda_n = \{x + \mathbb{Z} \in T : |x| < 1/(3n)\}$, and if $a \in T$ satisfies $na \in \Lambda_1$ for all $1 \leqslant n < \omega$, then $a = 0$. Let $\phi \in \mathbf{K}(T^\kappa, T)$; then $\Lambda_1 \phi^{-1}$ is open in T^κ and hence there exist $\gamma_0, \ldots, \gamma_{k-1} < \kappa$ and $m < \omega$ such that $\bigcap(\langle \gamma_j; \Lambda_m \rangle : j < k) \subseteq \Lambda_1 \phi^{-1}$. If $\langle a_\gamma \rangle, \langle b_\gamma \rangle \in T^\kappa$ with $a_{\gamma_j} = b_{\gamma_j}$ for all $j < k$, then $a_{\gamma_j} - b_{\gamma_j} = 0$ for all $j < k$, and hence

$$n(\langle a_\gamma \rangle - \langle b_\gamma \rangle) \in \bigcap(\langle \gamma_j; \Lambda_m \rangle : j < k) \qquad \text{for all} \quad 1 \leqslant n < \omega.$$

Thus $n(\langle a_\gamma \rangle \phi - \langle b_\gamma \rangle \phi) \in \Lambda_1$ for all $1 \leqslant n < \omega$ and so $\langle a_\gamma \rangle \phi - \langle b_\gamma \rangle \phi = 0$. Consequently $\{\gamma_j : j < k\}$ is a finite support for ϕ.

Hence by Theorem 3.7, $\langle \mathbf{Ab}(\text{-}, T), \mathbf{K}(\text{-}, T) \rangle$ is a duality and by Theorem 3.8 it remains to prove that $\epsilon_X : X \to \mathbf{Ab}(\mathbf{K}(X, T), T)$ is a surjection for all $X \in \mathbf{K}$. If ϵ_X is not onto, then there is a nonzero character of the topological factor group $\mathbf{Ab}(\mathbf{K}(X, T), T)/X\epsilon_X$ and hence there is a nonzero character of $\mathbf{Ab}(\mathbf{K}(X, T), T)$ whose restriction to $X\epsilon_X$ is the zero character. Thus it is sufficient to show that each character $\phi : \mathbf{Ab}(\mathbf{K}(X, T), T) \to T$ that is zero on $X\epsilon_X$ is identically zero.

Since $\langle \mathbf{Ab}(\text{-}, T), \mathbf{K}(\text{-}, T) \rangle$ is a duality there exists $\psi \in \mathbf{K}(X, T)$ with $\psi\eta_{\mathbf{K}(X,T)} = \Gamma_\psi = \phi$. Now, denoting the constant map onto $\{0\}$ by $\hat{0}$, we have

$$\begin{aligned} \phi | X\epsilon_X = \hat{0} &\Leftrightarrow \Gamma_\psi | X\epsilon_X = \hat{0} \Leftrightarrow x\epsilon_X \Gamma_\psi = 0 \qquad \text{for all} \quad x \in X \\ &\Leftrightarrow \psi(x\epsilon_X) = 0 \qquad \text{for all} \quad x \in X \Leftrightarrow x\psi = 0 \qquad \text{for all} \quad x \in X \\ &\Leftrightarrow \psi = \hat{0} \Leftrightarrow \phi = \Gamma_\psi = \hat{0}, \end{aligned}$$

as required. ▌

An application of Proposition 2.23 yields the **Comp**-free functor for **K** and an application of Theorem 3.14 yields the Bohr compactification functor.

4.1.2 PROPOSITION. (i) $\mathbf{Ab}(\mathscr{C}(\text{-}, T), T)$: **Comp** $\to$ **K** *is left adjoint to the forgetful functor from* **K** *into* **Comp**.

(ii) $\mathbf{Ab}(T^{(\text{-})}, T)$: **Set** $\to$ **K** *is left adjoint to the forgetful functor from* **K** *into* **Set**.

(iii) $\mathbf{Ab}(G\mathbf{Ab}(\text{-}, T), T)$: **Ab** $\to$ **K** *is left adjoint to the forgetful functor* $G : \mathbf{K} \to \mathbf{Ab}$.

Little need be said about the applications of this duality; one need only refer to the texts by Pontryagin [39] and Hewitt and Ross [22].

4.2 *Duality for Meet-Semilattices*

Although this duality dates back to Austin [4], the most recent and thorough exposition is certainly the monograph [24] by Hofmann *et al.* The proof given here is quite different from the proof given in [24], which uses the concepts of density and codensity in categories.

Let **Sl** be the category of meet-semilattices with zero and unit, let **Z** be the category of totally disconnected compact topological meet-semilattices, and let **2** denote the two element chain.

4.2.1 THEOREM. *$\langle$*$\mathbf{Sl}(\text{-}, \mathbf{2}), \mathbf{Z}(\text{-}, \mathbf{2})$*$\rangle$ is a full duality between* **Sl** *and* **Z**.

To prove this result we require two elementary facts about **Z**.

4.2.2 LEMMA. *For all $X \in \mathbf{Z}$, each $x \in X$ has a neighborhood basis of clopen subsemilattices.*

Proof (L. B. Schneperman [44]). Let U be a clopen neighborhood of x and by an application of Zorn's lemma let S be a subsemilatice of X that contains x, is a subset of U, and is maximal with respect to these properties. Since the closure of a subsemilattice is a subsemilattice and since U is closed, it follows that S is closed; it remains to prove that S is open.

Let $y \in S$; then $y \wedge S \subseteq S \subseteq U$ and hence, by a simple compactness argument using the continuity of the meet operation, there exists an open neighborhood V of y with $V \subseteq U$ and $V \wedge S \subseteq U$. We claim that $V \subseteq S$, whence S is open. If $z \in V - S$, then $T = S \cup (z \wedge S) \cup \{z\}$ is a subsemilattice of X which is strictly larger than S. But $S \subseteq U$, $z \wedge S \subseteq V \wedge S \subseteq U$, and $z \in V \subseteq U$, and thus $T \subseteq U$, contradicting the maximality of S. ∎

Although every $X \in \mathbf{Z}$ is necessarily a meet-complete lattice, we need only the fact that X has a zero.

For each $x \in X$ define the *translation* $t_x : X \to X$ by $yt_x = y \wedge x$; clearly t_x is continuous and hence $(x] = \mathrm{Im}(t_x)$ is closed in X.

4.2.3 LEMMA. *Every $X \in \mathbf{Z}$ has a zero.*

Proof. Consider X as a downward directed net on itself. Since X is compact, X has an accumulation point, say x. If x is not the zero of X, then there exists $y \in X$ with $y < x$. But then $U = X - (y]$ is an open neighborhood of x such that for all $z \leqslant y$, we have $z \notin U$, contradicting the fact that x is an accumulation point of X. ∎

Clearly **Sl**$(B, \mathbf{2})$ separates the points of B for all $B \in \mathbf{Sl}$, and hence $\mathbf{Sl} = \mathbf{ISP}(\{\mathbf{2}\})$. If we let $\mathscr{Q} = \{\wedge\}$, then since $(a \wedge b) \wedge (c \wedge d) = (a \wedge c) \wedge (b \wedge d)$, $0 \wedge 0 = 0$, and $1 \wedge 1 = 1$, it follows that $\wedge$ is a full operation on **Sl**$(B, \mathbf{2})$ for all $B \in \mathbf{Sl}$ (Proposition 3.11), and hence $\mathbf{X}_{\mathscr{Q}}$ is a subcategory of $\mathbf{Z}$ and **Sl**(-, **2**): $\mathbf{Sl} \to \mathbf{Z}$ is well defined. We now show that $\mathbf{X}_{\mathscr{Q}} = \mathbf{Z}$; that is, every totally disconnected compact meet-semilattice is iseomorphic to a closed subsemilattice of a power of **2**.

Clearly it is sufficient to show that $\mathbf{Z}(X, \mathbf{2})$ separates pairs of comparable elements of X. Let $x < y$; since $U = X - (x]$ is an open neighborhood of y, by Lemma 4.2.2 there is a clopen subsemilattice S of X with $x \in S \subseteq U$. Since S is closed in X it follows that $S \in \mathbf{Z}$ and hence S has a zero, say θ, by Lemma 4.2.3. Now $[\theta) = \{z \in X : z \wedge \theta \in S\} = St_{\theta}^{-1}$ is a clopen filter of X containing y but not containing x, and hence the characteristic function of $[\theta)$ is an element of $\mathbf{Z}(X, \mathbf{2})$ that separates x and y.

4.2.4 *Remark.* An alternative, though not particularly intuitive, proof may be obtained by showing that if $x < y$ and U is a clopen neighborhood of y with $x \notin U$, then

$$F = \{z \in X : z \wedge w \wedge y \in U \Leftrightarrow w \wedge y \in U, \text{ for all } w \in X\}$$

is a clopen filter with $y \in F$ and $x \notin F$; this can be proved without resource to Lemma 4.2.2 and Lemma 4.2.3.

To prove that **2** is injective in **Z** it is clearly sufficient to prove that if Y is a closed subsemilattice of X and U is a clopen filter of Y, then there exists a clopen filter V of X with $V \cap Y = U$. Let $x \in U$ and for all $y \in Y - U$ let $V_{x,y}$ be a clopen filter of X with $x \in V_{x,y}$ and $y \notin V_{x,y}$. Then $\{X - V_{x,y} : y \in Y - U\}$ is an open cover of $Y - U$ and hence has a finite subcover, say $\{X - V_{x,y_j} : j < n\}$. Let $V_x = \bigcap(V_{x,y_j} : j < n)$; then V_x is a clopen filter of X that contains x and satisfies $V_x \cap Y \subseteq U$. Since U is closed in Y it follows that $U \in \mathbf{Z}$ and hence U has a zero, say θ, by Lemma 4.2.3. Clearly $V = V_{\theta}$ is the required clopen filter.

That each $\phi \in \mathbf{Z}(\mathbf{2}^n, \mathbf{2})$ is a polynomial function is easily seen. If $1\phi^{-1}$ is empty, then $\phi = 0$, and if $1\phi^{-1} = \mathbf{2}^n$, then $\phi = 1$. Otherwise there exists $a \in \mathbf{2}^n$ with $1\phi^{-1} = [a)$ and it is clear that the polynomial $p(x_0, \ldots, x_{n-1}) = \bigwedge(x_j : a_j = 1)$ satisfies $b\phi = p(b)$ for all $b \in \mathbf{2}^n$.

Thus by Theorem 3.7, $\langle$**Sl**(-, **2**), **Z**(-, **2**)$\rangle$ is a duality and by Theorem 3.8 it remains to prove that $\epsilon_X : X \to \mathbf{Sl}(\mathbf{Z}(X, \mathbf{2}), \mathbf{2})$ is a surjection for all $X \in \mathbf{Z}$.

Let $g \in \mathbf{Sl}(\mathbf{Z}(X, \mathbf{2}), \mathbf{2})$; then $\{1\phi^{-1} : \phi \in 1g^{-1}\}$ has the finite intersection property and consequently $\bigcap(1\phi^{-1} : \phi \in 1g^{-1})$ is a nonempty closed subsemilattice of X that by Lemma 4.2.3 has a zero, say x. We shall prove that $x\epsilon_X = \Gamma_x = g$. Clearly $g \leqslant \Gamma_x$ since $\psi g = 1$ implies $x\psi = 1$, which in turn implies $\psi\Gamma_x = 1$. Now assume that $\psi\Gamma_x = x\psi = 1$. Then

$$1\psi^{-1} \supseteq [x) = \bigcap(1\phi^{-1}:\phi \in 1g^{-1}).$$

By compactness there exist $\phi_0, \ldots, \phi_{n-1} \in 1g^{-1}$ with $1\psi^{-1} \supseteq \bigcap(1\phi_j^{-1}:j<n)$. Hence $\psi \geqslant \bigwedge(\phi_j:j<n) \in 1g^{-1}$, and so $\psi g = 1$. Thus $\Gamma_x \leqslant g$.

This completes the proof of Theorem 4.2.1. ∎

Applying Proposition 2.23 we obtain the **ZComp**-free functor for **Z**. It is easily proved that for all $X \in$ **ZComp**, **Sl**($\mathscr{C}(X, \mathbf{2}), \mathbf{2}$) is iseomorphic to the hyperspace of X regarded as a meet-semilattice with set union as the operation (see [24, Proposition 2.6, p. 15]).

4.2.5 PROPOSITION. (i) **Sl**($\mathscr{C}$(-, **2**), **2**): **ZComp** → **Z** *is left adjoint to the forgetful functor from* **Z** *into* **ZComp**.

(ii) **Sl**($\mathbf{2}^{(-)}$, **2**): **Set** → **Z** *is left adjoint to the forgetful functor from* **Z** *into* **Set**.

A similar duality theorem may be proved for the category $\mathbf{Sl}_1$ of meet-semilattices with unit and the category $\mathbf{Z}_1$ of totally disconnected compact topological meet-semilattices with unit. In this case we could let $\mathscr{Q} = \{\wedge, 1\}$, prove that $\mathbf{Z}_1 = \mathbf{X}_{\mathscr{Q}}$, and then apply Theorem 3.7 and Theorem 3.8 as we did above; but there is no need, since the duality for $\mathbf{Sl}_1$ and $\mathbf{Z}_1$ can be obtained as a corollary of Theorem 4.2.1.

4.2.6 THEOREM. $\langle \mathbf{Sl}_1(\text{-}, \mathbf{2}), \mathbf{Z}_1(\text{-}, \mathbf{2})\rangle$ *is a full duality between* $\mathbf{Sl}_1$ *and* $\mathbf{Z}_1$.

Proof. For any $B \in \mathbf{Sl}_1$ let ${}_0B$ denote the bounded semilattice obtained by adjoining a new zero to B. If $g \in \mathbf{Sl}_1(B, \mathbf{2})$, then ${}_0g \in \mathbf{Sl}({}_0B, \mathbf{2})$ will denote the obvious extension of g. Clearly $g \to {}_0g$ is an iseomorphism of $\mathbf{Sl}_1(B, \mathbf{2})$ onto $\mathbf{Sl}({}_0B, \mathbf{2})$. It is also clear that for all $X \in \mathbf{Z}_1$,

$$\mathbf{Z}(X, \mathbf{2}) = \mathbf{Z}_1(X, \mathbf{2}) \cup \{\hat{0}\} \simeq {}_0(\mathbf{Z}_1(X, \mathbf{2})).$$

It follows at once that the functors $\mathbf{Sl}_1(\text{-}, \mathbf{2})$ and $\mathbf{Z}_1(\text{-}, \mathbf{2})$ are well defined, and furthermore

$$\mathbf{Z}_1(\mathbf{Sl}_1(B, \mathbf{2}), \mathbf{2}) \simeq \mathbf{Z}(\mathbf{Sl}({}_0B, \mathbf{2}), \mathbf{2}) - \{\hat{0}\} \simeq {}_0B - \{0\} = B$$

and

$$\mathbf{Sl}_1(\mathbf{Z}_1(X, \mathbf{2}), \mathbf{2}) \simeq \mathbf{Sl}({}_0(\mathbf{Z}_1(X, \mathbf{2})), \mathbf{2}) \simeq \mathbf{Sl}(\mathbf{Z}(X, \mathbf{2}), \mathbf{2}) \simeq X. \quad ∎$$

There is an obvious analog of Proposition 4.2.5 for $\mathbf{Sl}_1$ and $\mathbf{Z}_1$. Furthermore, since the algebras in both categories are of the same type, Theorem 3.14 describes the left adjoint to the forgetful functor from $\mathbf{Z}_1$ into $\mathbf{Sl}_1$.

4.2.7 PROPOSITION. (i) $\mathbf{Sl}_1(\mathscr{C}(\text{-}, \mathbf{2}), \mathbf{2})\colon \mathbf{ZComp} \to \mathbf{Z}_1$ *is left adjoint to the forgetful functor from* $\mathbf{Z}_1$ *into* **ZComp**.

(ii) $\mathbf{Sl}_1(\mathbf{2}^{(\text{-})}, \mathbf{2})\colon \mathbf{Set} \to \mathbf{Z}_1$ *is left adjoint to the forgetful functor from* $\mathbf{Z}_1$ *into* **Set**.

(iii) $\mathbf{Sl}_1(G\mathbf{Sl}_1(\text{-}, \mathbf{2}), \mathbf{2})\colon \mathbf{Sl}_1 \to \mathbf{Z}_1$ *is left adjoint to the forgetful functor* $G\colon \mathbf{Z}_1 \to \mathbf{Sl}_1$.

For applications of these dualities one need not look beyond the monograph [24]. Proposition 3.15 shows that the approach given here cannot be extended to develop a duality for meet-semilattices in general.

4.3 *Duality for Equational Classes Generated by Primal Algebras*

Hu [25] develops a duality theory for classes of algebras generated by locally primal algebras. The methods developed in Section 3 are well suited to proving the duality in the particular case where the locally primal algebra is finite and therefore primal.

4.3.1 DEFINITION. A finite nontrivial algebra A is said to be *primal* if for all $1 \leqslant n < \omega$, every map $\phi\colon A^n \to A$ is a polynomial function.

A discussion of primal algebras more than adequate for our purposes may be found in Pixley [36].

Let A be a primal algebra and let $\mathbf{A}$ be the equational class generated by A. Let $\mathscr{Q}$ be the empty set of polynomials. Then $\mathbf{X}_{\mathscr{Q}}$ is simply the category **ZComp** of Boolean spaces.

4.3.2 THEOREM. $\langle \mathbf{A}(\text{-}, A), \mathscr{C}(\text{-}, A)\rangle$ *is a full duality between* $\mathbf{A}$ *and* **ZComp**.

Proof. We have already seen that the functors are well defined (Lemma 2.3 and Proposition 2.4). Since a primal algebra is simple and has no proper subalgebras, and every algebra in A has distributive congruences, it follows by Jónsson's lemma (see [29, Corollary 3.4, p. 115]) that $\mathbf{A} = \mathbf{ISPHS}(\{A\}) = \mathbf{ISP}(\{A\})$. It is well known that any finite discrete space is injective in **ZComp**, and, by the definition of primality, every $\phi \in \mathscr{C}(A^n, A)$ is a polynomial function. Hence by Theorem 2.7 $\langle \mathbf{A}(\text{-}, A), \mathscr{C}(\text{-}, A)\rangle$ is a duality.

We could now prove that the duality is full by appealing to Jonsson's lemma again. Instead, we use the following results, which will have further applications anon. As usual, if B is a subalgebra of A^X, then for all $x \in X$, $\Gamma_x\colon B \to A$ is defined by $\phi\Gamma_x = x\phi$ for all $\phi \in B$. For all $a \in A$, $\hat{a}\colon X \to A$ denotes the constant map onto $\{a\}$. The monoid of endomorphisms of A is denoted by End(A), and the constant endomorphism onto a one-element subalgebra $\{a\}$ of A is denoted by $\bar{a}$.

4.3.3 PROPOSITION (Davey [8]). *Let A be a nontrivial finite algebra all of whose nontrivial subalgebras are subdirectly irreducible, and assume that every algebra in* $\mathbf{A} = \mathbf{ISP}(\{A\})$ *has distributive congruences. If X is a Boolean space and B is a subalgebra of $\mathscr{C}(X, A)$ containing the constant maps, then every homomorphism $g \in \mathbf{A}(B, A)$ is of the form $\Gamma_x e$ for some $x \in X$ and some* $e \in \mathrm{End}(A)$.

4.3.4 COROLLARY. *Assume that the conditions of the proposition hold and that $\{a_0, \ldots, a_{n-1}\}$ is the set of pairwise distinct elements that form one-element subalgebras of A. Let*

$$\mathscr{F}(X) = (X \times (\mathrm{End}(A) - \{\bar{a}_0, \ldots, \bar{a}_{n-1}\})) \cup \{\bar{a}_0, \ldots, \bar{a}_{n-1}\}$$

and define $\mu_X : \mathscr{F}(X) \to \mathbf{A}(\mathscr{C}(X, A), A)$ by $\langle x, e\rangle \mu_X = \Gamma_x e$ and $\bar{a}_j \mu_X = \hat{a}_j \; (j < n)$. Then μ_X is a homeomorphism of $\mathscr{F}(X)$ onto $\mathbf{A}(\mathscr{C}(X, A), A)$.

Since a primal algebra has no proper subalgebras and no proper endomorphisms it follows that for all $X \in \mathbf{ZComp}$ we have $X \simeq \mathscr{F}(X)$ and $\epsilon_X \simeq \mu_X$, whence Corollary 4.3.4 implies that ϵ_X is a homeomorphism. Thus the duality between $\mathbf{A}$ and $\mathbf{ZComp}$ is full. ∎

Stone's duality [45] for the category $\mathbf{B}$ of Boolean algebras follows at once.

4.3.5 THEOREM $\langle \mathbf{B}(\text{-}, \mathbf{2}), \mathscr{C}(\text{-}, \mathbf{2})\rangle$ *is a full duality between* $\mathbf{B}$ *and* $\mathbf{ZComp}$.

Proof. It is well known that the two-element Boolean algebra is primal; indeed, if $\phi : \mathbf{2}^n \to \mathbf{2}$, then $a\phi = p(a)$ for all $a \in \mathbf{2}^n$, where

$$p(x_0, \ldots, x_{n-1}) = \bigvee(\bigwedge(x_j : a_j = 1) \wedge \bigwedge(x_j' : a_j = 0) : a \in 1\phi^{-1}). \quad ∎$$

4.3.6 *Remark.* Keimel and Werner [30] have generalized Theorem 4.3.2 to equational classes generated by quasi-primal algebras. Since a quasi-primal algebra may have many one-element subalgebras, in fact a quasi-primal algebra may be idempotent, Proposition 3.15 shows that in general a duality for an equational class $\mathbf{A}$ generated by a quasi-primal algebra A cannot be obtained via the approach in Section 3. In Keimel and Werner's duality, the dual $\mathbf{A}(B, A)$ of an algebra $B \in \mathbf{A}$ is a Boolean space endowed with an action, by partial homeomorphisms, of the semigroup H of inner isomorphisms of A. If one attempts to apply Theorem 2.14 to prove this duality one finds that condition (D_0) is easily checked and that condition (D_1) is just the definition of quasi-primality. A proof for the injectivity condition (D_1) may be obtained by compiling the necessary facts from the proof of the duality given in [30], but there is no significant saving.

Applications of Stone's duality abound in both algebra and topology. Some applications of Hu's duality may be found in [25] and [26].

4.4 *Duality for Distributive Lattices*

The most famous duality for distributive lattices is certainly that of Stone [46]. His duality, which utilizes the concept of a spectral space, is purely topological.

That a finite distributive lattice is completely determined by its poset of join-irreducibles (and therefore by its poset of prime filters) is part of the folklore of the theory. In retrospect, it is surprising that a general duality theory for distributive lattices, utilizing the natural partial order of the prime filters, was not developed until the late 1960s.

Let **2** denote the two-element distributive lattice with the zero and unit as nullary operations. Then by the prime ideal theorem, $\mathbf{D} = \mathbf{ISP}(\{\mathbf{2}\})$ is the category of bounded distributive lattices. Let $\mathcal{Q} = \{\wedge\}$; then the objects of the category $\mathbf{R} = \mathbf{X}_{\mathcal{Q}}$ are (at least) totally disconnected compact topological partial meet-semilattices. The structure of the partial algebra $\mathbf{2}^{\kappa}$ is easily described; recall that whenever meet is defined in $\mathbf{2}^{\kappa}$ it is defined pointwise.

4.4.1 LEMMA. *For all $a, b \in \mathbf{2}^{\kappa}$, $a \wedge b$ is defined if and only if $a \leqslant b$ or $b \leqslant a$ in the pointwise order.*

Proof. For $\kappa = 1$ this is trivial, hence assume that $\kappa > 1$. If $a \leqslant b$, then for every κ-ary polynomial q, we have $q(a) \wedge q(b) = q(a) = q(a \wedge b)$ since polynomial functions are order-preserving. Conversely, assume that a and b are incomparable. Then there exist $\gamma, \lambda < \kappa$ such that $0 = a_{\gamma} < b_{\gamma} = 1$ and $1 = a_{\lambda} > b_{\lambda} = 0$. Put $q(x) = x_{\gamma} \vee x_{\lambda}$; then $q(a) \wedge q(b) = (0 \vee 1) \wedge (1 \vee 0) = 1 \neq 0 = (0 \wedge 1) \vee (1 \wedge 0) = q(a \wedge b)$, whence $a \wedge b$ is undefined. ∎

Recall that a subset U of a poset X is *increasing* (*decreasing*) if $x \in U$ and $y \geqslant x$ ($y \leqslant x$) imply that $y \in U$.

4.4.2 DEFINITION. A partially ordered topological space X is *totally order-disconnected* if for all $x, y \in X$ with $x \nleqslant y$, there exists a clopen increasing subset U of X such that $x \in U$ and $y \notin U$. Let **P** be the category whose objects are totally order-disconnected compact spaces and whose morphisms are continuous order-preserving maps.

4.4.3 PROPOSITION. (i) *Every object X of* **R** *becomes an object of* **P** *when partially ordered by*

$$x \leqslant y \Leftrightarrow (x \wedge y \text{ exists and equals } x).$$

(ii) *Every object* X *of* **P** *becomes an object of* **R** *when a partial meet operation is defined on* X *by*

$$(x \wedge y \text{ exists and equals } x) \Leftrightarrow x \leqslant y.$$

(iii) **R** *is isomorphic to the category* **P**.

Proof. (i) Let X be a closed subalgebra of $\mathbf{2}^\kappa$ and assume that $a, b \in X$ satisfy $a \nleqslant b$. Then there exists $\gamma < \kappa$ such that $a_\gamma = 1$ and $b_\gamma = 0$, and thus $U = \{c \in X : c_\gamma = 1\}$ is a clopen subset of X with $a \in U$ and $b \notin U$; finally, U is increasing by Lemma 4.4.1.

(ii) Let $X \in \mathbf{P}$ and endow X with its partial meet operation. The set $\mathbf{P}(X, \mathbf{2})$ separates the points of X, whence X is homeomorphic and order-isomorphic to a closed subspace of the totally order-disconnected space $\mathbf{2}^\kappa$, where κ is the cardinality of $\mathbf{P}(X, \mathbf{2})$. But any closed subspace of the partial algebra $\mathbf{2}^\kappa$ is a closed subalgebra, by Lemma 4.4.1, and hence by the definition of meet in X, any continuous order-isomorphism of X into $\mathbf{2}^\kappa$ is an iseomorphism of the partial algebra X onto a closed subalgebra. Thus $X \in \mathbf{R}$.

(iii) Clearly this a corollary of (i) and (ii). ▮

Henceforth we work with **P** rather than **R**. Note that if we were to choose $\mathcal{Q}$ to be the set of all operations in the type of **2**, that is $\mathcal{Q} = \{\wedge, \vee, 0, 1\}$, then $\mathbf{R} = \mathbf{X}_{\mathcal{Q}}$ would still be isomorphic to **P**; indeed, for all $X \in \mathbf{R}$, we have $\mathcal{D}(\vee, X) = \mathcal{D}(\wedge, X)$, and hence join induces the same partial order as meet, and $\mathcal{D}(0, X)$ and $\mathcal{D}(1, X)$ are both empty (see Remark 3.1(iii)).

4.4.4 Theorem (Priestly [40, 41]). $\langle \mathbf{D}(\text{-}, \mathbf{2}), \mathbf{P}(\text{-}, \mathbf{2})\rangle$ *is a full duality between* **D** *and* **P**.

Proof. To show that **2** is injective in **P** it is sufficient to show that for all $X, Y \in \mathbf{P}$, if Y is a closed subspace of X and U is a clopen increasing subset of Y, then there is a clopen increasing subset V of X with $V \cap Y = U$. As in the proof of the injectivity of **2** in **Z** (immediately following Remark 4.2.4), for all $x \in U$ let V_x be a clopen increasing subset of X with $x \in V_x$ and $V_x \cap Y \subseteq U$. Let $\{V_{x_j} : j < m\}$ be a finite subcover of the cover $\{V_x : x \in U\}$ of U. Then $V = \bigcup(V_{x_j} : j < m)$ satisfies $V \cap Y = U$, as required.

A map $\phi \in \mathbf{P}(\mathbf{2}^n, \mathbf{2})$ if and only if it is order-preserving, and hence $F = 1\phi^{-1}$ is increasing. If F is empty, then $\phi = 0$, and if $F = \mathbf{2}^n$, then $\phi = 1$. Otherwise there exist distinct, nonzero elements $a^0, \ldots, a^{m-1} \in \mathbf{2}^n$ with $F = \bigcup([a^j) : j < m)$. Define

$$p(x_0, \ldots, x_{n-1}) = \bigvee(\bigwedge(x_i : a_i^j = 1) : j < m);$$

clearly $b\phi = p(b)$ for all $b \in \mathbf{2}^n$.

Observe that **2** satisfies the conditions of Proposition 4.3.3 and that **2** has no proper endomorphisms. Hence for all $X \in \mathbf{P}$, $\epsilon_X : X \to \mathbf{D}(\mathbf{P}(X, \mathbf{2}), \mathbf{2})$ is onto by Proposition 4.3.3 with $B = \mathbf{P}(X, \mathbf{2})$.

Theorem 3.7 and Theorem 3.8 now combine to yield the result. ∎

4.4.5 *Remark.* Products in the category **R** are a little unorthodox since the ordered-space structure on $X \times Y$ is just the pointwise structure, while the partial algebra structure on $X \times Y$ need not be the pointwise one: if $x_1 < x_2$ in X and $y_1 > y_2$ in Y, then in the pointwise structure $\langle x_1, y_1 \rangle \wedge \langle x_2, y_2 \rangle$ is defined, but it is not defined in the product of X and Y in **R** since $\langle x_1, y_1 \rangle$ and $\langle x_2, y_2 \rangle$ are noncomparable.

There is a natural embedding of **ZComp** into **P** that endows a Boolean space with the discrete partial order. The Stone-Čech compactification functor may be lifted to a functor β: **Set** → **P** in the same way. Proposition 2.23 becomes rather trivial in the present setting.

4.4.6 Proposition. (i) *The natural embedding of* **ZComp** *into* **P** *is left adjoint to the forgetful functor from* **P** *into* **ZComp**.

(ii) β: **Set** → **P** *is left adjoint to the forgetful functor from* **P** *into* **Set**.

A duality for distributive lattices with unit may now be obtained very simply. Since the proof is almost identical to the proof of the duality between $\mathbf{Sl}_1$ and $\mathbf{Z}_1$ (Theorem 4.2.6), it is omitted.

Let **2** now denote the two-element distributive lattice with the unit as a nullary operation. By the prime ideal theorem, $\mathbf{D}_1 = \mathbf{ISP}(\{\mathbf{2}\})$ is the category of distributive lattices with unit. Let $\mathcal{Q} = \{\wedge, 1\}$; then the objects of the category $\mathbf{P}_1$ (which is isomorphic to $\mathbf{R}_1 = \mathbf{X}_{\mathcal{Q}}$) are totally order-disconnected compact spaces with unit and $\psi \in \mathbf{P}_1(X, Y)$ if and only if ϕ is continuous, order-preserving, and unit-preserving.

4.4.7 Theorem. $\langle \mathbf{D}_1(\text{-}, \mathbf{2}), \mathbf{P}_1(\text{-}, \mathbf{2}) \rangle$ *is a full duality between* $\mathbf{D}_1$ *and* $\mathbf{P}_1$.

Again, Proposition 3.15 shows that the natural extension of this approach will not yield a duality theory for distributive lattices in general.

Many applications have been found for these dualities. See, for example, [1, 2, 3, 6, 7, 8, 41, 42, 43].

4.5 *Duality for Stone Algebras*

For a thorough discussion of Stone algebras we refer the reader to [17]. The duality presented here is new but is very similar to the duality for Stone algebras developed independently by Priestley [42]. The main difference between the two approaches lies in the fact that here the category **St** of

Stone algebras is treated independently of the category **D** of bounded distributive lattices, and hence the three-element chain plays an important role, while in [42] the category **St** is treated as a subcategory of **D** and hence the two-element chain has the upper hand. (See Remark 4.5.4.)

Let **3** denote the three-element Stone algebra, $0 < a < 1$. Then **3** is an algebra of type $\langle 2, 2, 1, 0, 0\rangle$ with operations $\langle \wedge, \vee, *, 0, 1\rangle$, where $0^* = 1$, $a^* = 0$, and $1^* = 0$. Since **2** and **3** are the only subdirectly irreducible Stone algebras (see [16, 31, 33]) it follows that $\mathbf{S} = \mathbf{ISP}(\{\mathbf{3}\})$ is the category of Stone algebras. Let $\mathbf{W} = \mathbf{X}_{\mathscr{Q}}$, where $\mathscr{Q} = \{\wedge, **\}$. Recall that whenever the operations $\wedge$ and $**$ are defined in $\mathbf{3}^\kappa$, they are defined pointwise.

4.5.1 LEMMA. (i) *Define a partial order* $\leqslant^*$ *on the set* $\mathbf{3}^\kappa$ *as follows*:

$$b \leqslant^* c \Leftrightarrow \textit{for all } \gamma < \kappa, \qquad b_\gamma = c_\gamma \qquad \textit{or} \qquad (b_\gamma = a \textit{ and } c_\gamma = 1);$$

or equivalently,

$$b \leqslant^* c \Leftrightarrow b \leqslant c \textit{ in } \mathbf{3}^\kappa \qquad \textit{and} \qquad (b_\gamma = 0 \textit{ implies } c_\gamma = 0).$$

Then $b \wedge c$ *is defined in* $\mathbf{3}^\kappa$ *if and only if* $b \leqslant^* c$ *or* $c \leqslant^* b$.

(ii) b^{**} *is defined for all* $b \in \mathbf{3}^\kappa$, *and* b^{**} *is the unique maximal element of the poset* $\langle \mathbf{3}^\kappa; \leqslant^*\rangle$ *that dominates* b.

Proof. Throughout the following we denote the natural partial order on **3**, and its pointwise extension to $\mathbf{3}^\kappa$, by $\leqslant$, and $\leqslant^*$ will denote the partial order defined above.

(i) Assume that $b \wedge c$ is defined. Since lattice polynomials are *a fortori* Stone algebra polynomials, it follows that $b \leqslant c$ or $c \leqslant b$ as in the proof of Lemma 4.4.1. Without loss of generality assume that $b \leqslant c$. If $b_\gamma < c_\gamma$, then set $q(x) = x_\gamma{}^*$. Since $b \wedge c$ is defined we have

$$b_\gamma{}^* = q(b) = q(b \wedge c) = q(b) \wedge q(c) = b_\gamma{}^* \wedge c_\gamma{}^* = c_\gamma{}^*,$$

whence $b_\gamma = a$ and $c_\gamma = 1$.

Conversely, assume that $b \leqslant^* c$. We claim that for every κ-ary polynomial q, $q(b \wedge c) = q(b) \wedge q(c)$, that is $q(b) \leqslant q(c)$. If $q(x) \in \{x_\gamma, x_\gamma \wedge x_\lambda, x_\gamma \vee x_\lambda, x_\gamma{}^*\}$, then it is clear that $q(b) \leqslant q(c)$. A simple induction on the rank of q now shows that $q(b) \leqslant q(c)$ for every κ-ary polynomial q.

(ii) The identities $0^{**} = 1$, $1^{**} = 0$, $(x_\gamma \wedge x_\lambda)^{**} = x_\gamma^{**} \wedge x_\lambda^{**}$, and $(x_\gamma{}^*)^{**}$, $= (x_\gamma^{**})^*$ hold in every pseudocomplemented lattice. Since the identity $(x_\gamma \vee x_\lambda)^{**} = x_\gamma^{**} \vee x_\lambda^{**}$ is characteristic of Stone algebras (see [14]) it follows by Proposition 3.11 that $**$ is a full operation on $\mathbf{3}^\kappa$. Since $c \in \mathbf{3}^\kappa$ is maximal with respect to $\leqslant^*$ if and only if $c_\gamma \in \{0, 1\}$ for all $\gamma < \kappa$, and $0^{**} = 0$ and $a^{**} = 1^{**} = 1$, it is obvious that b^{**} is the unique maximal element that dominates b. ∎

4.5.2 DEFINITION. Let **V** be the category given by: X is an object of **V** if it is a totally order-disconnected compact space, every $x \in X$ is dominated by a unique maximal element x^{∞}, and the map ${}^{\infty}: X \to X$ is continuous; ϕ is a morphism of **V** if it is continuous, order-preserving, and ${}^{\infty}$-preserving.

The partial order on an object in **V** will be denoted by $\leqslant^*$. The partial order on **3**, as an object of **V**, is determined by

$$x <^* y \Leftrightarrow x = a \qquad \text{and} \qquad y = 1,$$

and on **3** the map ${}^{\infty}$ is given by $0^{\infty} = 0$, $a^{\infty} = 1^{\infty} = 1$.

4.5.3 PROPOSITION. (i) *Every object X of* **W** *becomes an object of* **V** *when partially ordered by*

$$x \leqslant^* y \Leftrightarrow (x \wedge y \text{ exists and equals } x);$$

the map ${}^{\infty}: X \to X$ *is then given by* $x^{\infty} = x^{**}$.

(ii) *Every object X of* **V** *becomes an object of* **W** *when a partial meet operation is defined on X by*

$$(x \wedge y \text{ exists and equals } x) \Leftrightarrow x \leqslant^* y,$$

and a unary operation ${}^{**}: X \to X$ *is defined by* $x^{**} = x^{\infty}$.

(iii) **W** *is isomorphic to the category* **V**.

Proof. (i) This follows from Lemma 4.5.1 since every object of **W** is iseomorphic to a subalgebra of $\mathbf{3}^{\kappa}$ for some κ.

(ii) Let $X \in \mathbf{V}$ and define $\wedge$ and ** on X as indicated. We show that $\mathbf{V}(X, \mathbf{3})$ separates the points of X, from which (ii) follows easily. If x and y are distinct points of X, then either $x \nleqslant^* y$ or $y \nleqslant^* x$, say $x \nleqslant^* y$. Thus there exists a clopen increasing subset U of X with $x \in U$ and $y \notin U$. Let $U_0 = \{z \in X : z^{\infty} \in U\}$; since ${}^{\infty}$ is continuous and order-preserving, U_0 is clopen and increasing, and since $U_0 = (U]$, U_0 is also decreasing. Define $\phi: X \to \mathbf{3}$ by

$$z\phi = \begin{cases} 1 & \text{if } z \in U, \\ a & \text{if } z \in U_0 - U, \\ 0 & \text{if } z \notin U_0. \end{cases}$$

It is easily seen that ϕ is an element of $\mathbf{V}(X, \mathbf{3})$ that separates x and y.

(iii) This follows from (i) and (ii). ∎

Had we chosen $\mathcal{Q} = \{\wedge, \vee, 0, 1, {}^*, {}^{**}\}$, then **W** and **V** would still be isomorphic, since, for all $X \in \mathbf{W}$, $\mathcal{D}(\vee, X) = \mathcal{D}(\wedge, X)$, and hence join and meet induce the same partial order, and $\mathcal{D}(0, X)$, $\mathcal{D}(1, X)$, and $\mathcal{D}({}^*, X)$ are all empty. The polynomial $p(x) = x_\gamma^{**}$ is an example of a polynomial that gives rise to a full operation while all its nontrivial proper subpolynomials (namely $x_\gamma{}^*$) give rise to partial operations with empty domain.

4.5.4 *Remark.* For the remainder of this section we work with **V** rather than **W**. Note that for all $B \in \mathbf{St}$, the partial order $\leqslant^*$ on $\mathbf{St}(B, \mathbf{3})$ is just the natural pointwise order, and for all $g \in \mathbf{St}(B, \mathbf{3})$, the map $g^\infty \in \mathbf{St}(B, \mathbf{3})$ is defined by $bg^\infty = (bg)^{**}$. It is well known (see [17, 19]) that in a Stone algebra B every prime filter is contained in a unique maximal filter, and it is easily shown (see [5, 17]) that $g \in \mathbf{St}(B, \mathbf{3})$ if and only if $1g^{-1}$ is a prime filter and $\{a, 1\}g^{-1}$ is the unique maximal filter containing $1g^{-1}$. Define $\Upsilon : \mathbf{St}(B, \mathbf{3}) \to \mathbf{D}(B, \mathbf{2})$ by $g\Upsilon = g_1$, where g_1 is the characteristic function of $1g^{-1}$. It is easily seen that Υ is a homeomorphism and an order-isomorphism. This remark provides the link between our approach and the approach in [42].

4.5.5 THEOREM. $\langle \mathbf{St}(\text{-}, \mathbf{3}), \mathbf{V}(\text{-}, \mathbf{3})\rangle$ *is a full duality between* **St** *and* **V**.

Proof. By Theorem 3.7, to show that we have a duality it is sufficient to prove that **3** is injective in **V** and that for all $1 \leqslant n < \omega$ every $\phi \in \mathbf{V}(\mathbf{3}^n, \mathbf{3})$ is a polynomial function; but this is precisely the content of the next two lemmas. Since **3** is injective in **St** (see [5, 17]), to show that the duality is full it is sufficient to prove that $\mathbf{St}(\mathscr{C}(\text{-}, \mathbf{3}), \mathbf{3}) : \mathbf{ZComp} \to \mathbf{V}$ is a **ZComp**-free functor for **V** (see Theorem 3.9); this is proved below in Proposition 4.5.8. ▮

4.5.6 LEMMA. **3** *is injective in* **V**.

Proof. Let X be a closed subalgebra of Y, let $\phi \in \mathbf{V}(X, \mathbf{3})$, and let $J_0 = 0\phi^{-1}$, $J_a = a\phi^{-1}$, and $J_1 = 1\phi^{-1}$. Since $\{a\}$ is decreasing in **3** and $\{1\}$ is increasing in **3**, J_a is clopen decreasing in X and J_1 is clopen increasing in X. Let U_a be a clopen decreasing subset of Y with $U_a \cap X = J_a$, and let U_1 be a clopen increasing subset of Y with $U_1 \cap X = J_1$; apply the injectivity of **2** in **P**. Set $V_1 = \{y \in Y : y^\infty \in U_1\}$ and $V_a = U_a - U_1$. Define $\phi' : Y \to \mathbf{3}$ by

$$y\phi' = \begin{cases} 1 & \text{if } \ y \in V_1 - V_a, \\ a & \text{if } \ y \in V_1 \cap V_a, \\ 0 & \text{if } \ y \in Y - V_1. \end{cases}$$

Then ϕ' is continuous since V_1 and V_a are clopen. The set V_1 is both increasing and decreasing, and V_a is decreasing. Hence $V_1 - V_a$ is increasing, $V_1 \cap V_a$ is decreasing, and $Y - V_1$ is both increasing and decreasing; whence ϕ' is order-preserving. Since ϕ' is order-preserving, to prove that ϕ' is $^{\infty}$-preserving it is sufficient to show that if $y\phi' = a$, then $y^{\infty}\phi' = 1$. If $y\phi' = a$, then $y \in V_1 \cap V_a$ and so $y^{\infty} \in U_1$. Thus $(y^{\infty})^{\infty} = y^{\infty} \in U_1$ and hence $y^{\infty} \in V_1$. Also, $y^{\infty} \in U_1$ implies that $y^{\infty} \notin V_a = U_a - U_1$. Hence $y^{\infty} \in V_1 - V_a$; giving $y^{\infty}\phi' = 1$, as required. It is easily proved that $J_0 \subseteq Y - V_1$, $J_a \subseteq V_1 \cap V_a$, and $J_1 \subseteq V_1 - V_a$; whence $\phi'|X = \phi$. ∎

4.5.7 LEMMA. *For all* $1 \leqslant n < \omega$, *every* $\phi \in \mathbf{V}(\mathbf{3}^n, \mathbf{3})$ *is a polynomial function.*

Proof. Throughout this proof we make the usual convention that the meet of an empty set of variables is the nullary polynomial 1, and the join of an empty set of variables is the nullary polynomial 0.

Set $X = 3^n$, $X^{\infty} = \{b \in X : b_j \in \{0, 1\} \text{ for all } j < n\}$, $X_0{}^{\infty} = 0\phi^{-1} \cap X^{\infty}$ and $X_1{}^{\infty} = 1\phi^{-1} \cap X^{\infty}$; clearly $X^{\infty} = X_0{}^{\infty} \cup X_1{}^{\infty}$. Define $p^0(x)$ by

$$p^0(x) = \bigwedge(\bigvee(x_j^{**} : b_j = 0) \vee \bigvee(x_j{}^{*} : b_j = 1) : b \in X_0{}^{\infty}).$$

Since $(c\phi = 0 \Leftrightarrow c^{\infty} \in X_0{}^{\infty})$ and $(c\phi \neq 0 \Leftrightarrow c^{\infty} \in X_1{}^{\infty})$, it follows that $(c\phi = 0 \Leftrightarrow p^0(c) = 0)$ and $(c\phi \neq 0 \Leftrightarrow p^0(c) = 1)$.

Let $(z]_* = \{y \in X : y \leqslant^* z\}$. If $X_1{}^{\infty}$ is nonempty, then for all $z \in X_1{}^{\infty}$, $1\phi^{-1} \cap (z]_*$ is a nonempty increasing subset of $(z]_*$. Let M_z be the set of minimal elements of $1\phi^{-1} \cap (z]_*$, and define $p^z(x)$ by

$$p^z(x) = \bigvee(\bigwedge(x_j : b_j = 1) \wedge \bigwedge(x_j^{**} : b_j = a) \wedge \bigwedge(x_j{}^{*} : b_j = 0) : b \in M_z).$$

We claim that the polynomial

$$p(x) = p^0(x) \wedge \bigvee(p^z(x) : z \in X_1{}^{\infty})$$

is the required polynomial. Clearly it is sufficient to prove that $c\phi \neq 0$ and $c^{\infty} = z$ imply that $p^z(c) = c\phi$ and, for $z \neq w \in X_1{}^{\infty}$, $p^w(c) = 0$. We shall constantly, and without specific reference, use the fact that $b \leqslant^* c$ and $b_j = 0$ imply that $c_j = 0$.

If $c\phi = 1$, then $b \leqslant^* c$ for some $b \in M_z$. Thus

$$\bigwedge(c_j : b_j = 1) = 1, \qquad \bigwedge(c_j^{**} : b_j = a) = 1, \qquad \text{and} \qquad \bigwedge(c_j{}^{*} : b_j = 0) = 1,$$

and so $p^z(c) = 1$.

If $c\phi = a$, then for all $b \in M_z$, $b \nleqslant^* c$. Thus for all $b \in M_z$ there exists $i < n$ such that $c_i = a$ and $b_i = 1$. It follows that for all $b \in M_z$,

$$\bigwedge(c_j : b_j = 1) = a, \qquad \bigwedge(c_j^{**} : b_j = a) = 1, \qquad \text{and} \qquad \bigwedge(c_j{}^{*} : b_j = 0) = 1,$$

and so $p^z(c) = a$.

If $z \neq w \in X_1{}^{\infty}$, then there exists $i < n$ such that either ($z_i = 0$ and $w_i = 1$) or ($z_i = 1$ and $w_i = 0$). If $z_i = 0$ and $w_i = 1$, then for all $b \in M_w$ either $b_i = a$

or $b_i = 1$. Thus for all $b \in M_w$ either

$$\wedge(c_j : b_j = 1) = 0 \qquad \text{or} \qquad \wedge(c_j : b_j = a) = 0,$$

giving $p^w(c) = 0$. If $z_i = 1$ and $w_i = 0$, then for all $b \in M_w$, we have $b_i = 0$, and hence $\wedge(c_j^* : b_j = 0) = 0$, giving $p^w(c) = 0$. ∎

If $X \in \mathbf{ZComp}$, then let $\mathscr{F}(X) = X \times \mathbf{2} \in \mathbf{V}$ be determined by $\langle x, 0\rangle^\infty = \langle x, 1\rangle^\infty = \langle x, 1\rangle$, and $\langle x, a\rangle <^* \langle y, b\rangle \Leftrightarrow (x = y,\ a = 0 \text{ and } b = 1)$. If $\psi \in \mathscr{C}(X, Y)$, then define $\mathscr{F}(\psi) \in \mathbf{V}(\mathscr{F}(X), \mathscr{F}(Y))$ by $\langle x, a\rangle\mathscr{F}(\psi) = \langle x\psi, a\rangle$. Clearly $\mathscr{F} : \mathbf{ZComp} \to \mathbf{V}$ is a well-defined functor.

4.5.8 Proposition. (i) *$\mathscr{F}$: $\mathbf{ZComp} \to \mathbf{V}$ is naturally isomorphic to $\mathbf{St}(\mathscr{C}(-, \mathbf{3}), \mathbf{3}) : \mathbf{ZComp} \to \mathbf{V}$ and is left adjoint to the forgetful functor from $\mathbf{V}$ into $\mathbf{ZComp}$.*

(ii) *$\mathscr{F}(\beta(-)) : \mathbf{Set} \to \mathbf{V}$ is left adjoint to the forgetful functor from $\mathbf{V}$ into $\mathbf{Set}$.*

Proof. Let $\mathrm{End}(\mathbf{3}) = \{e_0, e_1\}$, where e_0 is the identity map and e_1 is given by $ce_1 = c^{**}$. Define $\mu_X : \mathscr{F}(X) \to \mathbf{St}(\mathscr{C}(X, \mathbf{3}), \mathbf{3})$ by $\langle x, d\rangle\mu_X = \Gamma_x e_d$. Since $\mathbf{3}$ satisfies the conditions of Proposition 4.3.3, Corollary 4.3.4 implies that μ_X is a homeomorphism. Since $e_0 \leqslant e_1$ (pointwise), it follows that μ_X is an order-isomorphism, and since $e_0{}^\infty = e_1{}^\infty = e_1$, μ_X is ${}^\infty$-preserving. Thus μ_X is an iseomorphism. A simple calculation shows that μ is a natural transformation, and hence $\mathscr{F}$ and $\mathbf{St}(\mathscr{C}(-, \mathbf{3}), \mathbf{3})$ are naturally isomorphic.

The unit $\zeta : \mathrm{id}_{\mathbf{ZComp}} \to \mathbf{V}$ of the adjunction from $\mathbf{ZComp}$ to $\mathbf{V}$ is defined by $x\zeta_X = \langle x, 0\rangle$. If $Y \in \mathbf{V}$ and $\alpha : X \to Y$ is continuous, then define $\beta : \mathscr{F}(X) \to Y$ by $\langle x, 0\rangle\beta = x\alpha$ and $\langle x, 1\rangle\beta = (x\alpha)^\infty$; clearly $\beta \in \mathbf{V}(\mathscr{F}(X), Y)$. Since $x\zeta_X\beta = \langle x, 0\rangle\beta = x\alpha$, we have $\zeta_X\beta = \alpha$, and the uniqueness of β is immediate. This establishes (i), and (ii) follows as an obvious corollary. ∎

The duality can be applied to show that if $(B_\delta : \delta \in \Delta)$ is a family of Stone algebras, then their free product in $\mathbf{D}$ is a Stone algebra and is, in fact, their free product in $\mathbf{St}$. The description of the finitely generated free algebras also follows easily (see [17, 18]). We close the section by using the duality to obtain Balbes and Grätzer's [5] description of the injectives in $\mathbf{St}$; our proof illustrates well the use of the $\mathbf{ZComp}$-free functor. (The $\mathbf{ZComp}$-free functor is used similarly in Davey [8], although there the dualities considered are not full.)

4.5.9 Theorem. *The following are equivalent:*

(i) *P is sur-projective in $\mathbf{V}$;*

(ii) *P is a retract of $\mathscr{F}(X) = X \times \mathbf{2}$ for some extremally disconnected compact space X;*

(iii) *there are extremally disconnected compact spaces* X_0 *and* X_1 *such that* P *is iseomorphic to* $\mathscr{F}(X_0) \dot{\cup} X_1$, *where* X_1 *is endowed with the discrete structure* ($x \leqslant^* y \Leftrightarrow x = y$, *and* $x^\infty = x$).

Proof. By Proposition 2.26, only the equivalence of (ii) and (iii) remains to be proved.

(ii) $\Rightarrow$ (iii). For convenience we assume that P is a subalgebra of $X \times \mathbf{2}$ and that $\tau: X \times \mathbf{2} \to P$ is a retraction. If $\langle x, 0\rangle \in P$, then $\langle x, 1\rangle = \langle x, 0\rangle^\infty \in P$, but it may happen that $\langle x, 1\rangle \in P$ while $\langle x, 0\rangle \notin P$. Hence, setting

$$X_0 = \{x \in X : \langle x, 0\rangle \in P\} \qquad \text{and} \qquad X_1 = \{x \in X : \langle x, 1\rangle \in P; x \notin X_0\},$$

we have $P = (X_0 \times \mathbf{2}) \dot{\cup} (X_1 \times \{1\})$. We claim that (a) $X_0 \cup X_1$ is a retract of X, and (b) X_0 and X_1 are clopen in $X_0 \cup X_1$. The result then follows since (a) and (b) imply that X_0 and X_1 are retracts of X and it is well known that a retract of an extremally disconnected compact space is itself extremally disconnected and compact (see [20]).

(a) Define $\rho: X \to X_0 \cup X_1$ by $x\rho = \langle x, 1\rangle\tau\pi$, where $\pi: X \times \mathbf{2} \to X$ is the natural projection. Let U be open in X; then

$$U\rho^{-1} = \{x \in X : \langle x, 1\rangle\tau\pi \in U\} = [(X \times \{1\}) \cap (U \times \mathbf{2})\tau^{-1}]\pi,$$

which is open since τ is continuous and π is open.

(b) We construct a set V which is clopen in X and satisfies $V \cap (X_0 \cup X_1) = X_0$. In fact, let

$$V = [(X \times \{0\}) \cap (X \times \{0\})\tau^{-1}]\pi.$$

Then V is clopen since τ is continuous and π, being a projection parallel to a compact factor, is both open and closed. It is easily verified that $X_0 \subseteq V$ and that $X_1 \cap V$ is empty, whence $V \cap (X_0 \cup X_1) = X_0$.

(iii) $\Rightarrow$ (ii). If P is iseomorphic to $(X_0 \times \mathbf{2}) \dot{\cup} X_1$, then it is also iseomorphic to the retract $(X_0 \times \mathbf{2}) \dot{\cup} (X_1 \times \{1\})$ of $(X_0 \dot{\cup} X_1) \times \mathbf{2}$. Since both X_0 and X_1 are extremally disconnected and compact, so is $X_0 \dot{\cup} X_1$. ∎

If B is a Boolean algebra, then let $B^{[2]} = \{\langle b_0, b_1\rangle \in B^2 : b_0 \leqslant b_1\}$. It is easily seen that $B^{[2]}$ is a Stone algebra in which $\langle b_0, b_1\rangle^* = \langle b_1', b_1'\rangle$. It follows from the results of Davey [7], and it is easily verified directly, that $B^{[2]}$ is isomorphic to $\mathscr{C}(X, \mathbf{3})$, where X is the Stone space of B.

4.5.10 (Balbes and Grätzer [5]). *The following are equivalent*:

(i) *I is injective in* **St**;

(ii) *I is isomorphic to* $\mathscr{C}(X_0, \mathbf{3}) \times \mathscr{C}(X_1, \mathbf{2})$ *for some extremally disconnect compact spaces* X_0 *and* X_1;

(iii) *I is isomorphic to $B_0^{[2]} \times B_1$ for some complete Boolean algebras B_0 and B_1.*

Proof. (i)$\Leftrightarrow$(ii). As we have already noted, **3** is injective in **St**. Hence by Proposition 2.27, I is injective in **St** if and only if there is a sur-projective P in **V** with $I \simeq \mathbf{V}(P, \mathbf{3})$. Hence I is injective in **St** if and only if there are extremally disconnected compact spaces X_0 and X_1 with

$$I \simeq \mathbf{V}(\mathscr{F}(X_0) \dot{\cup} X_1, \mathbf{3}) \simeq \mathbf{V}(\mathscr{F}(X_0), \mathbf{3}) \times \mathbf{V}(X_1, \mathbf{3}).$$

By the duality,

$$\mathbf{V}(\mathscr{F}(X_0), \mathbf{3}) \simeq \mathbf{V}(\mathbf{St}(\mathscr{C}(X_0, \mathbf{3}), \mathbf{3}) \simeq \mathscr{C}(X_0, \mathbf{3}),$$

and since $x^\infty = x$ for all $x \in X_1$, $\mathbf{V}(X_1, \mathbf{3}) \simeq \mathscr{C}(X_1, \mathbf{2})$, as required.

(ii)$\Leftrightarrow$(iii). This is immediate since extremally disconnected compact spaces are precisely the Stone spaces of complete Boolean algebras (see [20]). ∎

References

1. M. E. Adams, The structure of distributive lattices. Ph.D. Thesis, Univ. of Bristol, 1973.
2. M. E. Adams, The Frattini sublattice of a distributive lattice, *Algebra Universalis* **3** (1973), 216–228.
3. M. E. Adams, A problem of A. Monteiro concerning relative complementation of lattices, *Colloq. Math.* **30** (1974), 61–67.
4. C. W. Austin, Duality theorems for some commutative semigroups, *Trans. Amer. Math. Soc.* **109** (1963), 245–256.
5. R. Balbes and G. Grätzer, Injective and projective Stone algebras, *Duke Math. J.* **38** (1971), 339–347.
6. B. A. Davey, A note on representable posets, *Algebra Universalis* **3** (1973), 345–347.
7. B. A. Davey, Free products of bounded distributive lattices, *Algebra Universalis* **4** (1974), 106–107.
8. B. A. Davey, Dualities for equational classes of Brouwerian algebras and Heyting algebras, *Trans. Amer. Math. Soc.* **221** (1976), 119–146.
9. J. Dugundji, "Topology," Allyn & Bacon, Rockleigh, New Jersey, 1966.
10. T. Evans, Endomorphisms of abstract algebras, *Proc. Roy. Soc. Edinburgh Sect. A* **66** (1962), 53–64.
11. S. Fajtlowicz, Duality in universal algebra I (abstract #691-08-7), *Notices Amer. Math. Soc.* **19** (1972), A-45.
12. S. Fajtlowicz, Duality in universal algebra II (abstract #711-08-3), *Notices Amer. Math. Soc.* **21** (1974), A-47.
13. S. Fajtlowicz, Duality for algebras, *in* "Contributions to Universal Algebra" (B. Csakany, ed.), Coll. Math. Soc. János Bolyai Vol. 17, pp. 101–112, Budapest, 1977.
14. O. Frink, Pseudocomplements in semilattices, *Duke Math. J.* **29** (1962), 505–514.
15. G. Grätzer, "Universal Algebra," Van Nostrand–Reinhold, Princeton, New Jersey, 1968.

16. G. GRÄTZER, Stone algebras form an equational class (notes on lattice theory III), *J. Austral. Math. Soc.* **9** (1969), 308–309.
17. G. GRÄTZER, "Lattice Theory. First Concepts and Distributive Lattices," Freeman, San Francisco, California, 1971.
18. G. GRÄTZER AND H. LAKSER, Some applications of free distributive products (abstract #69T-A27), *Notices Amer. Math. Soc.* **16** (1969), 405.
19. G. GRÄTZER AND E. T. SCHMIDT, On a problem of M. H. Stone, *Acta Math. Acad. Sci. Hungar.* **8** (1957), 455–460.
20. P. R. HALMOS, "Lecture on Boolean Algebras," Mathematical Studies No. 1, Van Nostrand Reinhold, Princeton, New Jersey, 1963.
21. Z. HENDRIL AND A. PULTR, On categorical embedding of topological structures into algebraic, *Comment. Math. Univ. Carolinae* **7** (1966), 377–400.
22. E. HEWITT AND K. A. ROSS, "Abstract Harmonic Analysis I," Academic Press, New York, 1963.
23. K. H. HOFMANN AND K. KEIMEL, A general character theory for partially ordered sets and lattices, *Mem. Amer. Math. Soc.* **122** (1972).
24. K. H. HOFMANN, M. MISLOVE, AND A. STRALKA, "The Pontryagin Duality of Compact 0-Dimensional Semilattices and its Applications," Lecture Notes in Mathematics No. 396, Springer–Verlag, Berlin and New York, 1974.
25. T. K. HU, Stone duality for primal algebra theory, *Math. Z.* **110** (1969), 180–198.
26. T. K. HU, Characterization of algebraic functions in equational classes generated by independent primal algebras, *Algebra Universalis* **1** (1971), 187–191.
27. J. R. ISBELL, Top and its adjoint relatives, *Proceedings of the Kanpur Topological Conference (1968)*, pp. 143–154, Academia, Prague, 1971.
28. J. R. ISBELL, General functorial semantics, I, *Amer. J. Math.* **44** (1972), 535–596.
29. B. JÓNSSON, Algebras whose congruence lattices are distributive, *Math. Scand.* **21** (1967), 110–121.
30. K. KEIMEL AND H. WERNER, Stone duality for varieties generated by quasi-primal algebras, *Mem. Amer. Math. Soc.* **148** (1974), 59–85.
31. H. LAKSER, The structure of pseudocomplemented distributive lattices. I: subdirect decomposition, *Trans. Amer. Math. Soc.* **156** (1971), 335–342.
32. J. LAMBEK AND B. A. RATTRAY, Localization and duality in additive categories, *Houston J. Math.* **1** (1975), 87–100.
33. K. B. LEE, Equational classes of distributive pseudocomplemented lattices, *Canad. J. Math.* **22** (1970), 881–891.
34. F. E. J. LINTON, Applied functorial semantics. I, *Ann. Mat. Pura Appl.* **86** (IV) (1970), 1–14.
35. S. MACLANE, "Categories for the Working Mathematician," Springer–Verlag, Berlin and New York, 1971.
36. A. F. PIXLEY, The ternary discriminator function in universal algebra, *Math. Ann.* **191** (1971), 167–180.
37. L. S. PONTRYAGIN, Sur les groupes abéliens continus, *C. R. Acad. Sci. Paris Sér. A-B* **198** (1934), 328–330.
38. L. S. PONTRYAGIN, The theory of topological commutative groups, *Ann. Math.* **35** (1934), 361–388.
39. L. S. PONTRYAGIN, "Topological Groups," 2nd ed., Gordon & Breach, New York, 1966.
40. H. A. PRIESTLEY, Representations of distributive lattices by means of ordered Stone spaces, *Bull. London Math. Soc.* **2** (1970), 186–190.
41. H. A. PRIESTLEY, Ordered topological spaces and the representation of distributive lattices, *Proc. London Math. Soc.* (3) **24** (1972), 507–530.
42. H. A. PRIESTLEY, Stone lattices: a topological approach, *Fund. Math.* **84** (1974), 127–143.

43. H. A. PRIESTLEY, The construction of spaces dual to pseudocomplemented distributive lattices, *Quart. J. Math. Oxford Ser.* **26**(2) (1975), 215–228.
44. L. B. SCHNEPERMAN, On the theory of characters of locally bicompact topological semigroups, *Mat. Sb.* **77** (1968), 508–532. (*Math. USSR-Sb.* **6** (1968), 471–492.)
45. M. H. STONE, The theory of representations for Boolean algebras, *Trans. Amer. Math. Soc.* **40** (1936), 37–111.
46. M. H. STONE, Topological representations for distributive lattices and Brouwerian logics, *Časopis Pěst. Mat.* **67** (1937), 1–35.

AMS (MOS) subject classifications: 54H10, 18A40, 08A15.

Up–Down and Down–Up Partitions[†]

L. CARLITZ

Department of Mathematics
Duke University
Durham, North Carolina

Let $A(n, k)$ denote the number of solutions in positive integers $a_1, a_2, \ldots, a_k$ of $n = a_1 + a_2 + \cdots + a_k$, where

$$a_1 \leqslant a_2, \qquad a_2 \geqslant a_3, \qquad a_3 \leqslant a_4, \qquad \ldots .$$

Similarly let $B(n, k)$ denote the number of solutions in positive integers $a_1, a_2, \ldots, a_k$ of $n = a_1 + a_2 + \cdots + a_k$, where

$$a_1 \geqslant a_2, \qquad a_2 \leqslant a_3, \qquad a_3 \geqslant a_4, \qquad \ldots .$$

Generating functions are obtained for these as well as certain more general enumerants. For example, if

$$G_k(x) = \sum_{n=k}^{\infty} B(n, k)x^n,$$

then

$$1 + \sum_{k=1}^{\infty} G_{2k}(x)z^{2k} = \frac{1}{C(x, z)}, \qquad \sum_{k=0}^{\infty} G_{2k+1}(x)z^{2k+1} = \frac{S(x, z)}{C(x, z)},$$

where

$$C(x, z) = \sum_{k=0}^{\infty} (-1)^k \frac{x^{k(k+1)}z^{2k}}{(x)_{2k}},$$

$$S(x, z) = \sum_{k=0}^{\infty} (-1)^k \frac{x^{k^2+3k+1}z^{2k+1}}{(x)_{2k+1}},$$

and

$$(x)_k = (1 - x)(1 - x^2) \cdots (1 - x^k), \qquad (x)_0 = 1.$$

1. INTRODUCTION

A permutation $(a_1, a_2, \ldots, a_n)$ of $Z_n = \{1, 2, \ldots, n\}$ is called an *up–down* permutation if

$$a_1 < a_2, \qquad a_2 > a_3, \qquad a_3 < a_4, \qquad \ldots;$$

[†] Supported in part by NSF grant GP-37924 X1.

ISBN 0-12-599101-0

it is called a *down–up* permutation if

$$a_1 > a_2, \qquad a_2 < a_3, \qquad a_3 > a_4, \qquad \ldots .$$

Such permutations can be represented graphically in the following way:

It is clear that the transformation

$$b_i = n - a_i + 1 \qquad (i = 1, 2, \ldots, n)$$

establishes a one to one correspondence between up–down and down–up permutations of Z_n for $n \geqslant 2$. Let $A(n)$ denote the number of up–down permutations of Z_n. It is known [3, 105–112] that

$$\sum_0^\infty A(n)\frac{x^n}{n!} = \sec x + \tan x \qquad (A(0) = A(1) = 1). \tag{1.1}$$

This result has been generalized in the following manner. Let $k \geqslant 2$, $t \geqslant 0$. By a (k, t)-permutation of Z_{kn+t} is meant one of the following kind:

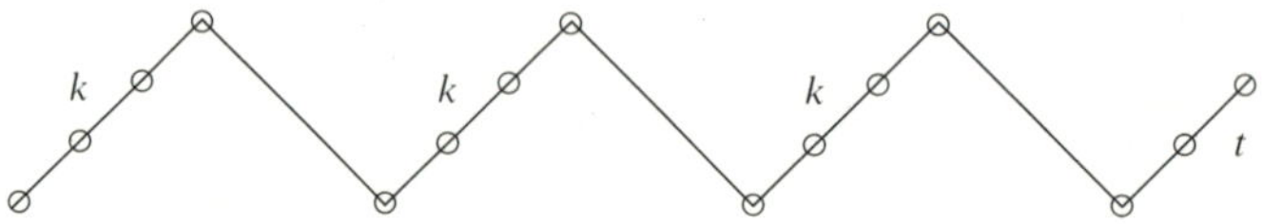

Let $A_{k,t}(kn + t)$ denote the number of (k, t)-permutations of Z_{kn+t}. It has been proved [1, 2] that

$$\sum_{n=0}^\infty A_{k,0}(kn)\frac{x^{kn}}{(kn)!} = \frac{1}{\phi_{k,0}(x)} \tag{1.2}$$

$$\sum_{n=0}^\infty A_{k,t}(kn)\frac{x^{kn+1}}{(kn+t)!} = \frac{\phi_{k,t}(x)}{\phi_{k,0}(x)} \qquad (t \geqslant 1), \tag{1.3}$$

where

$$\phi_{k,t}(x) = \sum_{n=0}^\infty (-1)^n \frac{x^{kn+t}}{(kn+t)!} \qquad (t = 0, 1, 2, \ldots). \tag{1.4}$$

For $0 \leqslant t \leqslant n - 1$, the $\phi_{k,t}(x)$ are the so-called Olivier functions [4].

Clearly, for $k = 2$, $t = 0, 1$, (1.2) and (1.3) reduce to (1.1).

Certain partition problems are suggested by the above. Let $A(n, k)$ denote the number of solutions in positive integers $a_1, a_2, \ldots, a_k$ of

$$n = a_1 + a_2 + \cdots + a_k, \tag{1.5}$$

where

$$a_1 \leqslant a_2, \qquad a_2 \geqslant a_3, \qquad a_3 \leqslant a_4, \qquad \ldots. \tag{1.6}$$

Similarly let $B(n, k)$ denote the number of positive solutions of (1.5) where now

$$a_1 \geqslant a_2, \qquad a_2 \leqslant a_3, \qquad a_3 \geqslant a_4, \qquad \ldots. \tag{1.7}$$

We may call such decompositions up–down and down–up partitions, respectively.

Clearly

$$A(n, 2k) = B(n, 2k). \tag{1.8}$$

Put

$$F_k(x) = \sum_{n=k}^{\infty} A(n, k)x^n, \qquad G_k(x) = \sum_{n=k}^{\infty} B(n, k)x^n.$$

We shall show that

$$1 + \sum_{k=1}^{\infty} G_{2k}(x)z^{2k} = \frac{1}{C(x, z)}, \tag{1.9}$$

$$\sum_{k=0}^{\infty} G_{2k+1}(x)z^{2k+1} = \frac{S(x, z)}{C(x, z)}, \tag{1.10}$$

where

$$C(x, z) = \sum_{k=0}^{\infty} (-1)^k \frac{x^{k(k+1)}z^{2k}}{(x)_{2k}}, \qquad S(x, z) = \sum_{k=0}^{\infty} (-1)^k \frac{x^{k^2+3k+1}z^{2k+1}}{(x)_{2k+1}}$$

and $(x)_k = (1 - x)(1 - x^2) \cdots (1 - x^k)$, $(x)_0 = 1$. Similar generating functions are obtained for $F_k(x)$.

Next we generalize these results by considering (k, t)-partitions of n into $sk + t$ parts with pattern

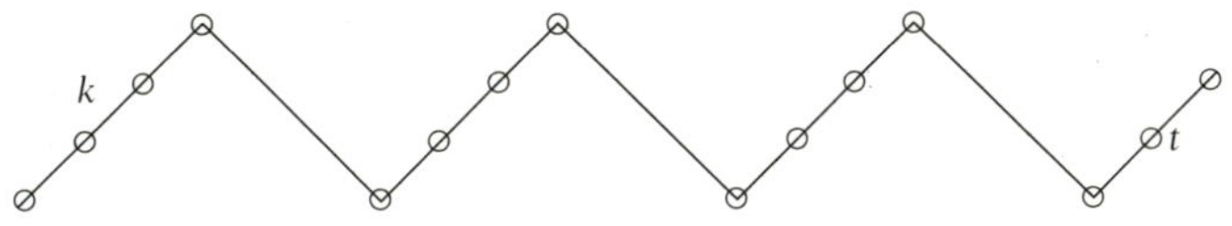

or

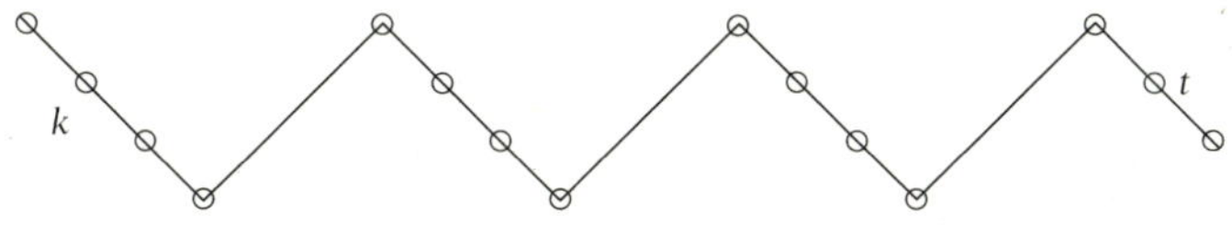

It is understood that all inequalities between adjacent elements of a partition are weak. We obtain generating functions for the corresponding enumerants that generalize (1.9) and (1.10) (see §§9, 10).

In §11 we extend these results further while in §12 we discuss several special cases of interest.

2. Notation

Let $A(n, k, a)$ denote the number of up–down partitions of n into k positive parts the first of which is equal to a; similarly, let $B(n, k, a)$ denote the number of down–up partitions of n into k positive parts the first of which is equal to a. For $k = 1$ we put

$$A(n, 1, a) = B(n, 1, a) = \delta_{n,a}.$$

Next we define, for $k \geqslant 1$,

$$F(x, k, a) = \sum_{n=1}^{\infty} A(n, k, a)x^n$$

$$F_k(x, y) = \sum_{a=1}^{\infty} F(x, k, a)y^a$$

$$G(x, k, a) = \sum_{n=1}^{\infty} B(n, k, a)x^n$$

$$G_k(x, y) = \sum_{a=1}^{\infty} G(x, k, a)y^a.$$

To complete the definitions we put

$$F_0(x, y) = G_0(x, y) = 1.$$

Clearly

$$A(n, k) = \sum_{a=1}^{n} A(n, k, a), \qquad B(n, k) = \sum_{a=1}^{n} B(n, k, a).$$

Also, since any partition of the type we are considering can be read in the opposite order it is evident that

$$A(n, 2k) = B(n, 2k). \tag{2.1}$$

However such a relation does not hold when the number of parts is odd.

It follows from the definitions that

$$F_k(x, 1) = \sum_{n=k}^{\infty} A(n, k)x^n, \qquad G_k(x, 1) = \sum_{n=k}^{\infty} B(n, k)x^n,$$

so that (2.1) may be replaced by

$$F_{2k}(x, 1) = G_{2k}(x, 1). \tag{2.2}$$

3. Down–Up Partitions

It is convenient to begin with down–up partitions. It is evident from the scheme

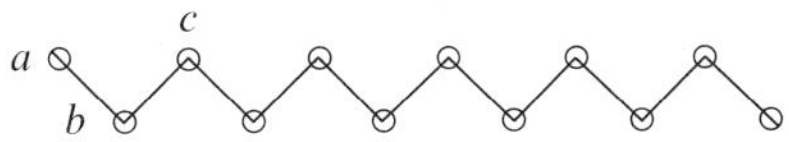

that, for $k > 2$,

$$B(n, k, a) = \sum_{\substack{b\leqslant a\\ b\leqslant c}} B(n-a-b, k-2, c).$$

Thus

$$\begin{aligned} G(x, k, a) &= \sum_{\substack{b\leqslant a\\ b\leqslant c}} x^{a+b} G(x, k-2, c) \\ &= \sum_{c=1}^{\infty} x^a G(x, k-2, c) \sum_{\substack{b\leqslant a\\ b\leqslant c}} x^b. \end{aligned}$$

It follows that

$$\begin{aligned} G_k(x, y) &= \sum_{a=1}^{\infty} y^a G(x, k, a) \\ &= \sum_{c=1}^{\infty} G(x, k-2, c) \sum_{b\leqslant c} x^b \sum_{a\geqslant b} (xy)^a \\ &= \frac{1}{1-xy} \sum_{c=1}^{\infty} G(x, k-2, c) \sum_{b=1}^{c} (x^2 y)^b \\ &= \frac{1}{1-xy} \sum_{c=1}^{\infty} G(x, k-2, c) \frac{x^2 y - (x^2 y)^{c+1}}{1-x^2 y}. \end{aligned}$$

Hence we get

$$G_k(x, y) = \frac{x^2 y}{(1-xy)(1-x^2 y)} [G_{k-2}(x, 1) - G_{k-2}(x, x^2 y)]. \tag{3.1}$$

Iteration of (3.1) gives

$$G_k(x, y) = \frac{x^2 y}{(xy)_2} G_{k-2}(x, 1) - \frac{x^6 y^2}{(xy)_4} [G_{k-4}(x, 1) - G_{k-4}(x, x^4 y)]$$

and generally, for $2s < k$,

$$G_k(x, y) = \sum_{j=1}^{s} (-1)^{j-1} \frac{x^{j(j+1)}y^j}{(xy)_{2j}} G_{k-2j}(x, 1) + (-1)^s \frac{x^{s(s+1)}y^s}{(xy)_{2s}} G_{k-2s}(x, x^{2s}y), \tag{3.2}$$

where

$$(z)_k = (1 - z)(1 - xz) \cdots (1 - x^{k-1}z), \qquad (z)_0 = 1.$$

In particular, replacing k by $2k$ and taking $s = k - 1$, (3.2) becomes

$$G_{2k}(x, y) = \sum_{j=1}^{k-1} (-1)^{j-1} \frac{x^{j(j+1)}y^j}{(xy)_{2j}} G_{2k-2j}(x, 1) + (-1)^{k-1} \frac{x^{k(k-1)}y^{k-1}}{(xy)_{2k-2}} G_2(x, x^{2k-2}y). \tag{3.3}$$

Now it is evident from the definition that

$$G_2(x, y) = \sum_{\substack{a,b \\ b \leqslant a}} x^{a+b}y^a = \sum_{b=1}^{\infty} \frac{(x^2y)^b}{1 - xy} = \frac{x^2y}{(xy)_2}.$$

Hence, since $G_0(x, y) = 1$, (3.3) becomes

$$G_{2k}(x, y) = \sum_{j=1}^{k} (-1)^{j-1} \frac{x^{j(j+1)}y^j}{(xy)_{2j}} G_{2k-2j}(x, 1). \tag{3.4}$$

For $y = 1$, (3.4) reduces to

$$\sum_{j=0}^{k} (-1)^j \frac{x^{j(j+1)}}{(x)_{2j}} G_{2k-2j}(x, 1) = \delta_{k,0}. \tag{3.5}$$

It follows that

$$\sum_{j=0}^{\infty} (-1)^j \frac{x^{j(j+1)}z^{2j}}{(x)_{2j}} \sum_{k=0}^{\infty} G_{2k}(x, 1)z^{2k} = 1$$

and therefore

$$\sum_{k=0}^{\infty} G_{2k}(x, 1)z^{2k} = \frac{1}{C(x, z)}, \tag{3.6}$$

where

$$C(x, z) = \sum_{k=0}^{\infty} (-1)^k \frac{x^{k(k+1)}z^{2k}}{(x)_{2k}}. \tag{3.7}$$

Returning to (3.4), we get

$$\sum_{k=0}^{\infty} G_{2k}(x, y)z^{2k} = 1 + \sum_{j=1}^{\infty} (-1)^{j-1} \frac{x^{j(j+1)}y^j z^{2j}}{(xy)_{2j}} \sum_{k=0}^{\infty} G_{2k}(x, 1)z^{2k}.$$

Therefore, by (3.6),

$$\sum_{k=0}^{\infty} G_{2k}(x, y)z^{2k} = 1 + \frac{1 - C(x, y, z)}{C(x, z)}, \tag{3.8}$$

where

$$C(x, y, z) = \sum_{j=0}^{\infty} (-1)^j \frac{x^{j(j+1)}y^j z^{2j}}{(xy)_{2j}}. \tag{3.9}$$

In the next place, replacing k by $2k + 1$ and taking $s = k$ in (3.4), we get

$$G_{2k+1}(x, y) = \sum_{j=1}^{k} (-1)^{j-1} \frac{x^{j(j+1)}y^j}{(xy)_{2j}} G_{2k-2j+1}(x, 1)$$
$$+ (-1)^k \frac{x^{k(k+1)}y^k}{(xy)_{2k}} G_1(x, x^{2k}y).$$

Since

$$G_1(x, y) = \sum_{a=1}^{\infty} (xy)^a = \frac{xy}{1 - xy},$$

we get

$$G_{2k+1}(x, y) = \sum_{j=1}^{k} (-1)^{j-1} \frac{x^{j(j+1)}y^j}{(xy)_{2j}} G_{2k-2j+1}(x, 1) + (-1)^k \frac{x^{k^2+3k+1}y^{k+1}}{(xy)_{2k+1}}. \tag{3.10}$$

In particular, for $y = 1$, (3.10) reduces to

$$\sum_{j=0}^{k} (-1)^j \frac{x^{j(j+1)}}{(x)_{2j}} G_{2k-2j+1}(x, 1) = (-1)^k \frac{x^{k^2+3k+1}}{(x)_{2k+1}}.$$

This gives

$$\sum_{k=0}^{\infty} G_{2k+1}(x, 1)z^{2k+1} = \frac{S(x, z)}{C(x, z)}, \tag{3.11}$$

where

$$S(x, z) = \sum_{k=0}^{\infty} (-1)^k \frac{x^{k^2+3k+1}z^{2k+1}}{(x)_{2k+1}}. \tag{3.12}$$

It follows from (3.10) that

$$\sum_{k=0}^{\infty} G_{2k+1}(x, y)z^{2k+1} = \sum_{j=1}^{\infty} (-1)^{j-1} \frac{x^{j(j+1)}y^j z^{2j+1}}{(xy)_{2j}} \sum_{k=0}^{\infty} G_{2k+1}(x, 1)z^{2k+1} + \sum_{k=0}^{\infty} (-1)^k \frac{x^{k^2+3k+1}y^{k+1}z^{2k+1}}{(xy)_{2k+1}},$$

so that

$$\sum_{k=0}^{\infty} G_{2k+1}(x, y)z^{2k+1} = [1 - C(x, y, z)] \frac{S(x, z)}{C(x, z)} + S(x, y, z), \tag{3.13}$$

where

$$S(x, y, z) = \sum_{k=0}^{\infty} (-1)^k \frac{x^{k^2+3k+1}y^{k+1}z^{2k+1}}{(xy)_{2k+1}}. \tag{3.14}$$

To sum up the results of this section, we state the following theorems.

THEOREM 3.1. *We have*

$$\sum_{k=0}^{\infty} G_{2k}(x, 1)z^{2k} = \frac{1}{C(x, z)}, \qquad \sum_{k=0}^{\infty} G_{2k+1}(x, 1)z^{2k+1} = \frac{S(x, z)}{C(x, z)},$$

where $C(x, z)$, $S(x, z)$ *are defined by* (3.7), (3.12), *respectively.*

THEOREM 3.2. *We have*

$$\sum_{k=0}^{\infty} G_{2k}(x, y)z^{2k} = 1 + \frac{1 - C(x, y, z)}{C(x, z)},$$

$$\sum_{k=0}^{\infty} G_{2k+1}(x, y)z^{2k+1} = (1 - C(x, y, z)) \frac{S(x, z)}{C(x, z)} + S(x, y, z),$$

where $C(x, y, z)$, $S(x, y, z)$ *are defined by* (3.9), (3.14), *respectively.*

4. AN IDENTITY

For later use we require the special case $y = x$ of both (3.8) and (3.11). Since, by (3.9),

$$C(x, x, z) = \sum_{j=0}^{\infty} (-1)^j \frac{x^{j(j+2)}z^{2j}}{(x^2)_{2j}},$$

it follows that

$$\frac{xz}{1-x} C(x, x, z) = \sum_{j=0}^{\infty} (-1)^j \frac{x^{(j+1)^2} z^{2j+1}}{(x)_{2j+1}} \equiv S_1(x, z). \tag{4.1}$$

Thus

$$\frac{xz}{1-x} \sum_{k=0}^{\infty} G_{2k}(x, x) z^{2k} = \frac{xz}{1-x} + \frac{[xz/(1-x)] - S_1(x, z)}{C(x, z)}. \tag{4.2}$$

As for (3.13), we have

$$S(x, x, z) = \sum_{k=0}^{\infty} (-1)^k \frac{x^{k^2+4k+2} z^{2k+1}}{(x^2)_{2k+1}},$$

so that

$$\frac{xz}{1-x} S(x, x, z) = 1 - \sum_{k=0}^{\infty} \frac{x^{k^2+2k} z^{2k}}{(x)_{2k}}.$$

Thus (3.13) yields

$$\frac{xz}{1-x} \sum_{k=0}^{\infty} G_{2k+1}(x, x) z^{2k+1} = 1 + \frac{xz}{1-x} \frac{S(x, z)}{C(x, z)} - \frac{C(x, z)C_1(x, z) + S(x, z)S_1(x, z)}{C(x, z)}, \tag{4.3}$$

where

$$\begin{aligned} C_1(x, z) &= \sum_{k=0}^{\infty} (-1)^k \frac{x^{k^2+2k} z^{2k}}{(x)_{2k}} \\ S_1(x, z) &= \sum_{k=0}^{\infty} (-1)^k \frac{x^{(k+1)^2} z^{2k+1}}{(x)_{2k+1}}. \end{aligned} \tag{4.4}$$

We shall now show that

$$C(x, z)C_1(x, z) + S(x, z)S_1(x, z) = 1. \tag{4.5}$$

We have

$$\begin{aligned} C(x, z)C_1(x, z) &= \sum_{j, k=0}^{\infty} (-1)^{j+k} \frac{x^{j(j+1)+k(k+2)}}{(x)_{2j}(x)_{2k}} z^{2j+2k} \\ &= \sum_{n=0}^{\infty} (-1)^n \frac{z^{2n}}{(x)_{2n}} \sum_{k=0}^{n} \begin{bmatrix} 2n \\ 2k \end{bmatrix} x^{k(k+2)+(n-k)(n-k+1)} \\ &= \sum_{n=0}^{\infty} (-1)^n \frac{x^{n^2+n} z^{2n}}{(x)_{2n}} \sum_{k=0}^{n} \begin{bmatrix} 2n \\ 2k \end{bmatrix} x^{k(2k-1)-2k(n-1)}, \end{aligned}$$

$$S(x,z)S_1(x,z) = \sum_{j,k=0}^{\infty} (-1)^{j+k} \frac{x^{j^2+3j+1+(k+1)^2}}{(x)_{2j+1}(x)_{2k+1}} z^{2j+2k+2}$$
$$= \sum_{n=1}^{\infty} (-1)^{n-1} \frac{z^{2n}}{(x)_{2n}} \sum_{j=0}^{n} \begin{bmatrix} 2n \\ 2j+1 \end{bmatrix} x^{j^2+3j+1+(n-j)^2}$$
$$= \sum_{n=1}^{\infty} (-1)^{n-1} \frac{x^{n^2+n}z^{2n}}{(x)_{2n}} \sum_{j=0}^{n} \begin{bmatrix} 2n \\ 2j+1 \end{bmatrix} x^{j(2j+1)-(2j+1)(n-1)}.$$

Thus

$$C(x,z)C_1(x,z) + S(x,z)S_1(x,z)$$
$$= 1 + \sum_{n=1}^{\infty} (-1)^n \frac{x^{n^2+n}z^{2n}}{(x)_{2n}} \sum_{k=0}^{2n} (-1)^k \begin{bmatrix} 2n \\ k \end{bmatrix} x^{\frac{1}{2}k(k-1)-k(n-1)},$$

where

$$\begin{bmatrix} n \\ k \end{bmatrix} = \frac{(x)_n}{(x)_k(x)_{n-k}}.$$

Since

$$(z)_n = \sum_{k=0}^{n} (-1)^k \begin{bmatrix} n \\ k \end{bmatrix} x^{\frac{1}{2}k(k-1)} z^k,$$

the inner sum on the right is equal to

$$\prod_{j=0}^{2n-1} (1 - x^{j-n+1}) = 0 \qquad (n > 0).$$

This completes the proof of

THEOREM 4.1. *The series* $C(x,z)$, $C_1(x,z)$, $S(x,z)$, $S_1(x,z)$ *satisfy the identity* (4.5).

Substituting from (4.5) in (4.3) we get

$$\frac{xz}{1-x} \sum_{k=0}^{\infty} G_{2k+1}(x,x) z^{2k+1} = 1 + \frac{xz}{1-x} \frac{S(x,z)}{C(x,z)} - \frac{1}{C(x,z)}. \tag{4.6}$$

It is easily verified that

$$C_1(x,z) = C(x, x^{\frac{1}{2}}z), \; S_1(x,z) = x^{\frac{1}{2}} S(x, x^{-\frac{1}{2}}z).$$

Thus (4.5) can be written in the form

$$C(x,z)C(x,x^{\frac{1}{2}}z) + x^{\frac{1}{2}} S(x,z) S(x, x^{-\frac{1}{2}}z) = 1. \tag{4.7}$$

5. Up–Down Partitions

To get generating functions for up–down partitions consider

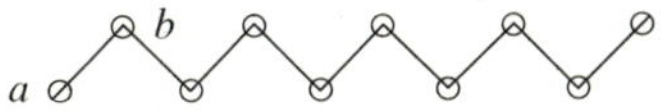

This gives

$$A(n, k, a) = \sum_{b \geqslant a} B(n - a, k - 1, b). \tag{5.1}$$

It follows from (5.1) that

$$F(x, k, a) = \sum_{b \geqslant a} x^a G(x, k - 1, b),$$

$$F_k(x, y) = \sum_{a=1}^{\infty} (xy)^a \sum_{b \geqslant a} G(x, k - 1, b)$$

$$= \sum_{b=1}^{\infty} G(x, k - 1, b) \sum_{a=1}^{b} (xy)^a,$$

so that

$$F_k(x, y) = \frac{xy}{1 - xy} [G_{k-1}(x, 1) - G_{k-1}(x, xy)] \qquad (k > 1). \tag{5.2}$$

In particular, for $y = 1$, (5.2) becomes

$$F_k(x, 1) = \frac{x}{1 - x} [G_{k-1}(x, 1) - G_{k-1}(x, x)]. \tag{5.3}$$

Replacing k by $2k$, (5.3) becomes

$$F_{2k}(x, 1) = \frac{x}{1 - x} [G_{2k-1}(x, 1) - G_{2k-1}(x, x)],$$

and

$$\sum_{k=0}^{\infty} F_{2k}(x, 1) z^{2k} = 1 + \frac{xz}{1 - x} \left[\sum_{k=0}^{\infty} G_{2k+1}(x, 1) z^{2k+1} - \sum_{k=0}^{\infty} G_{2k+1}(x, x) z^{2k+1} \right].$$

Making use of (3.11) and (4.6), we get

$$\sum_{k=0}^{\infty} F_{2k}(x, 1) z^{2k} = 1 + \frac{xz}{1 - x} \frac{S(x, z)}{C(x, z)} - \left[1 + \frac{xz}{1 - x} \frac{S(x, z)}{C(x, z)} - \frac{1}{C(x, z)} \right]$$

$$= \frac{1}{C(x, z)}.$$

This result was anticipated in (2.2) and thus furnishes a partial check for the general results found above.

Returning to (5.3) with k replaced by $2k+1$, we have

$$F_{2k+1}(x, 1) = \frac{x}{1-x}\,[G_{2k}(x, 1) - G_{2k}(x, x)] \qquad (k > 1).$$

Hence

$$\sum_{k=0}^{\infty} F_{2k+1}(x, 1)z^{2k+1} = \frac{xz}{1-x} + \frac{xz}{1-x}\left[\sum_{k=1}^{\infty} G_{2k}(x, 1)z^{2k} - \sum_{k=1}^{\infty} G_{2k}(x, x)z^{2k}\right].$$

Substituting from (3.6) and (4.2), we get

$$\sum_{k=0}^{\infty} F_{2k+1}(x, 1)z^{2k+1} = \frac{xz}{1-x}\,\frac{1}{C(x, z)} - \frac{[xz/(1-x)] - S_1(x, z)}{C(x, z)},$$

that is,

$$\sum_{k=0}^{\infty} F_{2k+1}(x, 1)z^{2k+1} = \frac{S_1(x, z)}{C(x, z)}. \tag{5.4}$$

More general results are implied by (5.2). It follows from

$$F_{2k}(x, y) = \frac{xy}{1-xy}\,[G_{2k-1}(x, 1) - G_{2k-1}(x, xy)] \qquad (k > 0)$$

together with (3.11) and (3.13) that

$$\sum_{k=0}^{\infty} F_{2k}(x, y)z^{2k}$$
$$= 1 + \frac{xyz}{1-xy}\,\frac{S(x, z)}{C(x, z)} - \frac{xyz}{1-xy}\left[1 - C(x, xy, z)\,\frac{S(x, z)}{C(x, z)} + S(x, xy, z)\right]. \tag{5.5}$$

If we put

$$S_1(x, y, z) = \frac{xyz}{1-xy}\,C(x, xy, z) = \sum_{k=0}^{\infty} (-1)^k\,\frac{x^{(j+1)^2}y^{j+1}z^{2j+1}}{(xy)_{2j+1}} \tag{5.6}$$

and

$$C_1(x, y, z) = 1 - \frac{xyz}{1-xyz}\,S(x, xy, z)$$
$$= \sum_{k=0}^{\infty} (-1)^k\,\frac{x^{k^2+2k}y^{k+1}z^{2k}}{(xy)_{2k}}, \tag{5.7}$$

then (5.5) becomes

$$\sum_{k=0}^{\infty} F_{2k}(x, y)z^{2k} = \frac{C(x, z)C_1(x, y, z) + S(x, z)S_1(x, y, z)}{C(x, z)}. \tag{5.8}$$

In the next place, if follows from

$$F_{2k+1}(x, y) = \frac{xy}{1 - xy}[G_{2k}(x, 1) - G_{2k}(x, xy)] \qquad (k > 1)$$

that

$$\begin{aligned}\sum_{k=0}^{\infty} F_{2k+1}(x, y)z^{2k+1} &= \frac{xyz}{1-xy} + \frac{xyz}{1-xy}\frac{1}{C(x, z)} - \frac{xyz}{1-xy}\left[1 + \frac{1 - C(x, xy, z)}{C(x, z)}\right] \\ &= \frac{xyz}{1-xy}\frac{C(x, xy, z)}{C(x, z)}.\end{aligned}$$

Since, by (5.6),

$$S_1(x, y, z) = \frac{xyz}{1 - xy} C(x, xy, z),$$

we therefore have

$$\sum_{k=0}^{\infty} F_{2k+1}(x, y)z^{2k+1} = \frac{S_1(x, y, z)}{C(x, z)}. \tag{5.9}$$

We now state

THEOREM 5.1. *We have*

$$\sum_{k=0}^{\infty} F_{2k}(x, 1)z^{2k} = \frac{1}{C(x, z)}, \qquad \sum_{k=0}^{\infty} F_{2k+1}(x, 1)z^{2k+1} = \frac{S_1(x, z)}{C(x, z)},$$

where $C(x, z)$, $S_1(x, z)$ *are defined by* (3.7), (4.4), *respectively.*

THEOREM 5.2. *We have*

$$\sum_{k=0}^{\infty} F_{2k}(x, y)z^{2k} = \frac{C(x, z)C_1(x, y, z) + S(x, z)S_1(x, y, z)}{C(x, z)},$$

$$\sum_{k=0}^{\infty} F_{2k+1}(x, y)z^{2k+1} = \frac{S_1(x, y, z)}{C(x, z)},$$

where $C_1(x, y, z)$, $S_1(x, y, z)$ *are defined by* (5.7), (5.6), *respectively.*

6. Corollaries

Put

$$A^{e}(n) = \sum_{2k \leqslant n} A(n, 2k), \qquad A^{o}(n) = \sum_{2k<n} A(n, 2k+1),$$

$$B^{e}(n) = \sum_{2k \leqslant n} B(n, 2k), \qquad B^{o}(n) = \sum_{2k<n} B(n, 2k+1).$$

Thus $A^{e}(n)$ is the number of up–down partitions of n into an even number of parts, $A^{o}(n)$ is the number of up–down partitions of n into an odd number of parts, and similarly for $B^{e}(n)$, $B^{o}(n)$.

Taking $z = 1$ in (3.6), (3.11) and (5.4), we get the following generating functions:

$$\sum_{n=0}^{\infty} A^{e}(n)x^n = \sum_{n=0}^{\infty} B^{e}(n)x^n = \frac{1}{\sum_{k=0}^{\infty} (-1)^k [x^{k(k+1)}/(x)_{2k}]}, \tag{6.1}$$

$$\sum_{n=0}^{\infty} A^{o}(n)x^n = \frac{\sum_{k=0}^{\infty} (-1)^k [x^{k^2+3k+1}/(x)_{2k+1}]}{\sum_{k=0}^{\infty} (-1)^k [x^{k(k+1)}/(x)_{2k}]}, \tag{6.2}$$

$$\sum_{n=0}^{\infty} B^{o}(n)x^n = \frac{\sum_{k=0}^{\infty} (-1)^k [x^{(k+1)^2}/(x)_{2k+1}]}{\sum_{k=0}^{\infty} (-1)^k [x^{k(k+1)}/(x)_{2k}]}. \tag{6.3}$$

7. Down–Up (k, t)-Partitions

Let $k \geqslant 2$, $t \geqslant 0$ be fixed integers. Let $B^{(k)}(n, sk + t)$ denote the number of down–up (k, t)-partitions with pattern

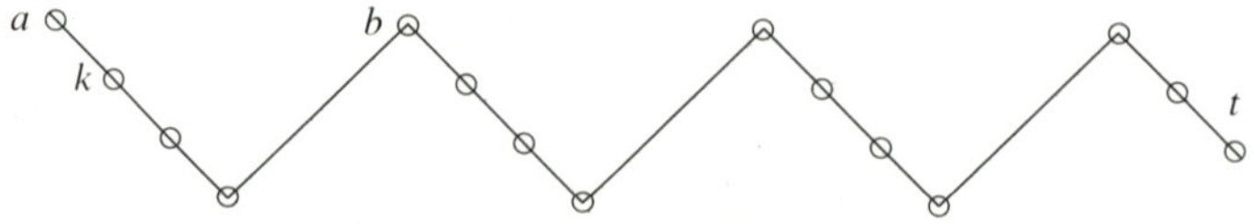

(7.1)

More precisely, $B^{(k)}(n, sk + t)$ is the number of solutions in positive integers $a_1, a_2, \ldots, a_{sk+t}$ of the equation

$$n = a_1 + a_2 + \cdots + a_{sk+t}$$

where

$$a_1 \geqslant a_2 \geqslant \cdots \geqslant a_k, \quad a_{k+1} \geqslant a_{k+2} \cdots \geqslant a_{2k}, \quad \ldots, \quad a_{sk+1} \geqslant \cdots \geqslant a_{sk+t}$$
$$a_k \leqslant a_{k+1}, \quad a_{2k} \leqslant a_{2k+1}, \quad \ldots . \tag{7.2}$$

Let $B^{(k)}(n, sk + t, a)$ denote the number of partitions satisfying these conditions and in addition $a_1 = a$. Clearly

$$B^{(k)}(n, sk + t) = \sum_{a=1}^{\infty} B^{(k)}(n, sk + t, a).$$

We first consider the case $t = 0$. It is clear from (7.1) that $B^{(k)}(n, sk, a)$ satisfies

$$B^{(k)}(n, sk, a) = \sum_{\substack{a=a_1 \geqslant \cdots \geqslant a_k \\ b \geqslant a_k}} B^{(k)}(n - a_1 - \cdots - a_k, (s-1)k, b) \qquad (s > 1). \tag{7.3}$$

Put

$$G^{(k)}(x, sk, a) = \sum_n B^{(k)}(n, sk, a)x^n$$

$$G_{sk}^{(k)}(x, y) = \sum_{a=1}^{\infty} G^{(k)}(x, sk, a)y^a.$$

Then, by (7.3),

$$G^{(k)}(x, sk, a) = \sum_n x^n \sum_{\substack{a=a_1 \geqslant \cdots \geqslant a_k \\ b \geqslant a_k}} B^{(k)}(n - a_1 - \cdots - a_k, (s-1)k, b)$$

$$= \sum_{\substack{a=a_1 \geqslant \cdots \geqslant a_k \\ b \geqslant a_k}} x^{a_1 + \cdots + a_k} G^{(k)}(x, (s-1)k, b),$$

$$G_{sk}^{(k)}(x, y) = \sum_{\substack{a_1 \geqslant \cdots \geqslant a_k \\ b \geqslant a_k}} x^{a_1 + \cdots + a_k} y^{a_1} G^{(k)}(x, (s-1)k, b)$$

$$= \sum_{b=1}^{\infty} G^{(k)}(x, (s-1)k, b) \sum_{\substack{a_1 \geqslant \cdots \geqslant a_k \\ a_k \leqslant b}} x^{a_1 + \cdots + a_k} y^{a_1}. \tag{7.4}$$

Put

$$\begin{aligned} a_1 &= c_1 + \cdots + c_{k-1} + a \\ a_2 &= c_2 + \cdots + c_{k-1} + a \\ &\vdots \\ a_{k-1} &= c_{k-1} + a, \qquad a_k = a, \end{aligned}$$

where $c_1, c_2, \ldots, c_{k-1} \geqslant 0$. Then the inner sum in (7.4) is equal to

$$\sum_{c_1, \ldots, c_{k-1}=0}^{\infty} \sum_{a=1}^{b} x^{c_1 + 2c_2 + \cdots + (k-1)c_{k-1} + ka} y^{c_1 + \cdots + c_{k-1} + a}$$

$$= \frac{1}{(xy)_{k-1}} \cdot x^k y \frac{1 - (x^k y)^b}{1 - x^k y}.$$

Substituting in (7.4), we get

$$G_{sk}^{(k)}(x, y) = \frac{x^k y}{(xy)_k} \sum_{b=1}^{\infty} G^{(k)}(x, (s-1)k, b)(1 - (x^k y)^b),$$

so that

$$G_{sk}^{(k)}(x, y) = \frac{x^k y}{(xy)_k} [G_{(s-1)k}^{(k)}(x, 1) - G_{(s-1)k}^{(k)}(x, x^k y)] \qquad (s \geqslant 2). \tag{7.5}$$

Iteration of (7.5) gives

$$G_{sk}^{(k)}(x, y) = \sum_{j=1}^{s-1} (-1)^{j-1} \frac{x^{\frac{1}{2}kj(j+1)} y^j}{(xy)_k} G_{(s-j)k}^{(k)}(x, 1)$$
$$+ (-1)^{s-1} \frac{x^{\frac{1}{2}ks(s-1)} y^{s-1}}{(xy)_{(s-1)k}} G_k^{(k)}(x, x^{(s-1)k} y).$$

Since

$$G_k^{(k)}(x, y) = \sum_{a_j \geqslant \cdots \geqslant a_k} x^{a_1 + \cdots + a_k} y^{a_1} = \frac{x^k y}{(xy)_k},$$

it follows that

$$G_{sk}^{(k)}(x, y) = \sum_{j=1}^{s} (-1)^{j-1} \frac{x^{\frac{1}{2}kj(j+1)} y^j}{(xy)_{jk}} G_{(s-j)k}^{(k)}(x, 1) \qquad (s \geqslant 1), \tag{7.6}$$

where we take

$$G_0^{(k)}(x, y) = 1.$$

In particular, for $y = 1$, (7.6) reduces to

$$\sum_{j=0}^{s} (-1)^j \frac{x^{\frac{1}{2}kj(j+1)}}{(x)_{jk}} G_{(s-j)k}^{(k)}(x, 1) = \delta_{s,0}. \tag{7.7}$$

This yields

$$\sum_{s=0}^{\infty} G_{sk}^{(k)}(x, 1) z^{sk} = \frac{1}{C^{(k)}(x, z)}, \tag{7.8}$$

where

$$C^{(k)}(x, z) = \sum_{j=0}^{\infty} (-1)^j \frac{x^{\frac{1}{2}kj(j+1)} z^{jk}}{(x)_{jk}}. \tag{7.9}$$

For $k = 2$ it is evident that (7.8) reduces to (3.6).

In the next place, it follows from (7.6) that

$$\sum_{s=0}^{\infty} G_{sk}^{(k)}(x, y)z^{sk} = 1 + \sum_{j=1}^{\infty} (-1)^{j-1} \frac{x^{\frac{1}{2}kj(j+1)}y^j}{(xy)_{jk}} \sum_{s=0}^{\infty} G_{sk}^{(k)}(x, 1)z^{sk}.$$

Therefore, by (7.8), we have

$$G_{sk}^{(k)}(x, y)z^{sk} = 1 + \frac{1 - C^{(k)}(x, y, z)}{C^{(k)}(x, z)}, \tag{7.10}$$

where

$$C^{(k)}(x, y, z) = \sum_{j=0}^{\infty} (-1)^j \frac{x^{\frac{1}{2}kj(j+1)}y^j z^{jk}}{(xy)_{jk}}. \tag{7.11}$$

8. Continuation

We now consider the general down–up (k, t)-partitions where $t \geqslant 1$:

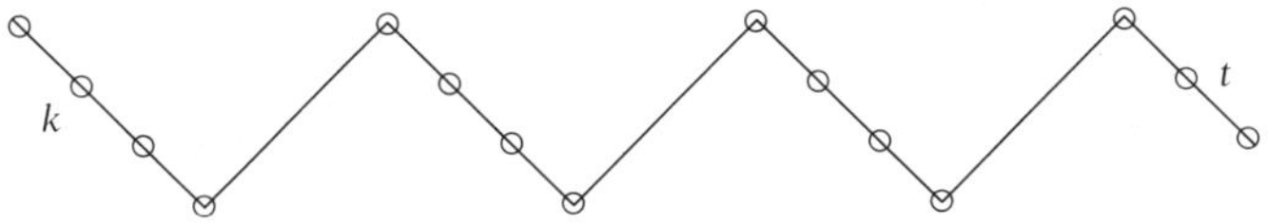

(8.1)

With $B^{(k)}(n, sk + t, a)$ as defined above, put

$$G^{(k)}(x, sk + t, a) = \sum_n x^n B^{(k)}(n, sk + t, a),$$

$$G_{sk+t}^{(k)}(x, y) = \sum_{a=1}^{\infty} y^a G^{(k)}(x, sk + t, a).$$

Then, exactly as in the proof of (7.5), we have

$$G_{sk+t}^{(k)}(x, y) = \frac{x^k y}{(xy)_k} [G_{(s-1)k+t}^{(k)}(x, 1) - G_{(s-1)k+t}^{(k)}(x, x^k y)] \qquad (s \geqslant 1). \tag{8.2}$$

Iteration of (8.2) leads to

$$G_{sk+t}^{(k)}(x, y) = \sum_{j=1}^{s} (-1)^{j-1} \frac{x^{\frac{1}{2}kj(j+1)}y^j}{(xy)_{jk}} G_{(s-j)k+t}^{(k)}(x, 1) + (-1)^s \frac{x^{\frac{1}{2}ks(s+1)}y^s}{(xy)_{sk}} G_t^{(k)}(x, x^{sk}y). \tag{8.3}$$

Since

$$G_t^{(k)}(x, y) = G_t^{(t)}(x, y) = \frac{x^t y}{(xy)_t},$$

(8.3) becomes

$$G_{sk+t}^{(k)}(x, y) = \sum_{j=1}^{s} (-1)^{j-1} \frac{x^{\frac{1}{2}kj(j+1)}y^j}{(xy)_{jk}} G_{(s-j)k+t}^{(k)}(x, 1) + (-1)^s \frac{x^{\frac{1}{2}ks(s+3)+t}y^{s+1}}{(xy)_{sk+t}} \qquad (s \geqslant 0). \tag{8.4}$$

In particular, for $y = 1$, we get

$$\sum_{j=0}^{s} (-1)^s \frac{x^{\frac{1}{2}kj(j+1)}}{(x)_{jk}} G_{(s-j)k+t}^{(k)}(x, 1) = (-1)^s \frac{x^{\frac{1}{2}ks(s+1)+t}}{(x)_{sk+t}}. \tag{8.5}$$

This yields

$$\sum_{j=0}^{\infty} (-1)^j \frac{x^{\frac{1}{2}kj(j+1)}z^{jk}}{(x)_{jk}} \sum_{s=0}^{\infty} G_{sk+t}^{(k)}(x, 1)z^{sk+t} = \sum_{s=0}^{\infty} (-1)^s \frac{x^{\frac{1}{2}ks(s+3)+t}z^{sk+t}}{(x)_{sk+t}}.$$

Therefore

$$\sum_{s=0}^{\infty} G_{sk+t}^{(k)}(x, 1)z^{sk+t} = \frac{S_t^{(k)}(x, z)}{C^{(k)}(x, z)}, \tag{8.6}$$

where

$$S_t^{(k)}(x, z) = \sum_{s=0}^{\infty} (-1)^s \frac{x^{\frac{1}{2}ks(s+3)+t}z^{sk+t}}{(x)_{sk+t}}. \tag{8.7}$$

For $k = 2$, $t = 1$, (8.6) reduces to (3.11).

Returning to (8.4), we have

$$\sum_{s=0}^{\infty} G_{sk+t}^{(k)}(x, y)z^{sk+t} = \sum_{j=1}^{\infty} (-1)^{j-1} \frac{x^{\frac{1}{2}kj(j+1)}y^j z^{jk}}{(xy)_{jk}} \sum_{s=0}^{\infty} G_{sk+t}^{(k)}(x, 1) + \sum_{s=0}^{\infty} (-1)^s \frac{x^{\frac{1}{2}ks(s+3)+t}y^{s+1}z^{sk+t}}{(xy)_{sk+t}}.$$

Hence, by (8.6) and (7.11),

$$\sum_{s=0}^{\infty} G_{sk+t}^{(k)}(x, y)z^{sk+t} = [1 - C^{(k)}(x, y, z)] \frac{S_t^{(k)}(x, z)}{C^{(k)}(x, z)} + S_t^{(k)}(x, y, z) \qquad (t \geqslant 1), \tag{8.8}$$

where

$$S_t^{(k)}(x, y, z) = \sum_{s=0}^{\infty} (-1)^s \frac{x^{\frac{1}{2}ks(s+3)+t} y^{s+1} z^{sk+t}}{(xy)_{sk+t}}. \tag{8.9}$$

For $k = 2, t = 1$, (8.9) reduces to (3.13). The results of §§7, 8 may be summed up in the following theorems.

THEOREM 8.1. *We have*

$$\sum_{s=0}^{\infty} G_{sk}^{(k)}(x, 1) z^{sk} = \frac{1}{C^{(k)}(x, z)}, \tag{8.10}$$

$$\sum_{s=0}^{\infty} G_{sk+t}^{(k)}(x, 1) z^{sk+t} = \frac{S_t^{(k)}(x, z)}{C^{(k)}(x, z)} \qquad (t \geqslant 1), \tag{8.11}$$

where $C^{(k)}(x, z)$, $S_t^{(k)}(x, z)$ are defined by (7.9), (8.7), *respectively.*

THEOREM 8.2. *We have*

$$\sum_{s=0}^{\infty} G_{sk}^{(k)}(x, y) z^{sk} = 1 + \frac{1 - C^{(k)}(x, y, z)}{C^{(k)}(x, z)}, \tag{8.12}$$

$$\sum_{s=0}^{\infty} G_{sk+t}^{(k)}(x, y) z^{sk+t} = \frac{[1 - C^{(k)}(x, y, z)] S_t^{(k)}(x, z) + C^{(k)}(x, z) S_t^{(k)}(x, y, z)}{C^{(k)}(x, z)} \qquad (t \geqslant 1), \tag{8.13}$$

where $C^{(k)}(x, y, z)$, $S_t^{(k)}(x, y, z)$ are defined by (7.11), (8.9), *respectively.*

Remark. For $t = k$, (8.11) becomes

$$\sum_{s=0}^{\infty} G_{sk}^{(k)}(x, 1) z^{sk} = 1 + \frac{S_k^{(k)}(x, z)}{C^{(k)}(x, z)}.$$

Since

$$C^{(k)}(x, z) + S_k^{(k)}(x, z) = \sum_{j=0}^{\infty} (-1)^j \frac{x^{\frac{1}{2}kj(j+1)} z^{jk}}{(x)_{jk}} + \sum_{j=1}^{\infty} (-1)^{j-1} \frac{x^{\frac{1}{2}kj(j+1)} z^{jk}}{(x)_{jk}} = 1,$$

it follows that (8.11) reduces to (8.10) in this case.

Similarly, for $t = k$, (8.13) becomes

$$\sum_{s=0}^{\infty} G_{sk}^{(s)}(x, y) z^{sk} = 1 + \frac{(1 - C^{(k)}(x, y, z)) S_k^{(k)}(x, z) + C^{(k)}(x, z) S_k^{(k)}(x, y, z)}{C^{(k)}(x, z)}$$

$$= \frac{1 - C^{(k)}(x, y, z)}{C^{(k)}(x, z)} + C^{(k)}(x, y, z) + S_k^{(k)}(x, y, z).$$

Since $C^{(k)}(x, y, z) + S_k^{(k)}(x, y, z) = 1$, it follows that (8.13) reduces to (8.12) when $t = k$.

9. Up–Down (k, t)-Partitions

Consider the up–down pattern

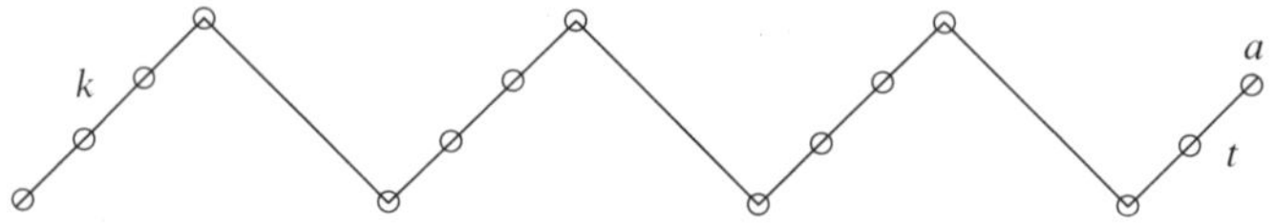

(9.1)

Let $A^{(k)}(n, sk + t, a)$ denote the number of up–down partitions of n into $sk + t$ parts with last part equal to a. More precisely, $A^{(k)}(n, sk + t, a)$ is the number of solutions in positive integers $a_1, a_2, \ldots, a_{sk+t}$ of the equation

$$n = a_1 + a_2 + \cdots + a_{sk+t},$$

where

$$\begin{gathered} a_1 \geqslant \cdots \geqslant a_k, \quad a_{k+1} \geqslant \cdots \geqslant a_{2k}, \quad \ldots, \quad a_{sk+1} \geqslant \cdots \geqslant a_{sk+t} \\ a_k \geqslant a_{k+1}, \quad a_{2k} \geqslant a_{2k+1}, \quad \ldots, \quad a_{sk} \geqslant a_{sk+1}. \end{gathered} \tag{9.2}$$

Put

$$A^{(k)}(n, sk + t) = \sum_a A^{(k)}(n, sk + t, a),$$

so that $A^{(k)}(n, sk + t)$ is the number of up–down (k, t)-partitions in which the last element is arbitrary.

We also define

$$\begin{aligned} F^{(k)}(x, sk + t, a) &= \sum_n A^{(k)}(n, sk + t, a)x^n, \\ F^{(k)}_{sk+t}(x, y) &= \sum_a F^{(k)}(x, sk + t, a)y^a. \end{aligned} \tag{9.3}$$

Reading (9.1) from right to left, we get

$$\begin{aligned} A^{(k)}(n, sk + t, a) &= \sum_{\substack{a_1 \leqslant \cdots \leqslant a_t = a \\ b \geqslant a_1}} B^{(k)}(n - a_1 - \cdots - a_t, sk, b), \\ F^{(k)}(x, sk + t, a) &= \sum_{\substack{a_1 \leqslant \cdots \leqslant a_t = a \\ b \geqslant a_1}} x^{a_1 + \cdots + a_t} G^{(k)}(x, sk, b), \\ F^{(k)}_{sk+t}(x, y) &= \sum_{\substack{a_1 \leqslant \cdots \leqslant a_t \\ b \geqslant a_1}} x^{a_1 + \cdots + a_t} y^{a_t} G^{(k)}(x, sk, b) \\ &= \sum_{b=1}^{\infty} G^{(k)}(x, sk, b) \sum_{\substack{a_1 \leqslant \cdots \leqslant a_t \\ a_1 \geqslant b}} x^{a_1 + \cdots + a_t} y^{a_t}. \end{aligned} \tag{9.4}$$

The inner sum is equal to

$$\sum_{a_1 \leqslant b} x^{a_1} \sum_{a_1 \leqslant \cdots \leqslant a_t} x^{a_2+\cdots+a_t} y^{a_t} = \sum_{a_1=1}^{b} x^{ta_1} y^{a_1} \sum_{0 \leqslant c_2 \leqslant \cdots \leqslant c_t} x^{c_2+\cdots+c_t} y^{c_t}$$

$$= \frac{x^t y - (x^t y)^{b+1}}{1 - x^t y} \frac{1}{(xy)_{t-1}} = x^t y \frac{1-(x^t y)^b}{(xy)_t}.$$

Substituting in (9.4), we get

$$F^{(k)}_{sk+t}(x, y) = \frac{x^t y}{(xy)_t} [G^{(k)}_{st}(x, 1) - G^{(k)}_{sk}(x, x^t y)] \qquad (s > 0). \tag{9.5}$$

Then by (7.8) and (7.10)

$$\sum_{s=0}^{\infty} F^{(k)}_{sk+t}(x, y) z^{sk+t} = F^{(k)}_t(x, y) z^t + \frac{x^t y}{(xy)_t} \frac{z^t}{C^{(k)}(x, z)}$$

$$- \frac{x^t y z^t}{(xy)_t} \left[1 + \frac{1 - C^{(k)}(x, x^t y, z)}{C^{(k)}(x, z)} \right]$$

$$= \left[F^{(k)}_t(x, y) - \frac{x^t y}{(xy)_t} \right] z^t + \frac{x^t y z^t}{(xy)_t} \frac{C^{(k)}(x, x^t y, z)}{C^{(k)}(x, z)}.$$

Since

$$\frac{x^t y z^t}{(xy)_t} C^{(k)}(x, x^t y, z) = \sum_{j=0}^{\infty} (-1)^j \frac{x^{\frac{1}{2}kj(j+1)+t(j+1)} y^{j+1} z^{jk+t}}{(xy)_{jk+t}}$$

and

$$F^{(k)}_t(x, y) = G^{(k)}_t(x, y) = G^{(t)}_t(x, y) = \frac{x^t y}{(xy)_t},$$

it follows that

$$\sum_{s=0}^{\infty} F^{(k)}_{sk+t}(x, y) z^{sk+t} = \frac{\bar{S}^{(k)}_t(x, y, z)}{C^{(k)}(x, z)} \qquad (t \geqslant 1), \tag{9.6}$$

where

$$\bar{S}^{(k)}_t(x, y, z) = \sum_{j=0}^{\infty} (-1)^j \frac{x^{\frac{1}{2}kj(j+1)+t(j+1)} y^{j+1} z^{jk+t}}{(xy)_{jk+t}}. \tag{9.7}$$

In particular, for $y = 1$, (9.7) becomes

$$\sum_{s=0}^{\infty} F^{(k)}_{sk+t}(x, 1) z^{sk+t} = \frac{\bar{S}^{(k)}_t(x, z)}{C^{(k)}(x, z)} \qquad (t \geqslant 1), \tag{9.8}$$

where

$$\bar{S}_t^{(k)}(x, z) = \sum_{j=0}^{\infty} (-1)^j \frac{x^{\frac{1}{2}kj(j+1)+t(j+1)} z^{jk+t}}{(x)_{jk+t}} \qquad (t \geqslant 1). \tag{9.9}$$

For $t = k$, (9.6) gives

$$\sum_{s=0}^{\infty} F_{sk}^{(k)}(x, y) z^{sk} = 1 + \frac{\bar{S}_k^{(k)}(x, y, z)}{C^{(k)}(x, z)}.$$

Since, by (9.7) and (7.11),

$$\bar{S}_k^{(k)}(x, y, z) = \sum_{j=1}^{\infty} (-1)^{j-1} \frac{x^{\frac{1}{2}kj(j+1)} y^j z^{kj}}{(xy)_{jk}} = 1 - C^{(k)}(x, y, z),$$

it follows that

$$\sum_{s=0}^{\infty} F_{sk}^{(k)}(x, y) z^{sk} = 1 + \frac{1 - C^{(k)}(x, y, z)}{C^{(k)}(x, z)}. \tag{9.10}$$

Thus, by comparison with (7.10),

$$\sum_{s=0}^{\infty} F_{sk}^{(k)}(x, y) z^{sk} = \sum_{s=0}^{\infty} G_{sk}^{(k)}(x, y) z^{sk}.$$

Indeed it follows from the definition that

$$F_{sk}^{(k)}(x, y) = G_{sk}^{(k)}(x, y) \qquad (s = 0, 1, 2, \ldots).$$

Thus again we have a partial check on the general results.
We now state the following theorems.

THEOREM 9.1. *We have*

$$\sum_{s=0}^{\infty} F_{sk}^{(k)}(x, 1) z^{sk} = \frac{1}{C^{(k)}(x, z)}, \tag{9.11}$$

$$\sum_{s=0}^{\infty} F_{sk+t}^{(k)}(x, 1) z^{sk+t} = \frac{\bar{S}_t^{(k)}(x, z)}{C^{(k)}(x, z)} \qquad (t \geqslant 1), \tag{9.12}$$

where $\bar{S}_t^{(k)}(x, z)$ *is defined by* (9.9).

THEOREM 9.2. *We have*

$$\sum_{s=0}^{\infty} F_{sk}^{(k)}(x, y) z^{sk} = 1 + \frac{1 - C^{(k)}(x, y, z)}{C^{(k)}(x, z)}, \tag{9.13}$$

$$\sum_{s=0}^{\infty} F_{sk+t}^{(k)}(x, y) z^{sk+t} = \frac{\bar{S}_t^{(k)}(x, y, z)}{C^{(k)}(x, z)} \qquad (t \geqslant 1), \tag{9.14}$$

where $C^{(k)}(x, y, z)$, $\bar{S}_t^{(k)}(x, y, z)$ *are defined by* (7.11), (9.7), *respectively.*

Remark (compare Remark at end of §8). It is easily verified that, for $t = k$, (9.12) reduces to (9.11) and (9.14) reduces to (9.13).

10. Corollaries

Put

$$A^{(k)}(n) = \sum_{a,s} A^{(k)}(n, sk, a),$$

$$A^{(k,t)}(n) = \sum_{a,s} A^{(k)}(n, sk + t, a) \qquad (t \geqslant 1),$$

$$B^{(k)}(n) = \sum_{a,s} B^{(k)}(n, sk, a)$$

$$B^{(k,t)}(n) = \sum_{a,s} B^{(k)}(n, sk + t, a) \qquad (t \geqslant 1).$$

Thus $A^{(k)}(n)$ is the number of up-down $(k, 0)$-partitions of n with sk parts, etc.

Taking $z = 1$ in (7.8) we get

$$\sum_{n=0}^{\infty} A^{(k)}(n) x^n = \sum_{n=0}^{\infty} B^{(k)}(n) x^n = \frac{1}{\sum_{j=0}^{\infty} (-1)^j [x^{\frac{1}{2}kj(j+1)}/(x)_{kj}]}. \tag{10.1}$$

Similarly, by (8.6) and (9.8), we get

$$\sum_{n=0}^{\infty} B^{(k,t)}(n) x^n = \frac{\sum_{s=0}^{\infty} (-1)^s [x^{\frac{1}{2}ks(s+3)+t}/(x)_{ks+t}]}{\sum_{j=0}^{\infty} (-1)^j [x^{\frac{1}{2}kj(j+1)}/(x)_{kj}]} \qquad (t \geqslant 1), \tag{10.2}$$

and

$$\sum_{n=0}^{\infty} A^{(k,t)}(n) x^n = \frac{\sum_{s=0}^{\infty} (-1)^s [x^{\frac{1}{2}kj(j+1)+t(j+1)}/(x)_{kj+t}]}{\sum_{j=0}^{\infty} (-1)^j [x^{\frac{1}{2}kj(j+1)}/(x)_{kj}]} \qquad (t \geqslant 1). \tag{10.3}$$

11. Partitions with Other Patterns

Fairly simple results can be obtained for certain more elaborate down–up patterns. Let $H(n, m, a)$ denote the number of partitions of n into m positive

parts associated with the pattern

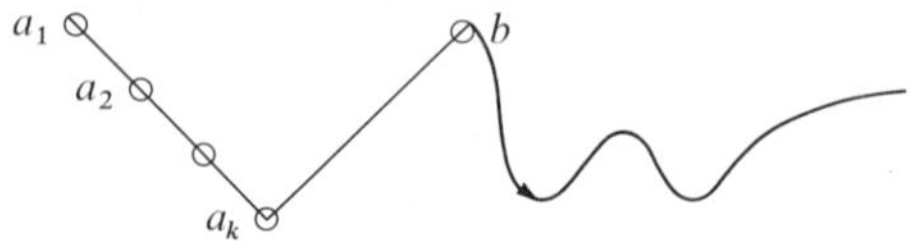

(11.1)

where the wavy line indicates an arbitrary pattern beginning with a descent at b. Let H' denote the enumerant associated with the wavy line pattern. Then we have

$$H(n, m, a) = \sum_{\substack{a_k \leqslant \cdots \leqslant a_1 = a \\ a_k \leqslant b}} H'(n - a_1 - \cdots - a_k, m - k, b).$$

If we put

$$\Phi(x, m, a) = \sum_n x^n H(n, m, a), \qquad \Phi_m(x, y) = \sum_a y^a \Phi(x, m, a),$$

$$\Phi'(x, m, a) = \sum_n x^n H'(n, m, a), \qquad \Phi_m'(x, y) = \sum_a y^a \Phi'(x, m, a),$$

it follows that

$$\Phi(x, m, a) = \sum_{\substack{a_k \leqslant \cdots \leqslant a_1 = a \\ a_k \leqslant b}} x^{a_1 + \cdots + a_k} \Phi'(x, m - k, b)$$

and

$$\begin{aligned}
\Phi_m(x, y) &= \sum_{\substack{a_k \leqslant \cdots \leqslant a_1 \\ a_k \leqslant b}} x^{a_1 + \cdots + a_k} y^{a_1} \Phi'(x, m - k, b) \\
&= \sum_{b=1}^{\infty} \Phi'(x, m - k, b) \sum_{a_k \leqslant b} x^{a_k} \sum_{a_1 \geqslant \cdots \geqslant a_k} x^{a_1 + \cdots + a_{k-1}} y^{a_1} \\
&= \sum_{b=1}^{\infty} \Phi'(x, m - k, b) \sum_{a=1}^{b} \sum_{c_1, \ldots, c_{k-1} = 0}^{\infty} x^{c_1 + 2c_2 + \cdots + (k-1)c_{k-1} + ka_k} \\
&\quad \cdot y^{c_1 + \cdots + c_{k-1} + a_k} \\
&= \frac{1}{(xy)_{k-1}} \sum_{b=1}^{\infty} \Phi'(x, m - k, b) \frac{x^k y - (x^k y)^{b+1}}{1 - x^k y} \\
&= \frac{x^k y}{(xy)_k} \sum_{b=1}^{\infty} \Phi'(x, m - k, b)(1 - (x^k y)^b).
\end{aligned}$$

We have therefore

$$\Phi_m(x, y) = \frac{x^k y}{(xy)^k} [\Phi'_{m-k}(x, 1) - \Phi'_{m-k}(x, x^k y)]. \tag{11.2}$$

This notation is somewhat inadequate; the $\Phi'_{m-k}(x, y)$ on the right is associated with the wavy line pattern in (11.1). Suppose now that the wavy line is taken as something similar to

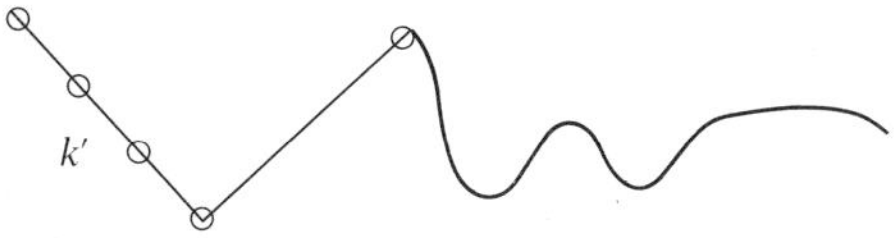

We may then iterate (11.2) to get

$$\Phi_m(x, y) = \frac{x^k y}{(xy)_k} = (xy)_k \Phi'_{m-k}(x, 1) - \frac{x^{2k+k'} y^2}{(xy)_{k+k'}} \Phi''_{m-k-k'}(x, 1)$$
$$+ \frac{x^{2k+k+k'}}{(xy)_{k+k'}} \cdot \Phi''_{m-k-k'}(x, x^{k+k'} y),$$

where $\Phi''_m(x, y)$ has the obvious meaning.

Continuing in this way we obtain the following result for the generalized down–up pattern

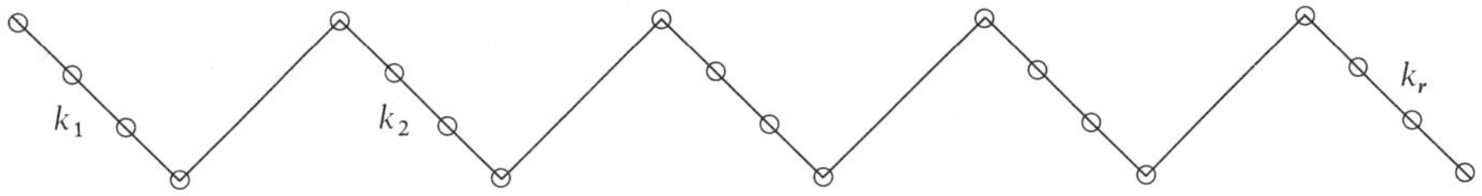

(11.3)

$$\Phi_m(x, y) = \sum_{j=1}^{r} (-1)^{j-1} \frac{x^{jk_1 + (j-1)k_2 + \cdots + k_j} y^j}{(xy)_{k_1 + \cdots + k_j}} \tag{11.4}$$
$$\cdot \Phi^{(j)}_{m-k_1-\cdots-k_j(x,1)} \qquad (m = k_1 + \cdots + k_r).$$

Since the various Φ's on the right apply to possibly different patterns, we cannot in general use (11.4) to get a generating function. However, if we assume that the pattern (11.3) is periodic with period s:

$$k_{i+js} = k_i \qquad (1 \leqslant i \leqslant s,\ 0 \leqslant j < t,\ r = st), \tag{11.5}$$

it follows that

$$\Phi^{(i+js)}_m(x, y) = \Phi^{(i)}_m(x, y).$$

For example let $t = 2$ and write k, k' in place of k_1, k_2. Then (11.4) becomes

$$\Phi_{sk+sk'}(x, y) = \sum_{j=1}^{s} \frac{x^{e_{2j-1}} y^{2j-1}}{(xy)_{jk+(j-1)k'}} \Phi'_{(s-j)k+(s-j+1)k'}(x, 1)$$
$$- \sum_{j=1}^{s} \frac{x^{e_{2j}} y^{2j}}{(xy)_{jk+jk'}} \Phi_{(s-j)k+(s-j)k'}(x, 1), \tag{11.6}$$

where

$$e_{2j-1} = j^2k + j(j-1)k', \qquad e_{2j} = j(j+1)k + j^2k'.$$

Similarly

$$\Phi_{sk+(s-1)k'}(x, y) = \sum_{j=1}^{s} \frac{x^{e_{2j-1}}y^{2j-1}}{(xy)_{jk+(j-1)k'}} \Phi'_{(s-j)k+(s-j)k'}(x, 1) - \sum_{j=1}^{s-1} \frac{x^{e_{2j}}y^{2j}}{(xy)_{jk+jk'}} \Phi_{(s-j)k+(s-j-1)k'}(x, 1). \tag{11.7}$$

Interchange of k and k' induces interchange of Φ and Φ'. Thus (11.6) and (11.7) may be replaced by

$$\Phi_{sk+sk'}(x, y) = \sum_{j=1}^{s} \frac{x^{e_{2j-1}}y^{2j-1}}{(xy)_{jk+(j-1)k'}} \Phi_{(s-j+1)k+(s-j)k'}(x, 1) - \sum_{j=1}^{s} \frac{x^{e_{2j}}y^{2j}}{(xy)_{jk+jk'}} \Phi_{(s-j)k+(s-j)k'}(x, 1) \qquad (s \geqslant 1), \tag{11.8}$$

$$\Phi_{sk+(s-1)k'}(x, y) = \sum_{j=1}^{s} \frac{x^{e_{2j-1}}y^{2j-1}}{(xy)_{jk+(j-1)k'}} \Phi_{(s-j)k+(s-j)k'}(x, 1) - \sum_{j=1}^{s-1} \frac{x^{e_{2j}}y^{2j}}{(xy)_{jk+jk'}} \Phi_{(s-j)k+(s-j-1)k'}(x, 1) \qquad (s \geqslant 1), \tag{11.9}$$

respectively.

In particular, for $y = 1$, it follows from (11.8) and (11.9) that

$$\sum_{j=0}^{\infty} \frac{x^{e_{2j}}z^{jk+jk'}}{(x)_{jk+jk'}} \sum_{s=0}^{\infty} \Phi_{sk+sk'}(x, 1)z^{sk+sk'} = 1 + \sum_{j=1}^{\infty} \frac{x^{e_{2j-1}}z^{jk+(j-1)k'}}{(x)_{jk+(j-1)k'}} \sum_{s=0}^{\infty} \Phi_{(s+1)k+sk'}(x, 1)z^{sk+(s-1)k'} \tag{11.10}$$

and

$$\sum_{j=0}^{\infty} \frac{x^{e_{2j}}z^{jk+jk'}}{(x)_{jk+jk'}} \sum_{s=1}^{\infty} \Phi_{sk+(s-1)k'}(x, 1)z^{sk+(s-1)k'} = \sum_{j=1}^{\infty} \frac{x^{e_{2j-1}}z^{jk+(j-1)k'}}{(x)_{jk+(j-1)k'}} \sum_{s=0}^{\infty} \Phi_{sk+sk'}(x, 1)z^{sk+sk'}. \tag{11.11}$$

We now put

$$C_{k,k'}(x, z) = \sum_{j=0}^{\infty} \frac{x^{e_{2j}} z^{jk+jk'}}{(x)_{jk+jk'}},$$

$$S_{k,k'}(x, z) = \sum_{j=1}^{\infty} \frac{x^{e_{2j-1}} z^{jk+(j-1)k'}}{(x)_{jk+(j-1)k'}},$$

$$\Phi_{k,k'}(x, z) = \sum_{s=0}^{\infty} \Phi_{sk+sk'}(x, 1) z^{sk+sk'},$$

$$\Psi_{k,k'}(x, z) = \sum_{s=1}^{\infty} \Phi_{sk+(s-1)k'}(x, 1) z^{sk+(s-1)k'}.$$

Then (11.10), (11.11) become

$$\begin{aligned} C_{k,k'}(x, z)\Phi_{k,k'}(x, z) &= 1 + z^{k'-k} S_{k,k'}(x, z)\Psi_{k,k'}(x, z) \\ S_{k,k'}(x, z)\Phi_{k,k'}(x, z) &= C_{k,k'}(x, z)\Psi_{k,k'}(x, z). \end{aligned} \tag{11.12}$$

Solving for Φ, Ψ, we obtain

THEOREM 11.1. *The generating functions $\Phi_{k,k'}(x, z)$, $\Psi_{k,k'}(x, z)$ satisfy*

$$\Phi_{k,k'}(x, z) = \frac{C_{k,k'}(x, z)}{C^2_{k,k'}(x, z) - z^{k'-k} S^2_{k,k'}(x, z)}, \tag{11.13}$$

$$\Psi_{k,k'}(x, z) = \frac{S_{k,k'}(x, z)}{C^2_{k,k'}(x, z) - z^{k'-k} S^2_{k,k'}(x, z)}. \tag{11.14}$$

Presumably formulas of this kind can be obtained for the more general situation defined by (11.5).

Remark. Formula 11.4 suggests the problem of finding a combinatorial interpretation of the coefficients of

$$\left[\sum_{j=0}^{\infty} (-1)^j \frac{x^{jk_1+(j-1)k_2+\cdots+k_j}}{(x)_{k_1+\cdots+k_j}} z^{k_1+\cdots+k_j}\right]^{-1}$$

for arbitrary positive k_j (or for $k_j > 1$).

12. SOME SPECIAL CASES

We have been tacitly assuming above that $k > 1$. This is, however, not necessary. For $k = 1$, (7.1) reduces to simply

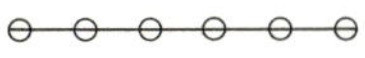

and the problem now is to enumerate the number of positive solutions of

$$n = a_1 + a_2 + \cdots + a_3$$

where

$$a_1 \geqslant a_2 \geqslant \cdots \geqslant a_3,$$

in other words, the number of (ordinary) partitions of n into k parts. Now, for $k = 1$, (7.8) becomes

$$\sum_{s=0}^{\infty} G_s(x, 1)z^s = \frac{1}{C(x, z)}, \tag{12.1}$$

where

$$C(x, z) = \sum_{j=0}^{\infty} (-1)^j \frac{x^{\frac{1}{2}j(j+1)}z^j}{(x)_j} = \prod_{j=1}^{\infty} (1 - x^j z)^{-1}.$$

Thus (12.1) is in agreement with a familiar result.

It is more interesting to examine the results of §11. If we take $k' = 1$ but $k > 1$, we get the patterns

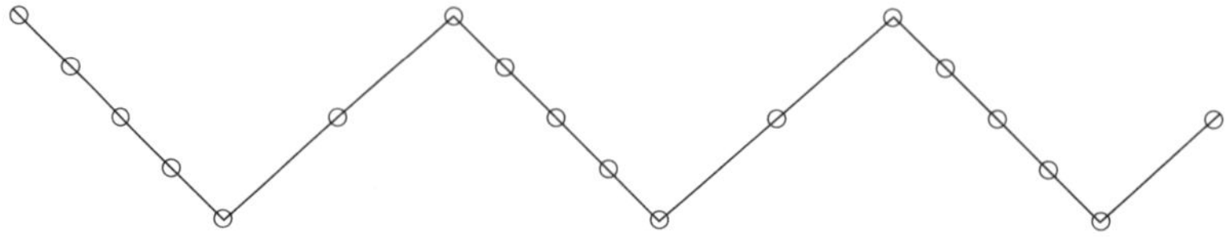

(12.2)

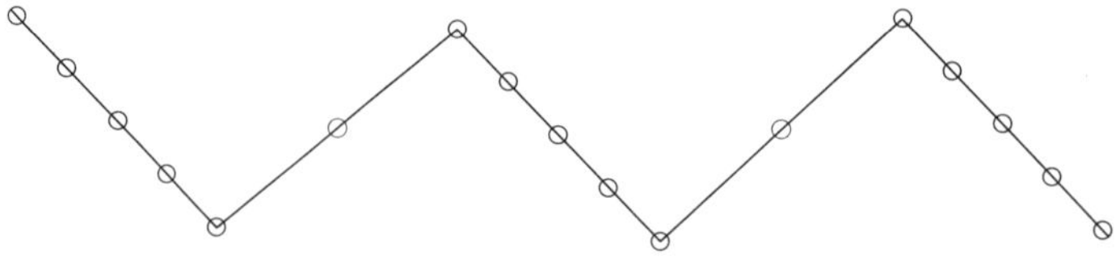

(12.3)

Formulas (11.13), (11.14) become

$$\Phi(x, z) = \frac{C(x, z)}{C^2(x, z) - z^{1-k}S^2(x, z)}, \tag{12.4}$$

$$\Psi(x, z) = \frac{S(x, z)}{C^2(x, z) - z^{1-k}S^2(x, z)}, \tag{12.5}$$

respectively, where

$$C(x, z) = \sum_{j=0}^{\infty} \frac{x^{j(j+1)k+j^2}z^{j(k+1)}}{(x)_{j(k+1)}},$$

$$S(x, z) = \sum_{j=1}^{\infty} \frac{x^{j^2k+j(j-1)}z^{j(k+1)-1}}{(x)_{j(k+1)-1}}.$$

The generating function $\Phi(x, z)$ corresponds to the pattern (12.2) while $\Psi(x, z)$ corresponds to (12.3).

The special case $k = 3$ is particular interesting; the patterns (12.2), (12.3) now become

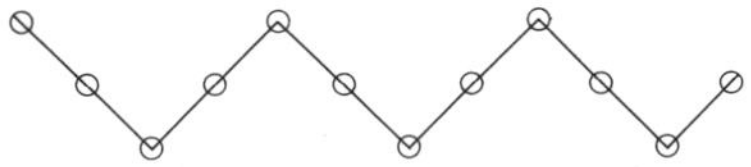

(12.6)

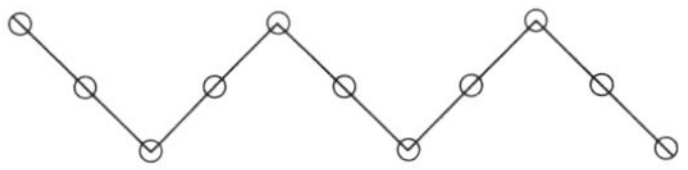

(12.7)

while $C(x, z)$, $S(x, z)$ become

$$C(x, z) = \sum_{j=0}^{\infty} \frac{x^{j(4j+3)} z^{4j}}{(x)_{4j}},$$

$$S(x, z) = \sum_{j=1}^{\infty} \frac{x^{j(4j-1)} z^{4j-1}}{(x)_{4j-1}}.$$

Specializing further we take $z = 1$. Also let $P^{\mathrm{e}}(n)$ denote the number of partitions of n with pattern (12.6), $P^{\mathrm{o}}(n)$ the number of partitions of n with pattern (12.7). We may state

THEOREM 12.1. *The enumerants $P^{\mathrm{e}}(n)$, $P^{\mathrm{o}}(n)$ satisfy*

$$\sum_{n=0}^{\infty} P^{\mathrm{e}}(n) x^n = \frac{C(x)}{C^2(x) - z^{-2} S^2(x)}, \qquad \sum_{n=0}^{\infty} P^{\mathrm{o}}(n) x^n = \frac{S(x)}{C^2(x) - z^{-2} S^2(x)},$$

where

$$C(x) = \sum_{j=0}^{\infty} \frac{x^{j(4j+1)}}{(x)_{4j}}, \qquad S(x) = \sum_{j=1}^{\infty} \frac{x^{j(4j-1)}}{(x)_{4j-1}}.$$

REFERENCES

1. L. CARLITZ, Generating functions for a special class of permutations, *Proc. Amer. Math. Soc.* **47**(1975), 251–256.
2. L CARLITZ, Permutations with prescribed pattern, *Math. Nachr.* **58**(1973), 31–53.
3. E. NETTO, "Lehrbuch der Combinatorik," Teubner, Leipzig und Berlin, 1927.
4. L. OLIVIER, Bermerkungen über eine Art von Funktionen, welche ähnliche Eigenschaften haben, wie die Cosinus und Sinus, *J. Reine Angew. Math.* **2** (1827), 243–251.

AMS (MOS) subject classifications: 05A17, 10A45, 33A70.

Plane Partitions (I): The MacMahon Conjecture†

George E. Andrews‡

Mathematics Research Center
University of Wisconsin
Madison, Wisconsin

1. Introduction

In 1898 MacMahon [7] presented his first study of symmetric higher dimensional partitions. The most interesting part of his paper is a conjecture concerning the generating function for $M(j, s; n)$, the number of plane partitions $\sum n_{ik}$ of n with the added conditions that (i) $n_{ik} = n_{ki}$, (ii) $n_{ik} = 0$ if $i > s$, (iii) $n_{11} \leqslant j$. Thus $M(j, s; n)$ is the number of symmetric plane partitions with at most s rows and with each part at most j. MacMahon [7, p. 153] conjectures

$$\sum_{N \geqslant 0} M(j, s; N)q^N = \prod_{i=1}^{s} \left[\frac{(1 - q^{j+2i-1})}{(1 - q^{2i-1})} \prod_{h=i+1}^{s} \frac{(1 - q^{2(j+i+h-1)})}{(1 - q^{2(i+h-1)})} \right]. \quad (1.1)$$

He demonstrates the truth of this conjecture in a few instances and remarks: "The proof of this formula, the truth of which seems unquestionable, is much to be desired."

Subsequently, in his monumental treatise "Combinatory Analysis," MacMahon [8, pp. 262–271] again discusses this conjecture at length, and again he asserts that, "The result has not been rigorously established." At the conclusion of his discussion [8, p. 270–271] MacMahon carefully examines the symmetry in the product in (1.1) and states that, "This property of the enumerating function is of great beauty and mathematical elegance."

In the late 1960s, Gordon [3, 4] proved MacMahon's conjecture when $s = \infty$. Gordon [4, p. 158] observes that Sylvester's mapping of self-conjugate partitions into partitions with distinct odd parts may be directly extended to plane partitions to show that $M(j, s; n)$ is also the number of plane partitions of n with strict decrease along rows where each part is odd and at most $2s - 1$ and there are at most j rows.

† Sponsored in part by the United States Army under Contract No. DAAG29-75-C-0024 and the National Science Foundation Grant MSP 74-07282.

‡ Present address: Department of Mathematics, Pennsylvania State University, University Park, Pennsylvania.

ISBN 0-12-599101-0

Also in the late 1960s, Bender and Knuth [2] developed a very powerful combinatorial method for treating many problems in plane partitions. They also extended much of the work of Gordon [3, 4] and Gordon and Houten [5, 6], and they showed [2, p. 50] that if

$$g_j(q) = \sum_{N \geqslant 0} M(j, m; N) q^N, \tag{1.2}$$

then

$$g_{2n}(q) = \det(C_{i-j} + C_{i+j-1})_{n \times n}, \tag{1.3}$$

and

$$g_{2n+1}(q) = \left[\prod_{i=1}^{m} (1 + q^{2i-1})\right] \det(C_{i-j} - C_{i+j})_{n \times n}, \tag{1.4}$$

where

$$C_k = q^{k^2} \binom{2m}{m+k}_2, \tag{1.5}$$

and $\binom{N}{M}_r$ is the Gaussian polynomial (or q-binomial coefficient) defined by

$$\binom{N}{M}_r = \begin{cases} \dfrac{(1-q^{Nr})(1-q^{(N-1)r}) \cdots (1-q^{(N-M+1)r})}{(1-q^{Mr})(1-q^{(M-1)r}) \cdots (1-q^r)}, & 0 < M \leqslant N \\ 1, & M = 0 \\ 0, & M < 0, \quad M > N. \end{cases} \tag{1.6}$$

However, Bender and Knuth [2, p. 50] go on to state that, "We have not been able to simplify these determinants any further, even for the limiting case as $q \to 1$ . . . But the known results, and calculations for small j give overwhelming empirical evidence that the answer has a simple form."

The object of this paper is to prove MacMahon's conjecture. In Section 2, we prove preliminary lemmas that are simply results from basic hypergeometric series. In Sections 3 and 4, we prove the conjecture by transforming the determinants in (1.3) and (1.4) three times each. In each case the third transformation produces a lower triangular determinant, and MacMahon's conjecture follows immediately.

In a subsequent paper we hope to treat a second conjecture of this nature due to Bender and Knuth [2, p. 50]. Presumably, our methods are adequate to treat it also. In [10, p. 265], Stanley mentions that Gordon possesses (unpublished) a proof of this latter conjecture; however, the implication from Stanley's comments is that Gordon's methods differ substantially from ours.

2. Summation Lemmas

First we require two results that are merely extensions of the q-analog of the Chu–Vandermonde summation [1, p. 469, Theorem 4.2]:

Lemma 1. *For integers $m \geqslant 0$, $i \geqslant 0$, $l \geqslant 1$,*

$$\sum_{j=-\infty}^{\infty} \binom{2m}{m+i-j}_2 q^{(i-j)^2} \binom{2l-1}{l-j}_2 \left(q^{(j-l)(j-l-2m)} + q^{(j+l-1)(j+l+2m-1)}\right)$$
$$= \binom{2m+2l-1}{m+i+l-1}_2 q^{(i-l)^2}\left(1 + q^{(2i-1)(2l-1)}\right).$$

Proof. We split the left sum into two parts:

$$\sum_{j=-\infty}^{\infty} \binom{2m}{m+i-j}_2 q^{(i-j)^2} \binom{2l-1}{l-j}_2 q^{(j-l)(j-l-2m)}$$
$$= \sum_{j=0}^{2l-1} \binom{2m}{m+i-l+j}_2 \binom{2l-1}{j}_2 q^{(i-l+1)^2+j(j+2m)}$$
$$= q^{(i-l)^2} \sum_{j=0}^{2l-1} \binom{2m}{m+i-l+j}_2 \binom{2l-1}{j}_2 q^{2j(j+m+i-l)}$$
$$= q^{(i-l)^2} \binom{2m+2l-1}{m+i+l-1}_2, \tag{2.1}$$

by the Chu–Vandermonde summation [1, p. 469, Theorem 4.2]. Next

$$\sum_{j=-\infty}^{\infty} \binom{2m}{m+i-j}_2 q^{(i-j)^2} \binom{2l-1}{l-j}_2 q^{(j+l-1)(j+l+2m-1)}$$
$$= \sum_{j=0}^{2l+1} \binom{2m}{m+i-j+l-1}_2 q^{(i-j+l-1)^2} \binom{2l-1}{2l-1-j}_2 q^{j(j+2m)}$$
$$= q^{(i+l-1)^2} \sum_{j=0}^{2l-1} \binom{2m}{m-i-l+1+j}_2 \binom{2l-1}{j}_2 q^{2j(j+m-i-l+1)}$$
$$= q^{(i+l-1)^2} \binom{2m+2l-1}{m+i+l-1}_2, \tag{2.2}$$

again by the Chu–Vandermonde summation [1, p. 469, Theorem 4.2].

If we now add together identities (2.1) and (2.2) we obtain Lemma 1. ∎

Lemma 2. *For integers $m \geqslant 0$, $i \geqslant 0$, $l \geqslant 0$,*

$$\sum_{j=-\infty}^{\infty} \binom{2m}{m+i-j}_2 q^{(i-j)^2} \binom{2l}{l+j}_2 \left(q^{(j-l)(j-l-2m)} - q^{(j+l)(j+l+2m)}\right)$$
$$= \binom{2m+2l}{m+i+l}_2 \left(q^{(i-l)^2} - q^{(i+l)^2}\right).$$

Proof. We split the left sum into two parts:

$$\sum_{j=-\infty}^{\infty} \binom{2m}{m+i-j}_2 q^{(i-j)^2} \binom{2l}{l+j}_2 q^{(j-l)(j-l-2m)}$$

$$= \sum_{j=0}^{2l} \binom{2m}{m+i-l+j}_2 q^{(i-l+j)^2} \binom{2l}{2l-j}_2 q^{j(j+2m)}$$

$$= q^{(i-l)^2} \sum_{j=0}^{2l} \binom{2m}{m+i-l+j}_2 \binom{2l}{j}_2 q^{2j(j+m+i-l)}$$

$$= q^{(i-l)^2} \binom{2m+2l}{m+i+l}_2, \tag{2.3}$$

by the Chu–Vandermonde summation [1, p. 469, Theorem 4.2]. Finally,

$$\sum \binom{2m}{m+i-j}_2 q^{(i-j)2} \binom{2l}{l+j}_2 q^{(j+l)(j+l+2m)}$$

$$= \sum_{j=0}^{2l} \binom{2m}{m+i+l-j}_2 q^{(i-j+l)^2} \binom{2l}{j}_2 q^{j(j+2m)}$$

$$= q^{(i+l)^2} \sum_{j=0}^{2l} \binom{2m}{m+i+l-j}_2 \binom{2l}{j}_2 q^{2j(j+m-i-l)}$$

$$= q^{(i+l)^2} \binom{2m+2l}{m+i+l}_2, \tag{2.4}$$

by the Chu–Vandermonde summation [1, p. 469, Theorem 4.2].

If we now subtract identity (2.4) from identity (2.3), we obtain Lemma 2. ∎

Our next result is a disguised form of the q-analog of the Pfaff–Saalschutz summation [9, p. 97, Eq. (3.3.2.2)].

LEMMA 3. *For integers* $n \geqslant k \geqslant 1$, $b = 0$ or 1, $i \geqslant 0$, $m \geqslant 0$,

$$\sum_{j=1}^{n} \binom{2m+2j-b}{m+i+j-b}_1 \frac{(-1)^{j+k}(q^{2m+2j+1-b})_{2k-2j} q^{\frac{1}{2}j(j+1)-kj}}{(q^{m+j+k-b})_{k-j}(q^{m+j})_{k-j}} \binom{k-1}{j-1}_1$$

$$= \frac{(-1)^{k+1} q^{1-k} (q)_{2m+2k-b} (q)_m (q^{1-i})_{k-1} (q^{b+1-i-k})_{k-1} (q)_{m+k-b}}{(q)_{m+2k-b-1} (q)_{m+k-1} (q)_{m+i-b+1} (q)_{m+k-i} (q^{b-i-m-k})_{k-1}},$$

where $(q)_n = (1-q)(1-q^2)\cdots(1-q^n)$.

Proof.

$$\sum_{j=1}^{n} \binom{2m+2j-b}{m+i+j-b}_1 \frac{(-1)^{j+k}(q^{2m+2j+1-b})_{2k-2j} q^{\frac{1}{2}j(j+1)-kj}}{(q^{m+j+k-b})_{k-j}(q^{m+j})_{k-j}} \binom{k-1}{j-1}_1$$

$$= \sum_{j=1}^{n} \frac{(q)_{2m+2k-b}(-1)^{j+k} q^{\frac{1}{2}j(j+1)-kj} (q)_{m+j+k-b-1} (q)_{m+j-1}}{(q)_{m+i+j-b} (q)_{m-i+j} (q)_{m+2k-b-1} (q)_{m+k-1}} \binom{k-1}{j-1}_1$$

$$= \frac{(-1)^{k+1}(q)_{2m+2k-b}(q)_{m+k-b}(q)_m}{(q)_{m+2k-b-1}(q)_{m+k-1}(q)_{m+i-b+1}(q)_{m-i+1}}$$

$$\times \sum_{j=0}^{k-1} \frac{(q^{-k+1})_j(q^{m+k-b+1})_j(q^{m+1})_j q^j}{(q)_j(q^{m+i-b+2})_j(q^{m-i+2})_j}$$

$$= \frac{(-1)^{k+1}q^{1-k}(q)_{2m+2k-b}(q)_{m+k-b}(q)_m(q^{1-i})_{k-1}(q^{b+1-i-k})_{k-1}}{(q)_{m+2k-b-1}(q)_{m+k-1}(q)_{m+i-b+1}(q)_{m+k-i}(q^{b-i-m-k})_{k-1}},$$

by the q-analog of the Pfaff–Saalschutz summation [9, p. 97, Eq. (3.3.2.2)]. We remark that the above is meaningful and valid provided $b \leqslant m + i + 1$; however the more restricted cases $b = 0$ or 1 are sufficient for us. ∎

Our next two lemmas treat q-identities that are really further formulas for sums of basic hypergeometric series. Since we were unable to find them or generalizations of them in the literature, we choose to prove them *de novo* using Dougall's method [10, pp. 55, 95].

LEMMA 4. *For integers* $s \geqslant j \geqslant 1$, $i \geqslant 1$, $m \geqslant 1$, *if*

$$a_i(r, s) = \frac{(-1)^{r+s}(q^2, q^2)_{m+r+s-1}(q^2, q^2)_{m+s-r}(q^{2i-2s}; q^2)_s(q^{2i}; q^2)_s}{(q^2; q^2)_{m+i+s-1}(q^2, q^2)_{m-i+s}(1 - q^{2i-2r})(1 - q^{2i+2r-2})}$$

$$\times \frac{(1 + q^{2i-1})(1 - q^{4r-2})q^{(i-s)^2+(s-r)}}{(1 + q^{2r-1})(q^2; q^2)_{s-1}(q^{2r}; q^2)_s} \binom{s-1}{r-1}_2, \tag{2.5}$$

then

$$\sum_{r=1}^{s} a_i(r, s)\binom{2m+2j-1}{m+r+j-1}_2 q^{(r-j)^2}(1 + q^{(2r-1)(2j-1)})$$

$$= \binom{2m+2j-1}{m+i+j-1}_2 q^{(i-j)^2}(1 + q^{(2i-1)(2j-1)}), \tag{2.6}$$

and

$$\sum_{r=1}^{s} a_i(r, s)\left[\binom{2m+2s+1}{m+r+s}_2 q^{(r-s-1)^2} \frac{(1 + q^{(2r-1)(2s+1)})}{(1 + q^{(2m+2s+1)(2s+1)})}\right.$$

$$\left. - \binom{2m+2s+1}{m+r+s}_2 q^{(r-1)^2-2ms-s^2} \frac{(1 + q^{2r-1})}{(1 + q^{2m+2s+1})}\right]$$

$$= \binom{2m+2s+1}{m+i+s}_2 q^{(i-s-1)^2} \frac{(1 + q^{(2i-1)(2s+1)})}{(1 + q^{(2m+2s+1)(2s+1)})}$$

$$- \binom{2m+2s+1}{m+i+s}_2 q^{(i-1)^2-2ms-s^2} \frac{(1 + q^{2i-1})}{(1 + q^{2m+2s+1})}, \tag{2.7}$$

where

$$(A;q)_n = \prod_{h=0}^{\infty} \frac{(1-Aq^h)}{(1-Aq^{h+n})}$$

$(= (1-A)(1-Aq)\cdots(1-Aq^{n-1})$ *when* n *is a positive integer*).

Remark. The apparent singularity of $a_i(r, s)$ for $i = r$ or $-r+1$ is removable since factors $(1-q^{2i-2r})$ and $(1-q^{2i+2r-2})$ appear in the numerator of $a_i(r, s)$. Note also that $1/(q^2;q^2)_{-n} = 0$ for any positive integer n.

Proof. If $i > m+s$ or $i \leqslant -m-s$, then (2.6) is trivial since both sides are identically zero. Multiply both sides of (2.6) by

$$q^{-(i-s)^2}(q^2;q^2)_{m+i+s-1}(q^2;q^2)_{m-i+s}(q^2;q^2)_{2m+2j-1}^{-1}(1+q^{2i-1})^{-1}.$$

Since we may assume $-m-s < i \leqslant m+s$, this quantity is neither infinite nor zero, and we have the equivalent identity:

$$\sum_{r=1}^{s} (-1)^{r+s}(q^{2m+2r+2j};q^2)_{s-j}(q^{2m+2j-2r+2};q^2)_{s-j}q^{(s-r)+(r-j)^2}$$
$$\times \frac{(1+q^{(2r-1)(2j-1)})(q^{2i-2s};q^2)_s(q^{2i};q^2)_s(1-q^{4r-2})}{(1+q^{2r-1})(1-q^{2i-2r})(1-q^{2i+2r-2})(q^2;q^2)_{s-1}(q^{2r};q^2)_s} \left({s-1 \atop r-1}\right)_s$$
$$= q^{(2i-j-s)(s-j)}\frac{(1+q^{(2i-1)(2j-1)})}{(1+q^{2i-1})}(q^{2m+2i+2j};q^2)_{s-j}(q^{2m+2j-2i+2};q^2)_{s-j}. \tag{2.8}$$

Now identity (2.8) can be viewed as a polynomial identity in q^{2i}, where the left polynomial has degree at most $2s-2$ and the right polynomial also has degree at most $(s-j)+(s-j)+(2j-2) = 2s-2$. If we can prove that the two sides of (2.8) are identical for at least $2s-1$ values of q^{2i}, then identity (2.8) must be valid for all i.

First let $i = t$ with $1 \leqslant t \leqslant s$. Then each term on the left of (2.8) vanishes with the exception of $r = t$; so in this case the left side of (2.8) is just

$$(q^{2m+2t+2j};q^2)_{s-j}(q^{2m+2j-2t+2};q^2)_{s-j}q^{-(s-t)^2+(t-j)^2}\frac{(1+q^{(2t-1)(2j-1)})}{(1+q^{2t-1})},$$

and this is the right side of (2.8) when $i = t$.

Next let $i = -t$ with $0 \leqslant t \leqslant s-1$. Then again each term on the left of (2.8) vanishes with the exception of $r = t+1$; so in this case the left side of

(2.8) is just

$$(q^{2m+2t+2j+2}; q^2)_{s-j}(q^{2m+2j-2t}; q^2)_{s-j} \frac{1+q^{(2t+1)(2j-1)}}{1+q^{2t+1}} q^{(j-t)^2-(s+t)^2+4t-2j+2},$$

and this is precisely the right side of (2.8) when $i = -t$. Consequently (2.8) is valid in general and thus (2.6) is valid in general.

We now pass to the proof of (2.7). If $i > m + s + 1$ or $i < -m - s$, then (2.7) is trivial since both sides are identically zero. Multiply both sides of (2.7) by

$$q^{2i-(i-s)^2}(q^2; q^2)_{m+i+s}(q^2; q^2)_{m-i+s+1}(q^2; q^2)_{2m+2s+1}^{-1}(1 + q^{2i-1})^{-1}.$$

Since we may assume $-m - s \leqslant i \leqslant m + s + 1$, this quantity is neither infinite nor zero, and we have the equivalent identity:

$$\begin{aligned} &q^{2i}(1 - q^{2m+2i+2s})(1 - q^{2m-2i+2s+2}) \\ &\quad\times \sum_{r=1}^{s} \frac{(-1)^{r+s}q^{(s-r)+(r-s-1)^2}(q^{2i-2s}; q^2)_s(q^{2i}, q^2)_s}{(1 - q^{2m+2r+2s})(1 - q^{2m+2s-2r+2})} \\ &\quad\times \frac{(1 - q^{4r-2})\binom{s-1}{r-1}_2}{(1 + q^{2r-1})(1 - q^{2i-2r})(1 - q^{2i+2r-2})(q^2; q^2)_{s-1}(q^{2r}; q^2)_s} \\ &\quad\times \left[\frac{(1 + q^{(2r-1)(2s+1)})}{(1 + q^{(2m+2s+1)(2s+1)})} - \frac{q^{-2s+2rs-2ms-2s^2}(1 + q^{2r-1})}{(1 + q^{2m+2s+1})}\right] \\ &= \frac{q^{2s+1}}{(1 + q^{(2m+2s+1)(2s+1)})} \\ &\quad\times \left[\frac{(1 + q^{(2i-1)(2s+1)})}{(1 + q^{2i-1})} - q^{2s(i-m-s-1)}\frac{(1 + q^{(2m+2s+1)(2s+1)})}{(1 + q^{2m+2s+1})}\right]. \end{aligned} \tag{2.9}$$

As before, identity (2.9) can be viewed as a polynomial identity in q^{2i}, where the left polynomial has degree at most $2s$ and the right polynomial has degree at most $2s$. We must now show that identity (2.9) is valid for $2s + 1$ values of i in order to establish it for all i. First of all we note that if $i = m + s + 1$ then both sides of (2.9) are identically zero.

Next let $i = t$ with $1 \leqslant t \leqslant s$. Then each term on the left of (2.9) vanishes with the exception of $r = t$; so in this case the left side of (2.9) is just

$$\frac{q^{2s+1}}{(1+q^{(2m+2s+1)(2s+1)})}\left[\frac{(1+q^{(2t-1)(2s+1)})}{(1+q^{2t-1})} - \frac{q^{2s(t-m-s-1)}(1+q^{(2m+2s+1)(2s+1)})}{(1+q^{2m+2s+1})}\right]$$

and this is precisely the right side of (2.9) when $i = t$.

Now let $i = -t$ with $0 \leqslant t \leqslant s-1$. As before each term on the left of (2.9) vanishes with the exception of $r = t+1$; hence for $i = -t$ the left side of (2.9) is just

$$\frac{q^{2s+1}}{(1+q^{(2m+2s+1)(2s+1)})} \times \left[\frac{q^{-2s(2t+1)}(1+q^{(2t+1)(2s+1)})}{(1+q^{2t+1})} - q^{-2s(t+m+s+1)}\frac{(1+q^{(2m+2s+1)(2s+1)})}{(1+q^{2m+2s+1})}\right],$$

and this is the right side of (2.9) when $i = -t$.

Hence both sides of (2.9) are identical for $i = m+s+1$ and $-s+1 \leqslant i \leqslant s$; hence (2.9) is valid for all i. Thus (2.7) is established. ∎

LEMMA 5. *For integers* $s \geqslant j \geqslant 1$, $i \geqslant 1$, $m \geqslant 1$, *if*

$$b_i(r,s) = \frac{(-1)^{r+s}(q^2;q^2)_{m+r+s}(q^2;q^2)_{m+s-r}(q^{2i-2s};q^2)_s(q^{2i+2};q^2)_s}{(q^2;q^2)_{m+i+s}(q^2;q^2)_{m-i+s}(1-q^{2i-2r})(1-q^{2i+2r})} \times \frac{(1-q^{4i})q^{(i-s)^2+(s-r)}}{(q^2;q^2)_{s-1}(q^{2r+2};q^2)_s}\binom{s-1}{r-1}_2, \tag{2.10}$$

then

$$\sum_{r=1}^{s} b_i(r,s)\binom{2m+2j}{m+r+j}_2 q^{(r-j)^2}(1-q^{4rj}) = \binom{2m+2j}{m+i+j}_2 q^{(i-j)^2}(1-q^{4ij}), \tag{2.11}$$

and

$$\sum_{r=1}^{s} b_i(r,s)\left\{\binom{2m+2s+2}{m+r+s+1}_2 q^{(r-s-1)^2}\frac{(1-q^{4r(s+1)})}{(1-q^{4(s+1)(s+m+1)})} - \binom{2m+2s+2}{m+r+s+1}_2 \frac{q^{(r-1)^2-2ms-s^2}(1-q^{4r})}{(1-q^{4m+4s+4})}\right\}$$
$$= \binom{2m+2s+2}{m+i+s+1}_2 \frac{q^{(i-s-1)^2}(1-q^{4i(s+1)})}{(1-q^{4(s+1)(s+m+1)})} - \binom{2m+2s+2}{m+i+s+1}_2 \frac{q^{(i-1)^2-2ms-s^2}(1-q^{4i})}{(1-q^{4m+4s+4})}. \tag{2.12}$$

Remark. The apparent singularity of $b_i(r,s)$ for $i = r$ or $-r$ is removable since factors $(1-q^{2i-2r})$ and $(1-q^{2i+2r})$ also appear in the numerator of $b_i(r,s)$.

Proof. If $i > m + s$ or $i < -m - s$ or $i = 0$, then (2.11) is trivial since both sides are identically zero. Multiply both sides of (2.11) by

$$q^{-(i-s)^2}(q^2;q^2)_{m+i+s}(q^2;q^2)_{m-i+s}(q^2;q^2)_{2m+2j}^{-1}(1-q^{4i})^{-1}.$$

Since we may assume $-m - s \leqslant i \leqslant m + s$ and $i \neq 0$, this quantity is neither infinite nor zero, and we have the equivalent identity:

$$\sum_{r=1}^{s}(-1)^{r+s}(q^{2m+2r+2j+2};q^2)_{s-j}(q^{2m+2j-2r+2};q^2)_{s-j}q^{(s-r)+(r-j)^2}$$
$$\times\frac{(1-q^{4rj})(q^{2i-2s};q^2)_s(q^{2i+2};q^2)_s}{(1-q^{2i-2r})(1-q^{2i+2r})(q^2;q^2)_{s-1}(q^{2r+2};q^2)_s}\binom{s-1}{r-1}_2$$
$$=q^{(2i-j-s)(s-j)}\frac{(1-q^{4ij})}{(1-q^{4i})}(q^{2m+2i+2j+2};q^2)_{s-j}(q^{2m+2j-2i+2};q^2)_{s-j}. \tag{2.13}$$

Identity (2.13) may be viewed as a polynomial identity in q^{2i}, where the left polynomial has degree at most $2s - 2$ and the right polynomial also has degree at most $2s - 2$. If we can prove that the two sides of (2.13) are identical for at least $2s - 1$ values of q^{2i}, then (2.13) is valid for all i.

First let $i = t$ with $1 \leqslant t \leqslant s$. Then each term on the left of (2.13) vanishes with the exception of $r = t$; so in this case the left side of (2.13) is just

$$(q^{2m+2t+2j+2};q^2)_{s-j}(q^{2m+2j-2t+2};q^2)_{s-j}q^{(2t-j-s)(s-j)}\frac{(1-q^{4tj})}{(1-q^{4t})},$$

and this is the right side of (2.13) when $i = t$.

Now let $i = -t$ with $1 \leqslant t \leqslant s$. In this case each term on the left of (2.13) vanishes with the exception of $r = t$; hence for $i = -t$ the left side of (2.13) is just

$$q^{-(2t+j+s)(s-j)}\frac{(1-q^{-4tj})}{(1-q^{-4t})}(q^{2m-2t+2j+2};q^2)_{s-j}(q^{2m+2j+2t+2};q^2)_{s-j},$$

and this is the right side of (2.13) when $i = -t$. Consequently (2.13) is valid for all i, and so (2.11) is also.

To conclude the proof of Lemma 5 we treat identity (2.12). If $i > m + s + 1$ or $i < -m - s - 1$, or $i = 0$, then (2.12) is trivial since both sides are identically zero. Multiply both sides of (2.12) by

$$q^{2i-(i-s)^2}(q^2;q^2)_{m+i+s+1}(q^2;q^2)_{m-i+s+1}(q^2;q^2)_{2m+2s+2}^{-1}(1-q^{4i})^{-1}.$$

Since we may assume $-m - s - 1 \leqslant i \leqslant m + s + 1$ and $i \neq 0$, this quantity is neither infinite nor zero and we have the equivalent identity:

$$q^{2i}(1-q^{2m+2i+2s+2})(1-q^{2m-2i+2s+2})$$
$$\times \sum_{r=1}^{s} \frac{(-1)^{r+s}q^{(s-r)}}{(1-q^{2m+2r+2s+2})(1-q^{2m-2r+2s+2})}$$
$$\times \frac{(q^{2i-2s};q^2)_s(q^{2i+2};q^2)_s\binom{s-1}{r-1}_2}{(1-q^{2i-2r})(1-q^{2i+2r})(q^2;q^2)_{s-1}(q^{2r+2};q^2)_s}$$
$$\times \left[\frac{q^{(r-s-1)^2}(1-q^{4r(s+1)})}{(1-q^{4(s+1)(s+m+1)})} - \frac{q^{(r-1)^2-2ms-s^2}(1-q^{4r})}{(1-q^{4m+4s+4})}\right]$$
$$= \frac{q^{2s+1}}{(1-q^{4(s+1)(s+m+1)})}\left[\frac{(1-q^{4i(s+1)})}{(1-q^{4i})} - \frac{q^{2s(i-m-s-1)}(1-q^{4(s+1)(s+m+1)})}{(1-q^{4m+4s+4})}\right]. \tag{2.14}$$

Identity (2.14) can be viewed as a polynomial identity in q^{2i}, where each side has degree at most $2s$. We must now show that identity (2.14) is valid for $2s+1$ values of i in order to establish it for all i.

If $i = m+s+1$, then both sides of (2.14) are identically zero.

If $i = t$ with $1 \leqslant t \leqslant s$, then each term on the left side of (2.14) vanishes with the exception of $r = t$; hence in this case the left side of (2.14) is just

$$\frac{q^{2s+1}}{(1-q^{4(s+1)(s+m+1)})}\left[\frac{(1-q^{4t(s+1)})}{(1-q^{4t})} - \frac{q^{2s(t-m-s-1)}(1-q^{4(s+1)(s+m+1)})}{(1-q^{4m+4s+4})}\right],$$

and this is, in fact, the right side of (2.14) when $i = t$.

Finally, let $i = -t$ with $1 \leqslant t \leqslant s$. As before, each term on the left of (2.14) vanishes with the exception of $r = t$; so for $i = -t$ the left side of (2.14) reduces to

$$\frac{q^{2s+1}}{(1-q^{4(s+1)(s+m+1)})}\left[\frac{(1-q^{-4t(s+1)})}{(1-q^{-4t})} - \frac{q^{-2s(t+m+s+1)}(1-q^{4(s+1)(s+m+1)})}{(1-q^{4m+4s+4})}\right],$$

and this is the right side of (2.14) when $i = -t$.

Thus identity (2.14) is valid for $i = m+s+1$, $1 \leqslant i \leqslant s$, and $-s \leqslant i \leqslant -1$; this implies that (2.14) is valid for all i. Identity (2.12) now follows. ∎

We now have all the necessary summations. These will be used in the next section for matrix multiplication lemmas.

3. Determinants and Matrices

To accomplish our goal, we must prove six lemmas that involve the following six matrices:

$$\alpha_n = \left(q^{(i-j)^2}\binom{2m}{m+i-j}_2 + q^{(i+j-1)^2}\binom{2m}{m+i+j-1}_2\right)_{n\times n}, \tag{3.1}$$

$$\alpha_n' = \left(q^{(i-j)^2}\binom{2m}{m+i-j}_2 - q^{(i+j)^2}\binom{2m}{m+i+j}_2\right)_{n\times n}, \tag{3.2}$$

$$\beta_n = \left[\binom{2m+2j-1}{m+i+j-1}_2 \frac{q^{(i-j)^2}(1+q^{(2i-1)(2j-1)})}{(1+q^{(2m+2j-1)(2j-1)})}\right]_{n\times n}, \tag{3.3}$$

$$\beta_n' = \left[\binom{2m+2j}{m+i+j}_2 \frac{q^{(i-j)^2}(1-q^{4ij})}{(1-q^{4j(m+j)})}\right]_{n\times n}, \tag{3.4}$$

$$\gamma_n = \left[\binom{2m+2j-1}{m+i+j-1}_2 \frac{q^{(i-1)^2-2m(j-1)-(j-1)^2}(1+q^{2i-1})}{(1+q^{2m+2j-1})}\right]_{n\times n}, \tag{3.5}$$

$$\gamma_n' = \left[\binom{2m+2j}{m+i+j}_2 \frac{q^{(i-1)^2-2m(j-1)-(j-1)^2}(1-q^{4i})}{(1-q^{4m+4j})}\right]_{n\times n}. \tag{3.6}$$

LEMMA 6. *Let*

$$\delta_n = \left[\binom{2j-1}{j-i}_2 \frac{(q^{(i-j)(i-j-2m)} + q^{(i+j-1)(i+j+2m-1)})}{(1+q^{(2j-1)(2j+2m-1)})}\right]_{n\times n}.$$

Then $\det\delta_n = 1$, *and* $\alpha_n \cdot \delta_n = \beta_n$.

Proof. We note immediately that δ_n is upper triangular with ones on the main diagonal. Hence $\det\delta_n = 1$.

The (i, j)th entry in $\alpha_n \cdot \delta_n$ is

$$\sum_{k=1}^{n} \left(q^{(i-k)^2}\binom{2m}{m+i-k}_2 + q^{(i+k-1)^2}\binom{2m}{m+i+k-1}_2\right)$$
$$\times \left[\binom{2j-1}{j-k}_2 \frac{(q^{(k-j)(k-j-2m)} + q^{(k+j-1)(k+j+2m-1)})}{(1+q^{(2j-1)(2j+2m-1)})}\right]$$
$$= \sum_{k=-\infty}^{\infty} \binom{2m}{m+i-k}_2 q^{(i-k)^2}\binom{2j-1}{j-k}_2 \frac{(q^{(k-j)(k-j-2m)} + q^{(k+j-1)(k+j+2m-1)})}{(1+q^{(2j-1)(2j+2m-1)})}$$
$$= \binom{2m+2j-1}{m+i+j-1}_2 \frac{q^{(i-j)^2}(1+q^{(2i-1)(2j-1)})}{(1+q^{(2j-1)(2j+2m-1)})},$$

by Lemma 1. Since this last expression is precisely the (i, j)th entry of β_n, Lemma 6 is established. ∎

LEMMA 7. *Let*

$$\delta_n' = \left(\binom{2j}{j+i}_2 \frac{(q^{(i-j)(i-j-2m)} - q^{(i+j)(i+j+2m)})}{(1-q^{4j(j+m)})}\right)_{n\times n}.$$

Then $\det\delta_n' = 1$, *and* $\alpha_n' \cdot \delta_n' = \beta_n'$.

Proof. The matrix δ_n' is upper triangular with ones on the main diagonal. Hence $\det \delta_n' = 1$.

The (i, j)th entry in $\alpha_n' \cdot \delta_n'$ is

$$\sum_{k=1}^{n} \left(q^{(i-k)^2}\binom{2m}{m+i-k}_2 - q^{(i+k)^2}\binom{2m}{m+i+k}_2\right)$$

$$\times \left(\binom{2j}{j+k}_2 \frac{(q^{(k-j)(k-j-2m)} - q^{(k+j)(k+j+2m)})}{(1-q^{4j(j+m)})}\right)$$

$$= \sum_{k=-\infty}^{\infty} \binom{2m}{m+i-k} q^{(i-k)^2}\binom{2j}{j+k}_2 \frac{(q^{(k-j)(k-j-2m)} - q^{(k+j)(k+j+2m)})}{(1-q^{4j(j+m)})}$$

$$= \binom{2m+2j}{m+i+j}_2 \frac{q^{(i-j)^2}(1-q^{4ij})}{(1-q^{4j(j+m)})},$$

by Lemma 2. Since this last expression is the (i, j)th entry of β_n', we have established Lemma 7. ▮

LEMMA 8.

$$\det \gamma_n = \prod_{i=1}^{m} \left[\frac{(1-q^{2n+2i-1})}{(1-q^{2i-1})} \prod_{h=i+1}^{m} \frac{(1-q^{2(2n+i+h-1)})}{(1-q^{2(i+h-1)})}\right].$$

Proof.

$$\det \gamma_n = \det\left[\binom{2m+2j-1}{m+i+j-1}_2 \frac{q^{(i-1)^2-2m(j-1)-(j-1)^2}(1+q^{2i-1})}{(1+q^{2m+2j-1})}\right]_{n\times n}$$

$$= \frac{q^{-mn(n-1)}(-q; q^2)_n}{(-q^{2m+1}; q^2)_n} \det\left[\binom{2m+2j-1}{m+i+j-1}_2\right]_{n\times n}.$$

Define

$$\epsilon_n = \left(\frac{(-1)^{i+j}(q^{4m+4i}; q^2)_{2j-2i}q^{i^2+i-2ij}}{(q^{2m+2i+2j-2}; q^2)_{j-i}(q^{2m+2i}; q^2)_{j-1}} \binom{j-1}{i-1}_2\right)_{n\times n}.$$

Then ϵ_n is upper triangular, and its (j, j)th entry is q^{-j^2+j}. Therefore $\det \epsilon_n = \prod_{j=1}^{n} q^{j-j^2}$. If we let

$$\gamma_n^* = \left[\binom{2m+2j-1}{m+i+j-1}_2\right]_{n\times n}$$

then the (i, j)th entry of $\gamma_n^* \cdot \epsilon_n$ is (by Lemma 3 with $b = 1$ and q replaced by q^2):

$$\frac{(-1)^{j+1}q^{2-2j}(q^2; q^2)_{2m+2j-1}(q^2; q^2)_m(q^{2-2i}; q^2)_{j-1}(q^{4-2i-2j}; q^2)_{j-1}}{(q^2; q^2)_{m+2j-2}(q^2; q^2)_{m+i}(q^2; q^2)_{m+j-i}(q^{2-2i-2m-2j}; q^2)_{j-1}}.$$

Now since this last expression is zero for $j > i$, we see that $\gamma_n^* \cdot \epsilon_n$ is lower triangular. Consequently,

$$\det \gamma_n = \frac{q^{-mn(n-1)}(-q;q^2)_n(\det \gamma_n^*)}{(-q^{2m+1};q^2)_n}$$

$$= \frac{q^{-mn(n-1)}(-q;q^2)_n}{(-q^{2m+1};q^2)_n(\det \epsilon_n)} \prod_{j=1}^{n}$$

$$\times \frac{(-1)^{j+1}q^{2-2j}(q^2;q^2)_{2m+2j-1}(q^{2-2j};q^2)_{j-1}(q^{4-4j};q^2)_{j-1}}{(q^2;q^2)_{m+2j-2}(q^2;q^2)_{m+j}(q^{2-2m-4j};q^2)_{j-1}}$$

$$= \frac{(-q;q^2)_n}{(-q^{2m+1};q^2)_n} \prod_{j=1}^{n} \frac{(q^2;q^2)_{2m+2j-1}(q^2;q^2)_{j-1}(q^{2j};q^2)_{j-1}}{(q^2;q^2)_{m+2j+2}(q^2;q^2)_{m+j}(q^{2m+2j+2};q^2)_{j-1}}$$

$$= \frac{(-q;q^2)_n}{(-q^{2m+1};q^2)_n} \prod_{j=1}^{n} \frac{(q^2;q^2)_{2m+2j-1}(q^2;q^2)_{2j-2}}{(q^2;q^2)_{m+2j-2}(q^2;q^2)_{m+2j-1}}$$

$$= \frac{(q^{2n+1};q^2)_m}{(q;q^2)_m} \prod_{j=1}^{n} \frac{(q^2;q^2)_{2m+2j-2}(q^2;q^2)_{2j-1}}{(q^2;q^2)_{m+2j-2}(q^2;q^2)_{m+2j-1}}$$

$$= \frac{(q^{2n+1};q^2)_m \prod_{j=1}^{n}(q^{4j-2};q^2)_{2m}}{(q;q^2)_m \prod_{j=1}^{2n}(q^{2j};q^2)_m}$$

$$= \frac{(q^{2n+1};q^2)_m \prod_{j=1}^{2n}(q^{2j};q^2)_{2m}}{(q;q^2)_m (\prod_{j=1}^{2n}(q^{2j};q^2)_m)(\prod_{j=1}^{n}(q^{4j};q^2)_{2m})}$$

$$= \frac{(q^{2n+1};q^2)_m \prod_{j=1}^{2n}(q^{2j+2m};q^2)_m}{(q;q^2)_m \prod_{j=1}^{n}(q^{4j};q^2)_{2m}}$$

$$= \frac{(q^{2n+1};q^2)_m \prod_{j=1}^{2n}(q^{2j+2m};q^2)_m}{(q;q^2)_m \prod_{j=1}^{n}(q^{4j};q^4)_m(q^{4j+2};q^4)_m}$$

$$= \frac{(q^{2n+1};q^2)_m}{(q;q^2)_m} \prod_{j=1}^{2n-1} \prod_{i=1}^{m} \frac{(1-q^{2m+2j+2i})}{(1-q^{2j+4i})}$$

$$= \frac{(q^{2n+1};q^2)_m}{(q;q^2)_m} \prod_{i=1}^{m} \frac{(q^{2m+2i};q^2)_{2n}}{(q^{4i};q^2)_{2n}}$$

$$= \frac{(q^{2n+1};q^2)_m}{(q;q^2)_m} \prod_{i=1}^{m} \frac{(q^2;q^2)_{2n+i+m-1}(q^2;q^2)_{2i-1}}{(q^2;q^2)_{m+i-1}(q^2;q^2)_{2n+2i-1}}$$

$$= \frac{(q^{2n+1};q^2)_m}{(q;q^2)_m} \prod_{i=1}^{m} \prod_{j=i+1}^{m} \frac{(1-q^{2(2n+i+j-1)})}{(1-q^{2i+2j-2})}$$

$$= \prod_{i=1}^{m} \left\{ \frac{(1-q^{2n+2i-1})}{(1-q^{2i-1})} \prod_{j=i+1}^{m} \frac{(1-q^{2(2n+i+j-1)})}{(1-q^{2(i+j-1)})} \right\}. \qquad \blacksquare$$

LEMMA 9.

$$(-q;q^2)_m \det \gamma_n' = \prod_{i=1}^{m}\left[\frac{(1-q^{2n+2i})}{(1-q^{2i-1})}\prod_{h=i+1}^{m}\frac{(1-q^{2(2n+i+h)})}{(1-q^{2(i+h-1)})}\right].$$

Proof.

$$\det\gamma_n' = \det\left[\binom{2m+2j}{m+i+j}_2 \frac{q^{(i-1)^2-2m(j-1)-(j-1)^2}(1-q^{4i})}{(1-q^{4m+4j})}\right]_{n\times n}$$

$$= \frac{q^{-mn(n-1)}(q^4;q^4)_n}{(q^{4m+4};q^4)_n}\det\left[\binom{2m+2j}{m+i+j}_2\right]_{n\times n}.$$

Define

$$\epsilon_n' = \left(\frac{(-1)^{i+j}(q^{4m+4i+2};q^2)_{2j-2i}q^{i^2+i-2ij}}{(q^{2m+2i+2j};q^2)_{j-i}(q^{2m+2i};q^2)_{j-i}}\binom{j-1}{i-1}_2\right)_{n\times n}.$$

Then ϵ_n' is upper triangular, and its (j,j)th entry is q^{-j^2+j}. Therefore $\det\epsilon_n' = \prod_{j=1}^{n} q^{j-j^2}$. If we let

$$\gamma_n^{\#} = \left(\binom{2m+2j}{m+i+j}_2\right)_{n\times n},$$

then the (i,j)th entry of $\gamma_n^{\#}\cdot\epsilon_n'$ is (by Lemma 3 with $b=0$ and q replaced by q^2):

$$\frac{(-1)^{j+1}q^{2-2j}(q^2;q^2)_{2m+2j}(q^2;q^2)_m(q^{2-2i};q^2)_{j-1}(q^{2-2i-2j};q^2)_{j-1}(1-q^{2m+2j})}{(q^2;q^2)_{m+2j-1}(q^2;q^2)_{m+i+1}(q^2;q^2)_{m+j-i}(q^{-2i-2m-2j};q^2)_{j-1}}.$$

This last expression is zero for $j>i$, so we see that $\gamma_n^{\#}\cdot\epsilon_n'$ is lower triangular. Consequently,

$$(-q;q^2)_m\det\gamma_n' = \frac{(-q;q^2)_m q^{-mn(n-1)}(q^4;q^4)_n(\det\gamma_n^{\#})}{(q^{4m+4};q^4)_n}$$

$$= \frac{(-q;q^2)_m q^{-mn(n-1)}(q^4;q^4)_n}{(q^{4m+4};q^4)_n(\det\epsilon_n')}$$

$$\times\prod_{j=1}^{n}\frac{(-1)^{j+1}q^{2-2j}(q^2;q^2)_{2m+2j}(q^{2-2j};q^2)_{j-1}(q^{2-4j};q^2)_{j-1}(1-q^{2m+2j})}{(q^2;q^2)_{m+2j-1}(q^2;q^2)_{m+j+1}(q^{-2m-4j};q^2)_{j-1}}$$

$$= \frac{(-q;q^2)_m(q^4;q^4)_n}{(-q^{2m+2};q^2)_n}\prod_{j=1}^{n}\frac{(q^2;q^2)_{2m+2j}(q^2;q^2)_{j-1}(q^{2j+2};q^2)_{j-1}}{(q^2;q^2)_{m+2j-1}(q^2;q^2)_{m+j+1}(q^{2m+2j+4};q^2)_{j-1}}$$

$$= \frac{(-q;q^2)_m(-q^2;q^2)_n}{(-q^{2m+2};q^2)_n}\prod_{j=1}^{n}\frac{(q^2;q^2)_{2m+2j}(q^2;q^2)_{2j-1}}{(q^2;q^2)_{m+2j-1}(q^2;q^2)_{m+2j}}$$

$$= \frac{(-q;q^2)_m(-q^2;q^2)_n\prod_{j=1}^{n}(q^{4j+2};q^2)_{2m}}{(-q^{2m+2};q^2)_n\prod_{j=1}^{2n}(q^{2j+2};q^2)_m}$$

$$= \frac{(-q;q^2)_m(-q^2;q^2)_n \prod_{j=1}^{2n} (q^{2j+2};q^2)_{2m}}{(-q^{2m+2};q^2)_n \left(\prod_{j=1}^{2n} (q^{2j+2};q^2)_m\right)\left(\prod_{j=1}^{n} (q^{4j};q^2)_{2m}\right)}$$

$$= \frac{(-q;q^2)_m(-q^2;q^2)_n \prod_{j=1}^{2n} (q^{2j+2m+2};q^2)_m}{(-q^{2m+2};q^2)_n \prod_{j=1}^{n} (q^{4j};q^2)_{2m}}$$

$$= \frac{(-q;q^2)_m(-q^2;q^2)_n \prod_{j=1}^{2n} (q^{2j+2m+2};q^2)_m}{(-q^{2m+2};q^2)_n \prod_{j=1}^{n} (q^{4j};q^4)_m(q^{4j+2};q^4)_m}$$

$$= \frac{(-q;q^2)_m(-q^2;q^2)_n}{(-q^{2m+2};q^2)_n} \prod_{j=0}^{2n-1} \prod_{i=1}^{m} \frac{(1-q^{2m+2j+2i+2})}{(1-q^{2j+4i})}$$

$$= \frac{(-q;q^2)_m(-q^2;q^2)_n}{(-q^{2m+2};q^2)_n} \prod_{i=1}^{m} \frac{(q^{2m+2i+2};q^2)_{2n}}{(q^{4i};q^2)_{2n}}$$

$$= \frac{(-q;q^2)_m(-q^2;q^2)_n}{(-q^{2m+2};q^2)_n} \prod_{i=1}^{m} \frac{(q^2;q^2)_{2n+m+i}(q^2;q^2)_{2i-1}}{(q^2;q^2)_{2n+2i-1}(q^2;q^2)_{m+i}}$$

$$= \frac{(-q;q^2)_m(-q^2;q^2)_n}{(-q^{2m+2};q^2)_n} \prod_{i=1}^{m} \left[\frac{(1-q^{4n+4i})}{(1-q^{2m+2i})} \prod_{j=i+1}^{m} \frac{(1-q^{4n+2i+2j})}{(1-q^{2(i+j-1)})}\right]$$

$$= \frac{(-q;q^2)_m(-q^2;q^2)_n(q^{4n+4};q^4)_m}{(-q^{2m+2};q^2)_n(q^{2m+2};q^2)_m} \prod_{i=1}^{m} \prod_{j=i+1}^{m} \frac{(1-q^{2(2n+i+j)})}{(1-q^{2(i+j-1)})}$$

$$= \frac{(-q;q^2)_m(q^4;q^4)_{m+n}(-q^2;q^2)_m(q^2;q^2)_m}{(q^2;q^2)_n(-q^2;q^2)_{m+n}(q^2;q^2)_{2m}} \prod_{i=1}^{m} \prod_{j=i+1}^{m} \frac{(1-q^{2(2n+i+j)})}{(1-q^{2(i+j-1)})}$$

$$= \frac{(-q;q^2)_m(q^2;q^2)_{m+n}(q^4;q^4)_m}{(q^2;q^2)_n(q^2;q^4)_m(q^4;q^4)_m} \prod_{i=1}^{m} \prod_{j=i+1}^{m} \frac{(1-q^{2(2n+i+j)})}{(1-q^{2(i+j-1)})}$$

$$= \frac{(q^{2n+2};q^2)_m}{(q;q^2)_m} \prod_{i=1}^{m} \prod_{j=i+1}^{m} \frac{(1-q^{2(2n+i+j)})}{(1-q^{2(i+j-1)})}$$

$$= \prod_{i=1}^{m} \left[\frac{(1-q^{2n+2i})}{(1-q^{2i-1})} \prod_{j=i+1}^{m} \frac{(1-q^{2(2n+i+j)})}{(1-q^{2(i+j-1)})}\right]. \quad \blacksquare$$

LEMMA 10. *There exist* c_{ij} *such that* $c_{ij} = 0$ *if* $i > j$, $c_{ii} = 1$, *and for all* n

$$\beta_n \cdot (c_{ij})_{n \times n} = \gamma_n.$$

Proof. We proceed by induction on n. In the case $n = 1$, the result is trivial since β_1 and γ_1 are the 1×1 matrix whose single entry is

$$\binom{2m+1}{m+1}_2 \frac{(1+q)}{(1+q^{2m+1})}.$$

Now we assume that the c_{ij} have been found for $i < n$ and $j < n$. We choose $c_{nj} = 0$ for $j < n$, $c_{nn} = 1$, and for $1 \leqslant h < n$ we choose c_{hn} so that the following system of equations is valid:

$$\sum_{j=1}^{n} \binom{2m+2j-1}{m+i+j-1}_2 \frac{q^{(i-j)^2}(1+q^{(2i-1)(2j-1)})}{(1+q^{(2m+2j-1)(2j-1)})} c_{jn} = \binom{2m+2n-1}{m+i+n-1}_2 \frac{q^{(i-1)^2-2m(n-1)-(n-1)^2}(1+q^{2i-1})}{(1+q^{2m+2n-1})}, \tag{3.7}$$

where $1 \leqslant i \leqslant n-1$. Now $c_{nn} = 1$, so that (3.7) is a system of $(n-1)$ equations in $(n-1)$ unknowns. The determinant of the system is $\det \beta_{n-1}$, and by the induction hypothesis

$$\det \beta_{n-1} = \frac{\det \gamma_{n-1}}{\det(c_{ij})_{n-1\times n-1}} = \det \gamma_{n-1} \neq 0$$

by Lemma 8. Therefore the c_{jn} exist and are unique.

I claim now that (3.7) is valid for every $i \geqslant 1$. This is because by Lemma 4,

$$\begin{aligned}
&\sum_{j=1}^{n} \binom{2m+2j-1}{m+i+j-1}_2 \frac{q^{(i-j)^2}(1+q^{(2i-1)(2j-1)})}{(1+q^{(2m+2j-1)(2j-1)})} c_{jn} \\
&\quad - \binom{2m+2n-1}{m+i+n-1}_2 \frac{q^{(i-1)^2-2m(n-1)-(n-1)^2}(1+q^{2i-1})}{(1-q^{2m+2n-1})} \\
&= \sum_{j=1}^{n-1} \sum_{r=1}^{n-1} a_i(r, n-1) \binom{2m+2j-1}{m+r+j-1}_2 \frac{q^{(r-j)^2}(1+q^{(2r-1)(2j-1)})}{(1+q^{(2m+2j-1)(2j-1)})} c_{jn} \\
&\quad + \sum_{r=1}^{n-1} a_i(r, n-1) \left[\binom{2m+2n-1}{m+r+n-1}_2 \frac{q^{(r-n)^2}(1+q^{(2r-1)(2n-1)})}{(1+q^{(2m+2n-1)(2n-1)})} \right. \\
&\quad \left. - \binom{2m+2n-1}{m+r+n-1}_2 \frac{q^{(r-1)^2-2m(n-1)-(n-1)^2}(1+q^{2r-1})}{(1+q^{2m+2n-1})} \right] \\
&= \sum_{r=1}^{n-1} a_i(r, n-1) \left[\sum_{j=1}^{n} \binom{2m+2j-1}{m+r+j-1}_2 \frac{q^{(r-j)^2}(1+q^{(2r-1)(2j-1)})}{(1+q^{(2m+2j-1)(2j-1)})} c_{jn} \right. \\
&\quad \left. - \binom{2m+2n-1}{m+r+n-1}_2 \frac{q^{(r-1)^2-2m(n-1)-(n-1)^2}(1+q^{2r-1})}{(1+q^{2m+2n-1})} \right] \\
&= 0,
\end{aligned}$$

since the internal expressions are zero by (3.7) with $1 \leqslant r \leqslant n-1$. Therefore in multiplying β_n on the right by $(c_{ij})_{n\times n}$ we see that the (i,j)th entry is the (i,j)th entry in γ_n if $j < n$ and $i < n$ by the induction hypothesis and if $j = n$ by (3.7). When $i = n$ and $j = j_0 < n$, we see that our above argument establishes (3.7) for all i with n replaced by j_0, and so again the resulting

(i, j)th entry is the (i, j_0)th entry of γ_n. Hence

$$\beta_n \cdot (c_{ij})_{n \times n} = \gamma_n. \quad \blacksquare$$

LEMMA 11. *There exist d_{ij} such that $d_{ij} = 0$ if $i > j$, $d_{ii} = 1$, and for all n*

$$\beta_n' \cdot (d_{ij})_{n \times n} = \gamma_n'.$$

Proof. We proceed by induction on n. In the case $n = 1$, the result is trivial since β_1' and γ_1' are the 1×1 matrix whose single entry is

$$\binom{2m+2}{m+2}_2 \frac{(1 - q^4)}{(1 - q^{4m+4})}.$$

Now we assume that the d_{ij} have been found for $i < n$ and $j < n$. We choose $d_{nj} = 0$ for $j < n$, $d_{nn} = 1$, and for $1 \leqslant h < n$, we choose d_{nn} so that the following system of equations is valid:

$$\sum_{j=1}^{n} \binom{2m+2j}{m+i+j}_2 \frac{q^{(i-j)^2}(1 - q^{4ij})}{(1 - q^{4j(m+j)})} d_{jn} = \binom{2m+2n}{m+i+n}_2 \frac{q^{(i-1)^2 - 2m(n-1) - (n-1)^2}(1 - q^{4i})}{(1 - q^{4m+4n})}, \tag{3.8}$$

where $1 \leqslant i \leqslant n - 1$. Since $d_{nn} = 1$, (3.8) is a system of $(n - 1)$ equations in $(n - 1)$ unknowns. The determinant of the system is $\det \beta'_{n-1}$, and by the induction hypothesis $\det \beta'_{n-1} = \det \gamma'_{n-1}/\det(d_{ij})_{n-1 \times n-1} = \det \gamma'_{n-1} \neq 0$ by Lemma 9. Therefore the d_{jn} exist and are unique. I claim now that (3.8) is valid for every $i \geqslant 1$. This is because by Lemma 5,

$$\begin{aligned}
&\sum_{j=1}^{n} \binom{2m+2j}{m+i+j}_2 \frac{q^{(i-j)^2}(1 - q^{4ij})}{(1 - q^{4j(m+j)})} d_{jn} - \binom{2m+2n}{m+i+n}_2 \frac{q^{(i-1)^2 - 2m(n-1) - (n-1)^2}(1 - q^{4i})}{(1 - q^{4m+4n})} \\
&= \sum_{j=1}^{n-1} \sum_{r=1}^{n-1} b_i(r, n-1) \binom{2m+2j}{m+r+j}_2 \frac{q^{(r-j)^2}(1 - q^{4rj})}{(1 - q^{4j(m+j)})} d_{jn} \\
&\quad + \sum_{r=1}^{n-1} b_i(r, n-1) \left[\binom{2m+2n}{m+r+n}_2 \frac{q^{(r-n)^2}(1 - q^{4rn})}{(1 - q^{4n(m+n)})} \right. \\
&\quad \left. - \binom{2m+2n}{m+r+n}_2 \frac{q^{(r-1)^2 - 2m(n-1) - (n-1)^2}(1 - q^{4r})}{(1 - q^{4m+4n})} \right] \\
&= \sum_{r=1}^{n-1} b_i(r, n-1) \left[\sum_{j=1}^{n} \binom{2m+2j}{m+r+j}_2 \frac{q^{(r-j)^2}(1 - q^{4rj})}{(1 - q^{4j(m+j)})} d_{jn} \right. \\
&\quad \left. - \binom{2m+2n}{m+r+n}_2 \frac{q^{(r-1)^2 - 2m(n-1) - (n-1)^2}(1 - q^{4r})}{(1 - q^{4m+4n})} \right] \\
&= 0,
\end{aligned}$$

since the internal expressions are zero by (3.8) with $1 \leqslant r \leqslant n-1$. Therefore in multiplying β_n' on the right by $(d_{ij})_{n\times n}$ we see that the (i,j)th entry is the (i,j)th entry in γ_n' if $j<n$ and $i<n$ by the induction hypothesis and if $j=n$ by (3.8). When $i=n$ and $j=j_0<n$, we see that our above argument establishes (3.8) for all i with n replaced by j_0, and so in any event the resulting (i,j_0)th entry is the (i,j_0)th entry of γ_n'. Therefore

$$\beta_n' \cdot (d_{ij})_{n\times n} = \gamma_n'. \quad ▮$$

4. MacMahon's Conjecture

Theorem 1 (MacMahon's conjecture).

$$\sum_{N\geqslant 0} M(n,m;N)q^N = \prod_{i=1}^{m}\left[\frac{(1-q^{n+2i-1})}{(1-q^{2i-1})}\prod_{h=i+1}^{m}\frac{(1-q^{2(n+i+h-1)})}{(1-q^{2(i+h-1)})}\right]. \tag{4.1}$$

Proof. We must treat separately the cases n even and n odd. First

$$\begin{aligned}
&\sum_{N\geqslant 0} M(2n,m;N)q^N \\
&= g_{2n}(q) && \text{(by (1.2))}\\
&= \det \alpha_n && \text{(by (1.3) and (3.1))}\\
&= \det \beta_n && \text{(by Lemma 6)}\\
&= \det \gamma_n && \text{(by Lemma 10)}\\
&= \prod_{i=1}^{m}\left[\frac{(1-q^{2n+2i-1})}{(1-q^{2i-1})}\prod_{h=i+1}^{m}\frac{(1-q^{2(2n+i+h-1)})}{(1-q^{2(i+h-1)})}\right] && \text{(by Lemma 8).}
\end{aligned}$$

Hence (4.1) is valid when n is even.

When $n=1$, the right side of (4.1) is

$$\begin{aligned}
\prod_{i=1}^{m}\frac{(1-q^{2i})}{(1-q^{2i-1})}\cdot\frac{(1-q^{2i+2m})}{(1-q^{4i})} &= \frac{(q^2;q^2)_m(q^{2m+2};q^2)_m}{(q;q^2)_m(q^4;q^4)_m}\\
&= \frac{(q^2;q^2)_{2m}}{(q;q^2)_m(q^4;q^4)_m}\\
&= \frac{(q^2;q^4)_m(q^4;q^4)_m}{(q;q^2)_m(q^4;q^4)_m}\\
&= (-q;q^2)_m.
\end{aligned}$$

On the other hand $M(1,m;N)$ is clearly just the number of linear partitions with distinct odd parts each at most $2m - 1$. Hence

$$\sum_{N \geqslant 0} M(1, m; N)q^N = (1 + q)(1 + q^3) \cdots (1 + q^{2m-1}) = (-q; q^2)_m.$$

Thus (4.1) is valid when $n = 1$.

$$\begin{aligned}
&\sum_{N \geqslant 0} M(2n + 1, m; N)q^N \\
&\quad = g_{2n+1}(q) && \text{(by (1.2))} \\
&\quad = (-q; q^2)_m \det \alpha_n' && \text{(by (1.4) and (3.2))} \\
&\quad = (-q; q^2)_m \det \beta_n' && \text{(by Lemma 7)} \\
&\quad = (-q; q^2)_m \det \gamma_n' && \text{(by Lemma 11)} \\
&\quad = \prod_{i=1}^{m} \left\{ \frac{(1 - q^{2n+2i})}{(1 - q^{2i-1})} \prod_{h=i+1}^{m} \frac{(1 - q^{2(2n+i+h)})}{(1 - q^{2(i+h-1)})} \right\} && \text{(by Lemma 9).}
\end{aligned}$$

Hence (4.1) is valid when n is odd and larger than 1. Therefore (4.1) is valid for all n. ∎

5. Conclusion

First of all we remark that the Bender–Knuth conjecture [2, p. 50] is so clearly of the same type as MacMahon's conjecture that it ought to be provable by our methods.

We also mention that the case $q = 1$ of MacMahon's conjecture can be treated much more easily in that Lemmas 4, 5, 10, and 11 are unnecessary.

Note added in proof. Ian Macdonald of Queen Mary College has independently obtained a proof of MacMahon's conjecture from group representation theory. The interrelationship of Lemmas 4 and 5 with reciprocal polynomials and basic hypergeometric series is explored in my paper Implications of the MacMahon conjecture, *in* "Combinatoire et réprésentation du groupe symetrique" (D. Foata, ed.), Lecture Notes in Mathematics, No. 579, Springer-Verlag, Berlin and New York pp. 287–296, 1977. The relationship between MacMahon's conjecture and the Bender–Knuth conjecture has been established in my paper Plane partitions (II): The equivalence of the Bender–Knuth and MacMahon conjectures, *Pac. J. Math* **72** (1977), 283–291.

References

1. G. E. Andrews, Applications of basic hypergeometric functions, *SIAM Rev.* **16** (1974), 441–484.
2. E. A. Bender and D. E. Knuth, Enumeration of plane partitions, *J. Combinatorial Theory Ser. B* **13** (1972), 40–54.

3. B. Gordon, Two new representations of the partition function, *Proc. Amer. Math. Soc.* **13** (1962), 869–873.
4. B. Gordon, Notes on plane partitions V, *J. Combinatorial Theory Ser. B* **11** (1971), 157–168.
5. B. Gordon and L. Houten, Notes on plane partitions I, *J. Combinatorial Theory Ser. B* **4** (1968), 72–80.
6. B. Gordon and L. Houten, Notes on plane partitions II, *J. Combinatorial Theory Ser. B* **4** (1968), 81–99.
7. P. A. MacMahon, Partitions of numbers whose graphs possess symmetry, *Trans. Cambridge Phil. Soc.* **17** (1898–1899), 149–170.
8. P. A. MacMahon, "Combinatory Analysis," Vol. 2, Cambridge Univ. Press, London and New York, 1916. (Reprinted: Chelsea, Bronx, New York, 1960).
9. L. J. Slater, "Generalized Hypergeometric Functions," Cambridge Univ. Press, London and New York, 1966.
10. R. P. Stanley, Theory and applications of plane partitions II, *Studies in Appl. Math.* **50** (1971), 259–279.

STUDIES IN FOUNDATIONS AND COMBINATORICS
ADVANCES IN MATHEMATICS SUPPLEMENTARY STUDIES, VOL. 1

Graph Theory in Statistical Physics†

F. Y. Wu

Department of Physics
Northeastern University
Botson, Massachusetts

This paper introduces the subject of graph theory in lattice statistics and reviews some of its recent developments. The vertex model is defined and its known properties are reviewed. It is shown that the Whitney polynomial is equivalent to the partition function of the Potts model in statistical physics and, for planar graphs, is also equivalent to the partition function of a vertex model on the medial graph. A graph-theoretical expansion of the Whitney polynomial due to Nagle and Temperley is also derived and generalized.

1. Introduction

Problems in statistical physics can often be formulated in graphical terms [1]. The connection between the Ising, percolation, and graph coloring problem in lattice statistics and the Whitney rank function in graph theory has been expounded by Essam [1]. More recently, it has been shown that the solutions of the Potts model and a certain vertex statistical model are also derivable from the Whitney function [2, 3]. In view of the important role played by these models in the modern theory of phase transitions [4], we present here a review of these latter developments. It is also our hope that this paper will stimulate some interest among graph-theoreticians to these outstanding problems in mathematical physics.

According to the principles of statistical mechanics, the thermodynamics of a physical system is completely determined by a certain partition function pertaining to the system. Thus the study of the physical properties is reduced to a mathematical problem of evaluating the partition function. It is in the latter problem that the graph-theoretical considerations arise.

Consider a physical system whose atomic constituents are arranged on a lattice. We shall think of the lattice as represented by a connected graph G specified by a vertex (site) set V and an edge set E. While in most physical considerations G is regular, i.e., periodic in some sense, we shall assume G arbitrary unless otherwise stated. The partition function Z_G is then a certain

† Supported in part by the National Science Foundation.

ISBN 0-12-599101-0

summation defined on G. A subgraph expansion of Z_G arises if a (1–1) correspondence can be made between the summands and the subgraphs of G. The partition function then becomes a graph generating function.

The procedure is best illustrated by considering the problem of the q-colorings of G. Let the vertices of G be colored with q colors such that no two adjacent vertices bear the same color. We wish to find a subgraph expansion of the number of such colorings, the chromatic polynomial $Z(G, q)$ (see, e.g., [5]).

The problem is unchanged if we color each site independently while at the same time introduce a factor

$$F(\xi_i, \xi_j) \begin{cases} = 1, & \text{if} \quad \xi_i = \xi_j, \\ = 0, & \text{if} \quad \xi_i \neq \xi_j, \end{cases} \tag{1}$$

to an edge connecting sites i and j. Here, $\xi_i = 1, 2, \ldots, q$ specifies the color of the ith vertex. Thus we write

$$Z(G, q) = \sum_{\xi_i = 1}^{q} \prod_{E} F(\xi_i, \xi_j). \tag{2}$$

The product in (2) is taken over all edges of G and is of the form

$$\prod_{E} [1 + f_{ij}] \tag{3}$$

with $f_{ij} = -\delta_{Kr}(\xi_i, \xi_j)$. There are $2^{|E|}$ terms when (3) is multiplied out. The terms differ from one another only in the f-factors contained therein. A natural graphical representation now arises if we associate with each term a subgraph $G' \subseteq G$ whose edge set E' corresponds to the f-factors contained in the term. Clearly, the correspondence between G' and the terms in (3) is (1–1). This type of graphical representation was first used by Mayer [6] in the theory of an imperfect gas many years ago.

We now carry out the summation in (2). It is clear that this gives rise to a factor q for each connected piece in G'. Furthermore, each edge in G' carries an additional factor -1. Thus we arrive at the expression

$$Z(G, q) = \sum_{G' \subseteq G} (-1)^{e(G')} q^{n(G')}, \tag{4}$$

where $e(G') = |E'|$ is the number of edges and $n(G')$ is the number of components in G'. This is the Birkhoff formula for $Z(G, q)$ [7–9].

2. The Whitney Rank Function

The Whitney rank function [8] of a graph G is defined by

$$W_G(x, y) = \sum_{G' \subseteq G} x^{r(G')} y^{c(G')}, \tag{5}$$

where

$$\begin{aligned} r(G) &= \text{rank of } G = e(G) - c(G), \\ c(G) &= \text{corank of } G = \text{number of independent circuits in } G \\ &= e(G) + n(G) - |V|. \end{aligned} \tag{6}$$

The last expression in (6) is Euler's relation. Comparing (4) with (5), we see that the chromatic polynomial is a special case of the Whitney rank function

$$Z(G, q) = q^{|V|} W_G(-1/q, -1). \tag{7}$$

A related definition is the Whitney polynomial

$$\bar{W}_G(x, y) = \sum_{G' \subseteq G} x^{e(G')} y^{c(G')} = W_G(x, xy). \tag{8}$$

The usefulness of the latter definition is that it can be extended to allow different weights for the edges. Let x_α denote the weight of the αth edge in E. Then (8) can be generalized to

$$\bar{W}_G(x_1, x_2, \ldots; y) = \sum_{G' \subseteq G} \left[\prod_{\alpha \in E'} x_\alpha \right] y^{c(G')}, \tag{9}$$

which reduces to (8) when all x_α's are equal:

$$\bar{W}_G(x, x, \ldots; y) = \bar{W}_G(x, y). \tag{10}$$

We now derive a duality relation for $\bar{W}_G$ for planar G. Let D be the dual of G. To each $G' \subseteq G$ introduce a $D' \subseteq D$ whose edge set E_D' complements those of E'. It is easy to see that each face of G' contains one and only one component of D' and vice versa. We then have

$$n(D') = c(G') + 1. \tag{11}$$

Substituting (6) and (11) into (9), we obtain

$$\begin{aligned} \bar{W}_G(x_1, x_2, \ldots; y) &= \left[\prod_{\alpha \in E} x_\alpha \right] \sum_{G' \subseteq G} \left[\prod_{\alpha \in E'} \frac{1}{x_\alpha} \right] y^{c(G')} \\ &= \left[\prod_{\alpha \in E} x_\alpha \right] \sum_{D' \subseteq D} \left[\prod_{\alpha \in E_D'} \frac{1}{x_\alpha} \right] y^{c(D') - e(D') + |V_D| - 1} \\ &= y^{|V_D| - 1} \left[\prod_{\alpha \in E} x_\alpha \right] \bar{W}_D\left(\frac{1}{x_1 y}, \frac{1}{x_2 y}, \ldots; y \right), \end{aligned} \tag{12}$$

where $|V_D|$ is the number of sites of D. Equation (12) is a duality relation for $\bar{W}_G$. When $x_1 = x_2 = \cdots = x$, this reduces to the duality relation for the Whitney rank function [10]:

$$W_G(x, y) = x^{|V| - 1} y^{|V_D| - 1} W_D(1/y, 1/x). \tag{13}$$

It now follows from (7) that the chromatic polynomial for a planar G can also be written as a Whitney rank function on D:

$$Z(G, q) = (-1)^{|V|+|V_D|} q W_D(-1, -q). \tag{14}$$

3. The Potts Model

The chromatic polynomial is a special case of the partition function of a physical system, the Potts model [11]. In a Potts model the vertices of a graph G are occupied by q kinds of atoms. There is one atom per site and no restriction is imposed. Between any two sites i and j connected by an edge, there is an interaction energy

$$E(\xi_i, \xi_j) = -\epsilon \delta_{Kr}(\xi_i, \xi_j), \tag{15}$$

where $\xi_i = 1, 2, \ldots, q$ specifies the type of atom at the site i. This defined the Potts model.

The physical reasoning that leads to the consideration of the Potts model is as follows: The basic ingredient of a magnetic system is that an "ordered" atomic arrangement is energetically favored. Since for $\epsilon > 0$ the Potts model favors a configuration in which all vertices are occupied by the same kind of atom, the model can be considered as a simulation of a magnetic substance.

The partition function of the Potts model is

$$Z_G^{\text{Potts}}(K, q) = \sum_{\substack{\xi_i = 1 \\ i \in V}}^{q} \prod_{E} \exp[K \delta_{Kr}(\xi_i, \xi_j)], \tag{16}$$

where $K = \beta\epsilon$, $\beta = 1/kT$, k being the Boltzmann constant and T the absolute temperature of the system. Once Z_G is evaluated, all thermodynamic properties of the model can be obtained from the derivatives of Z_G. For example, the energy and specific heat are, respectively,

$$E_G = -(d/d\beta) \ln Z_G, \tag{17}$$

$$c_G = dE_G/dT. \tag{18}$$

It is seen from (16) that the chromatic polynomial is

$$Z(G, q) = Z_q^{\text{Potts}}(-\infty, q). \tag{19}$$

Physically speaking, $K \to -\infty$ corresponds to the zero temperature limit of an $\epsilon < 0$ Potts model. In this limit, all configurations in (16) carry zero weight except those in which all neighboring atoms are different. The partition function then becomes the chromatic polynomial.

In real physical systems, G is usually regular and has $\sim 10^{23}$ sites. For large $|V|$ it is expected [12] that Z_G behaves as $\exp[f|V|]$, where f is independent of $|V|$. It is therefore appropriate to consider the "free energy" per site

$$f_G(K, q) = \lim_{|V| \to \infty} |V|^{-1} \ln Z_G^{\text{Potts}}(K, q) \tag{20}$$

for regular lattices G. Furthermore, the thermodynamic behavior of real magnetic substances exhibits some kind of anomaly, such as a divergent specific heat, at a specific temperature. Since for $|V|$ finite $Z_G^{\text{Potts}}(K, q)$ given by (16) is always analytic in K, or T, the anomaly, if any, can occur only after taking the thermodynamic limit $|V| \to \infty$, and is manifested as some kind of nonanalyticity of $f_G(K, q)$ in K. Thus, we are always interested in the limiting free energy (20) in physical considerations.

The evaluation of $f_G(K, q)$ for general q has proven to be a most formidable task. Except for the trivial case of a one-dimensional array of lattice points, the only known solution to this date is for $q = 2$, for which the Potts model is also known as the Ising model [13], and for regular planar G. For completeness, we give here the $q = 2$ solution for a square lattice [14]:

$$f_{\text{SQ}}(K, 2) = \ln(1 + e^{2K}) + (1/2\pi) \int_0^\pi d\theta \ln \tfrac{1}{2}[1 + (1 - k^2 \sin^2 \theta)^{1/2}], \tag{21}$$

where $k = 2 \sinh K/\cosh^2 K$. It is seen that $f_{\text{SQ}}(K, 2)$ is nonanalytic at $|K| = K_c = \frac{1}{2}\ln 2$ and that the specific heat (18) diverges as $\ln|K - K_c|$ at K_c. Other established results for $f_G(K, q)$ include the value $f_{\text{SQ}}(-\infty, 3) = \frac{3}{2}\ln\frac{4}{3}$ [4, 15], the value of K_c for some regular planar G [11, 16, 17], and the fact that the nature of the nonanalyticity of f_{SQ} is strongly q-dependent [18]. Bounds on $f_{\text{SQ}}(-\infty, q)$ have also been obtained [19].

We now turn to the problem of finding a subgraph expansion for $Z_G(K, q)$. This can be done in different ways. First, as in Section 1, we rewrite the exponential factor in (16) as

$$\exp[K\delta_{Kr}(\xi_i, \xi_j)] = 1 + f_{ij}, \tag{22}$$

where

$$f_{ij} = v\delta_{Kr}(\xi_i, \xi_j), \qquad v = e^K - 1. \tag{23}$$

Then we may carry out the expansion exactly as in Section 1. The only difference is that now each edge in G' carries a factor v, instead of -1. Thus, we arrive at the expression [3]

$$Z_G^{\text{Potts}}(K, q) = \sum_{G' \subseteq G} v^{e(G')} q^{n(G')} = q^{|V|} W_G\left(\frac{v}{q}, v\right). \tag{24}$$

The Potts partition function is now written as a Whitney rank function.

The expression (24) is readily extended to the case that the interaction parameter ϵ in (15) is different along different edges. Let ϵ_α be the interaction along the αth edge. The partition function of this Potts model is given by

$$Z_G^{\text{Potts}}(K_1, K_2, \ldots; q) = \sum_{\xi_i=1}^{q} \prod_{\alpha \in E} \exp[K_\alpha \delta_{Kr}(\xi_i, \xi_j)], \tag{25}$$

where $K_\alpha = \beta\epsilon_\alpha$. In place of (24), we now have

$$Z_G^{\text{Potts}}(K_1, K_2, \ldots; q) = q^{|V|}\bar{W}_G(v_1/q, v_2/q, \ldots; q), \tag{26}$$

where $v_\alpha = e^{K_\alpha} - 1$ and $\bar{W}_G$ is the Whitney polynomial defined in (9). The duality relation (12) for $\bar{W}_G$ now implies the following duality for Z_G^{Potts}:

$$Z_G^{\text{Potts}}(K_1, K_2, \ldots; q) = q^{1-|V_D|}\left[\prod_{\alpha \in E}(e^{K_\alpha} - 1)\right] Z_D^{\text{Potts}}(K_1^*, K_2^*, \ldots; q), \tag{27}$$

provided that

$$(e^{K_\alpha} - 1)(e^{K_\alpha^*} - 1) = q. \tag{28}$$

This result has been obtained by Wu and Wang [20] from a purely algebraic approach.

We can derive yet another subgraph expansion for Z_G^{Potts}. In place of (22) we write more generally for the αth edge which connects sites i and j,

$$\exp[K_\alpha \delta_{Kr}(\xi_i, \xi_j) = t_\alpha(1 + f_{ij}^\alpha). \tag{29}$$

Here,

$$f_{ij}^\alpha = t_\alpha^{-1} - 1 + v_\alpha t_\alpha^{-1}\delta_{Kr}(\xi_i, \xi_j) \tag{30}$$

and t_α is a parameter to be chosen at our disposal. Note that (22) corresponds to taking $t_\alpha = 1$. Substituting (29) into (25) and expanding the product as before, we obtain

$$Z_G^{\text{Potts}}(K_1, K_2, \ldots; q) = \left[\prod_{\alpha \in E} t_\alpha\right] \sum_{G' \subseteq G} w(G'), \tag{31}$$

where

$$w(G') = q^{|V|-|V'|} \sum_{\substack{\xi_i=1 \\ i \in V'}}^{q} \prod_{\alpha \in E'} f_{ij}^\alpha \tag{32}$$

is the weight of a subgraph $G' = \{V', E'\}$.

One choice of t_α is to require, for all i,

$$\sum_{\xi_j=1}^{q} f_{ij}^\alpha = 0 \tag{33}$$

so that $w(G') = 0$ whenever G' contains one or more vertices of degree one. This leads to

$$t_\alpha = q/(q + v_\alpha), \qquad f_{ij}^\alpha = \lambda^\alpha[-q^{-1} + \delta_{Kr}(\xi_i, \xi_j)], \tag{34}$$

where

$$\lambda^\alpha = v_\alpha q/(q + v_\alpha). \tag{35}$$

We can explicitly sum over all vertices of degree two in (32). Consider a link of r consecutive edges in G' connecting two sites m and n and having interactions $\epsilon_1, \epsilon_2, \ldots, \epsilon_r$. Let F^α be the $q \times q$ matrix whose ijth element is f_{ij}^α. It is easily seen that the eigenvalues of F^α are

$$\lambda_0{}^\alpha = 0, \qquad \lambda_1{}^\alpha = \lambda_2{}^\alpha = \cdots = \lambda_{q-1}^\alpha = \lambda^\alpha. \tag{36}$$

Then, the contribution in (32) due to this r-link is

$$\begin{aligned}\sum_{\xi_1=1}^{q} \cdots \sum_{\xi_{r-1}=1}^{q} f_{m1}^1 f_{12}^2 \cdots f_{r-1,n}^r &= [F^1F^2 \cdots F^r]_{mn} \\ &= \frac{1}{q}\sum_{k=1}^{q-1} e^{i2\pi k(\xi_m - \xi_n)/q}\lambda_k{}^1\lambda_k{}^2 \cdots \lambda_k{}^r \\ &= (\lambda^1\lambda^2 \cdots \lambda^r)[-q^{-1} + \delta_{Kr}(\xi_m, \xi_n)].\end{aligned} \tag{37}$$

Here, we have used

$$f_{ij}^\alpha = \frac{1}{q}\sum_{k=0}^{q-1} e^{i2\pi k(\xi_i - \xi_j)/q}\lambda_k{}^\alpha. \tag{38}$$

Thus, after summing over the $r - 1$ vertices of degree two in (37), the r-link is "contracted" into a single one with an effective edge factor (37). All vertices of degree two in G' can be contracted in this fashion. We are then led to consider the "skeleton" of G', or $G_s' = \{V_s', E_s'\}$, which is G' with all vertices of degree 2 contracted. The subgraph weight (32) now takes the form

$$w(G') = q^{|V|-|V'|}\left[\prod_{\alpha \in E'} \lambda^\alpha\right] \sum_{\substack{\xi_i = 1\\ i \in V_s'}}^{q} \prod_{E_s'} [-q^{-1} + \delta_{Kr}(\xi_i, \xi_j)]. \tag{39}$$

Again, we expand the product in (39) and obtain a subgraph expansion as in (3). For each connected piece in $G_s'' \subseteq G_s'$ the summations in (39) yield a factor q, and for each edge of G_s' *not* in G_s'' there is a factor $-q^{-1}$. Thus (39) becomes

$$w(G') = q^{|V|+|E'|-|V'|}\left[\prod_{\alpha \in E'}\left(1 + \frac{q}{v_\alpha}\right)\right]^{-1} \sum_{G_s'' \subseteq G_s'} (-q)^{e(G_s'')-|E_s'|}q^{n(G_s'')}. \tag{40}$$

Using (5) and the fact that

$$n(G_s'') - c(G_s'') = n(G') - c(G') = |V'| - |E'|, \tag{41}$$

we find from (40) and (31) the following subgraph expansion for Z_G^{Potts}:

$$Z_G^{\text{Potts}}(K_1, K_2, \ldots; q) = q^{|V|}\left[\prod_{\alpha \in E}\left(1 + \frac{v_\alpha}{q}\right)\right] \sum_{G' \subseteq G}^{(c)} \left[\prod_{\alpha \in E'}\left(1 + \frac{q}{v_\alpha}\right)\right]^{-1} \times (-1)^{|E_s'|} W_{G_s'}(-1, -q). \tag{42}$$

The superscript (c) in the summation denotes the fact that the summation is taken over only those subgraphs G' that are *closed*. The expansion (42) is valid for any G and can be used to generate series expansions for Z_G. For example, it is useful in writing down the partition function for some simple G such as the star topology [21].

For planar G, we may further use (14) to reduce (42) to

$$Z_G^{\text{Potts}}(K_1, K_2, \ldots; q) = q^{|V|-1}\left[\prod_{\alpha \in E}\left(1 + \frac{v_\alpha}{q}\right)\right] \sum_{G' \subseteq G}^{(c)} \left[\prod_{\alpha \in E'}\left(1 + \frac{q}{v_\alpha}\right)^{-1}\right] Z(D_s', q), \tag{43}$$

where $Z(D_s', q)$ is the chromatic polynomial for D_s', the dual of G_s'. Since G' and G_s' contain the same faces, $Z(D_s', q)$ is also the chromatic polynomial for q-coloring the *faces* of G'. The expansion (43) was first used by Nagle [22] to obtain a series expansion for the chromatic polynomial. The generalization to the Potts model and, in particular, the connection with the face colorings of G' have since been pointed out by Temperley [23]. Here, we have extended the results to the case of unequal Potts interactions. Using (26), (42) and (43) can also be regarded as expansions of the Whitney polynomial (9).

4. The Vertex Model

One of the most important developments in the mathematical theory of phase transitions in recent years is the solution of certain vertex model in lattice statistics [4]. The partition function of a vertex model is a graph generating function

$$Z_G = \sum_{G' \subseteq G} w(G') \tag{44}$$

whose subgraph weight $w(G')$ is given by a product of individual vertex weights. It is assumed that the individual vertex weights are determined by the configuration of the edges incident to the vertex. Associate to each edge of G a two-valued variable $s = \pm 1$ such that $s = 1$ (-1) corresponds to the edge being present (absent), and let $\{s\}_i$ denote the set of s-variable associated with the edges incident at the ith vertex in a given G'. Then the weight of the ith vertex is $w_i(\{s\}_i)$ and the subgraph weight for G' is

$$w(G') = \prod_{i \in V} w_i(\{s\}_i). \tag{45}$$

We may now rewrite (44) as

$$Z_G = \sum_{\{s\}} \prod_{i \in V} w_i(\{s\}_i), \tag{46}$$

where the summation is taken over the $|E|$ s-variables in G. This defines the vertex model.

We first derive a symmetry relation for Z_G known as the weak-graph transformation [9, 24]. In (46) a single s-variable is assigned to each edge. If we cut each edge into two halves, let each half carry its own variable, s and s', and introduce at the same time a factor $\delta_{Kr}(s, s')$ to the edge, the partition function (46) remains unchanged. Thus we have

$$Z_G = \sum_{\{s,s'\}} \left[\prod_{i \in V} w_i(\{s\}_i)\right]\left[\prod_{E} \delta_{Kr}(s, s')\right], \tag{47}$$

where the summation is taken over the set of $2|E|$ variables $\{s, s'\}$.

Let T be a 2×2 matrix satisfying

$$\tilde{T}T = E, \tag{48}$$

where $\tilde{T}$ is the transpose of T and E is the unit matrix. Next, we replace the edge factors in (47) by

$$\delta_{Kr}(s, s') = \sum_{s^* = \pm 1} T(s^*, s)T(s^*, s'), \tag{49}$$

where $T(s, s')$ are the elements of T. Here, $s^* = \pm 1$ is a variable associated with the edge. We may then independently carry out the summations over $\{s, s'\}$ in (47) and obtain

$$Z_G = \sum_{\{s^*\}} \prod_{i \in V} w_i^*(\{s^*\}_i), \tag{50}$$

where

$$w_i^*(\{s^*\}_i) = \sum_{\{s\}_i} \left[\prod_{j=1}^{\gamma_i} T(s_j^*, s_j)\right] w_i(\{s\}_i). \tag{51}$$

Here, γ_i is the degree of the ith vertex in G and $\{s\}_i = \{s_1, \ldots, s_{\gamma_i}\}$, $\{s^*\}_i = \{s_1{}^*, \ldots, s^*_{\gamma_i}\}$. Comparing (50) with (46), we see that Z_G is invariant under the transformation (51). This symmetry relation, which holds for any G, has been derived in the literature in several special instances [25–27].

In physical considerations we are again interested in the analytic properties of the limiting free energy per site f_G, as in (20). Unfortunately, very little progress has been made toward the exact evaluation of f_G. The following results are known:

A trivially soluble case is when the vertex weights are given by

$$w_i(\{s\}_i) = u^{n_i}, \tag{52}$$

where n_i is the number of incident edges. In this case we may instead assign a weight u^2 to each edge in G'. Since each edge can be independently present or absent, we have

$$Z_G = (1 + u^2)^{|E|}, \tag{53}$$

and

$$f_G = (|E|/|V|)\ln(1 + u^2), \tag{54}$$

which is analytic in u.

If

$$w_i(\{s\}_i)\begin{cases} = u^{n_i}, & n_i = 0, 1, \ldots, \gamma_i - 1, \\ = v, & n_i = \gamma_i, \end{cases} \tag{55}$$

it can be shown [28, 29] that, for $\gamma_i = \gamma$, this vertex model is equivalent to an Ising model in a nonzero magnetic field. The analytic properties of f_G can then be analyzed accordingly, even though it still cannot be evaluated. In particular, the nonanalytic point of f_G can be located for regular G (see, e.g., [30]). Another related situation is

$$w_i(\{s\}_i)\begin{cases} = u^{n_i}, & n_i = \text{even}, \\ = u^{n_i}v, & n_i = \text{odd}. \end{cases} \tag{56}$$

In this case, (46) can be identified as the high-temperature expansion of the Ising model in a nonzero magnetic field [30] and the analyticity of f_G can be similarly discussed. The case of $v = 0$ corresponds to a zero magnetic field and the vertex model (56) reduces to the $q = 2$ Potts model upon replacing $u = \tanh(K/2)$.

Most of the other known results are for square lattices. The most complete solution to this date is that of the eight-vertex model [4, 31]. The solution of the more general sixteen-vertex model is known only in a few isolated special cases [32–34]. Readers are referred to the literature for details of these solutions.

5. Equivalence of the Whitney Polynomial with a Vertex Model

A remarkable equivalence between the Whitney polynomial and the partition function of a vertex model has been obtained by Temperley and Lieb [2] for a square lattice. An extension of their result to arbitrary planar graph is as follows [3]:

On each edge of a planar graph G a midpoint is selected. The medial graph M of G is constructed by connecting the midpoints of consecutive edges on the boundary of a face of G (see, e.g., [35]). An example is shown in Fig. 1. It is seen that every site of G lies inside a face of M. In order for M to have straight edges only, it is necessary to add to M some "external" sites. These are the sites of degree 2 in Fig. 1. Other sites have degree 4 and are the "internal" sites. Note that the construction of the external sites is not unique, but the result will be independent of this construction.

The faces of M now fall into two color classes, each edge of M lying on the boundary of one face in each category:

α-faces, each lying within a single face of G;
β-faces, each enclosing a single vertex of G.

The β-faces are shaded in Fig. 1.

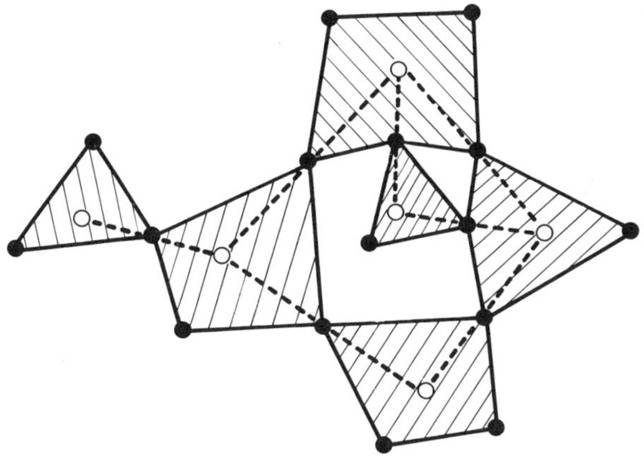

FIG. 1. A graph G (open circles and dashed lines) and its medial graph M (solid circles and lines). The β-faces of M are the shaded regions.

Now, we define a vertex model on M by assigning the following weights $w_i(\{s\}_i)$ to the ith vertex:

(a) At each internal site, only the six vertex configurations shown in Fig. 2 are allowed. Their weights are

$$\begin{aligned} &\omega_1 = z^{\alpha-\gamma}, \qquad \omega_2 = z^{\gamma-\alpha}, \qquad \omega_3 = x_i z^{\beta-\delta}, \\ &\omega_4 = x_i z^{\delta-\beta}, \qquad \omega_5 = z^{-\beta-\delta} + x_i z^{\alpha+\gamma}, \qquad \omega_6 = z^{\beta+\delta} + x_i z^{-\alpha-\gamma}, \end{aligned} \tag{57}$$

where α, β, γ, δ are the angles shown and x_i is a parameter, all of which may differ from vertex to vertex.

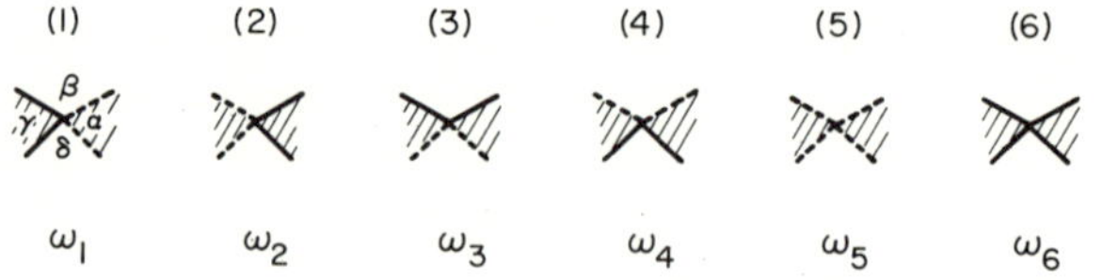

FIG. 2. The six allowed configurations at an internal site of M. The vertex weights are given in (57).

(b) At each external site, only the two vertex configurations shown in Fig. 3 are allowed. Their weights are

$$\omega_1{}' = z^{\pi-\alpha}, \qquad \omega_2{}' = z^{\alpha-\pi}, \tag{58}$$

where α is the angle shown. Note that α is the angle of a corner of a β-face.

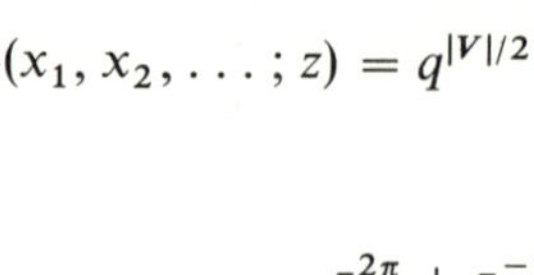

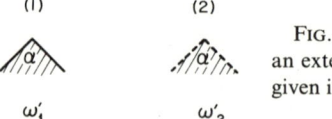

FIG. 3. The two allowed configurations at an external site of M. The vertex weights are given in (58).

The equivalence established in [3] can be stated as

$$Z_M(x_1, x_2, \ldots; z) = q^{|V|/2} \bar{W}_G(x_1/\sqrt{q}, x_2/\sqrt{q}, \ldots; q), \tag{59}$$

provided that

$$z^{2\pi} + z^{-2\pi} = \sqrt{q}. \tag{60}$$

Here Z_M is the partition function of the vertex model on M and $\bar{W}_G$ is the Whitney polynomial on G defined in (9). Note that Z_M is independent of the angles α, β, γ, δ. Using (25), we also equate the vertex model with a Potts model on G:

$$Z_M(x_1, x_2, \ldots; z) = q^{-|V|/2} Z_G^{\text{Potts}}(K_1, K_2, \ldots; q), \tag{61}$$

provided that (60) is satisfied and

$$x_i = (e^{K_i} - 1)/\sqrt{q}. \tag{62}$$

The duality relation (12) for $\bar{W}_G$ now implies a similar relation to hold for Z_M. We note that while $\bar{W}_G$ is a graph generating function whose subgraph weight $w(G')$ is determined by some "global" properties, such as the number of connected pieces, of G', (59) relates it to a graph generating function Z_M whose subgraph weight is a "local" property of G'. It is rather remarkable that they are equivalent.

The equivalence (59) is established as follows: First, we redraw the vertex configurations at the internal sites as in Fig. 4 so that the solid and dotted lines form the corners of a β-face. Note that this can be done in two different ways for vertices of types (5) and (6). To each of these we associate one term of the vertex weight as shown. The weights in Fig. 4 now have a simple geometric interpretation:

(a) To each corner of a β-face associate the weight (58).
(b) Associate a weight x_i if the two β-faces at the ith vertex are connected.

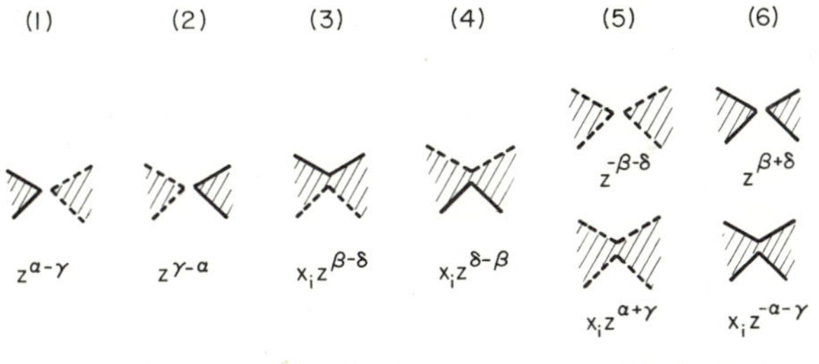

FIG. 4. The vertex configurations in Fig. 1 are redrawn so that both the solid and dashed lines form polygonal corners. Vertices of types (5) and (6) are redrawn in two different ways, each associated with a weight as shown.

It can easily be verified that all weights in Fig. 4 are generated by (a) and (b). For example,

$$\omega_1 = z^{\pi-\gamma} z^{\alpha-\pi} = z^{\alpha-\gamma}, \qquad \omega_3 = x_i z^{\pi-(2\pi-\beta)} z^{(2\pi-\delta)-\pi} = x_i z^{\beta-\delta}.$$

We also note from Fig. 4 that the solid and dotted lines in an allowed vertex configuration form nonintersecting polygons on M.

Next we observe that there exists a (1–1) correspondence between the subgraphs G' on G and the configurations of nonintersecting polygons on

M. This can be seen if we regard the β-faces enclosed within a polygon on M as covering a connected piece in G'. Thus, if we write ω_5 and ω_6 as the sum of two terms, multiply out the product involving ω_5 and ω_6 in (46), and consider the graphs thus generated, each nonintersecting polygonal configuration on M is generated in this process. In fact, each polygon is generated exactly twice, once with solid lines and once with dotted lines. Since the weight (a) can be evaluated for each polygon and the weight (b) is simply $\prod_{i \in E'} x_i$, where E' is the edge set of G', we obtain from (46)

$$Z_M = \sum_{G' \subseteq G} \left[\prod_{i \in E'} x_i \right] \prod_{\text{polygon}} w_{\text{p}}, \tag{63}$$

where w_{p} is the weight (a) of a polygon. Consider an n-polygon having interior angles $\alpha_1, \alpha_2, \ldots, \alpha_n$ inside β-faces. The weight (a) for this polygon is

$$z^{(\pi - \alpha_1) + \cdots + (\pi - \alpha_n)} = z^{2\pi}$$

if the polygon has a solid (line) boundary, and

$$z^{(\alpha_1 - \pi) + \cdots + (\alpha_n - \pi)} = z^{-2\pi}$$

if the polygon has a dotted boundary. Since each polygon can have solid or dotted boundary independently, we then have, for all polygons,

$$w_{\text{p}} = z^{2\pi} + z^{-2\pi}. \tag{64}$$

Thus

$$Z_M = \sum_{G' \subseteq G} \left[\prod_{i \in E'} x_i \right] (z^{2\pi} + z^{-2\pi})^{p(G')}, \tag{65}$$

where $p(G')$ is the number of nonintersecting polygons on M for a given G', or

$$p(G') = c(G') + n(G') = 2c(G') - e(G') + |V|. \tag{66}$$

Finally, we obtain (59) from (65) upon using (9), (60), and (66). Q.E.D.

References

1. J. W. Essam, Graph theory and statistical physics, *Discrete Math.* **1** (1971), 83–112.
2. H. N. V. Temperley and E. H. Lieb, Relation between the 'percolation' and 'colouring' problem and other graph-theoretical problems associated with regular planar lattices: some exact results for the 'percolation' problem, *Proc. Roy. Soc. Ser. A* **322** (1971), 251–280.
3. R. J. Baxter, S. B. Kelland, and F. Y. Wu, Equivalence of the Potts model or Whitney polynomial with an ice-type model, *J. Phys. A* **9** (1976), 397–406.

4. E. H. Lieb and F. Y. Wu, Two-dimensional ferroelectric models, *in* "Phase Transitions and Critical Phenomena" (C. Domb and M. S. Green, eds.), pp. 331–490, Academic Press, New York, 1972.
5. R. C. Read, An introduction to chromatic polynomials, *J. Combinatorial Theory* **4** (1968), 52–71.
6. J. E. Mayer, The statistical mechanics of condensed systems: I and II, *J. Chem. Phys.* **5** (1937), 67–83.
7. G. D. Birkhoff, A determinant formula for the number of ways of coloring a map, *Ann. of Math.* **14** (1912), 42–46; On the number of ways of coloring a map, *Proc. Edinburgh Math. Soc.* **2** (1930), 83–91.
8. H. Whitney, A logical expansion in mathematics, *Bull. Amer. Math. Soc.* **38** (1932), 572–579.
9. J. F. Nagle, Weak-graph method for obtaining formal series expansions for lattice statistical problems, *J. Math. Phys.* **9** (1968). 1007–1019.
10. H. Whitney, The coloring of graphs, *Ann. of Math.* **33** (1932), 688–718.
11. R. B. Potts, Some generalized order–disorder transformations, *Proc. Cambridge Phil. Soc.* **48** (1952), 106–109.
12. D. Ruelle, "Statistical Mechanics," Chapter 2, Benjamin, New York, 1969.
13. E. Ising, Beitrag zur Theorie des Ferromagnetismus, *Z. Phys.* **31** (1925), 253–258.
14. L. Onsager, Crystal statistics. I. A two-dimensional model with an order–disorder transition, *Phys. Rev.* **65** (1944), 117–149.
15. E. H. Lieb, Residual entropy of square ice, *Phys. Rev.* **162** (1967), 162–172.
16. D. Kim and R. J. Joseph, Exact transition temperature of the Potts model with q states per site for the triangular and honeycomb lattices, *J. Phys. C* **7** (1974), L167–170.
17. A. Hinterman, H. Kunz, and F. Y. Wu, Exact results for the Potts model in two dimensions, preprint.
18. R. J. Baxter, Potts model at the critical temperature, *J. Phys. C* **6** (1973), L445–448.
19. N. Biggs, Colouring square lattice graphs, preprint.
20. F. Y. Wu and Y. K. Wang, Duality transformation in a many-component spin model, *J. Math. Phys.* **17** (1976), 439–440.
21. C. Domb, Configurational studies of the Potts model, *J. Phys. A* **7** (1974), 1335–1348.
22. J. F. Nagle, A new subgraph expansion for obtaining coloring polynomials for graphs, *J. Combinatorial Theory Ser. B* **10** (1971), 42–59.
23. H. N. V. Temperley, Transformation of graph-theoretical problems into one another, preprint.
24. F. J. Wegner, A transformation including the weak-graph theorem and the duality transformation, *Physica* **68** (1973), 570–578.
25. C. Fan and F. Y. Wu, General lattic model of phase transitions, *Phys. Rev. B* **2** (1970), 723–733.
26. C. Thibaudier and J. Villian, On the three-dimensional eight-vertex model, *J. Phys. C* **5** (1972), 3429–3437.
27. B. Sutherland, Exact results for the eight-vertex model in three dimensions, *Phys. Rev. Lett.* **31** (1973), 1504–1506.
28. F. Y. Wu, Phase transition in a sixteen-vertex lattice model, *Phys. Rev. B* **5** (1972), 1810–1813.
29. F. Y. Wu, Phase transition in a vertex model in three dimensions, *Phys. Rev. Lett.* **32** (1974), 460–463.
30. G. F. Newell and E. W. Montroll, On the theory of the Ising model of ferromagnetism, *Rev. Mod. Phys.* **25** (1953), 353–389.
31. R. J. Baxter, Partition function of the eight-vertex lattice model, *Ann. Physics* **70** (1972), 193–228.

32. F. Y. WU, Exact solution of a general lattice statistical model, *Solid State Comm.* **10** (1972), 115–117.
33. J. RAE, Some simple solutions of the 16-vertex model, *J. Phys. A* **6** (1973), L140–143.
34. A. GAAFF, The two-dimensional Ising model in a magnetic field $\frac{1}{2}i\pi$ as a soluble case of the 16-vertex model, *Phys. Lett. A* **49** (1974), 103–105.
35. O. ORE, "The Four-Color Problem, p. 125," Academic Press, New York, (1967).

AMS (MOS) subject classification: 82A25.

STUDIES IN FOUNDATIONS AND COMBINATORICS
ADVANCES IN MATHEMATICS SUPPLEMENTARY STUDIES, VOL. 1

Secondary Structure of Single-Stranded Nucleic Acids†

MICHAEL S. WATERMAN

Los Alamos Scientific Laboratory
Los Alamos, New Mexico

DEDICATED TO JOHN R. KINNEY

The primary structure of a single-stranded nucleic acid, such as a tRNA, is the sequence of nucleotides or bases making up the molecule. Secondary structure of such a molecule is a class of graphs in the plane which preserves the bonds in the primary structure but allows helical regions. Prediction of the most stable secondary structure is an important problem at the most basic biological level. This paper gives the first graph theoretic definition of secondary structure and derives some associated properties. A classification of secondary structures is given and used as a basis for new and efficient algorithms to find the most stable secondary structure.

1. INTRODUCTION

The sequence of nucleotides or bases of a tRNA is known as its primary structure. When primary structure of a single-stranded tRNA is known, the question arises of which bases form pairs and allow the sequence to form helical regions in two dimensions. This latter structure is known as secondary structure and was proposed quite early [9, 10]. In fact, a cloverleaf model was proposed when the first tRNA primary structure was known [14]. Secondary structure has received much attention [1] and has, as at least part of its function, a role in the interactions of tRNAs with proteins [19], in stabilizing mRNA, in packing RNA into virus particles, and in recognition of specific sites by components of the translating system [23]. Prediction of the minimum free energy (i.e., most stable) secondary structure, then, is an important problem at the most basic biological level.

To approximate the Helmholtz free energy of a proposed secondary structure, much biochemical information must be obtained. Work by Uhlenbeck *et al.* [21], Gralla and Crothers [11, 12], DeLisi and Crothers [3], and DeLisi [4] has contributed to this task. Once this information is obtained, secondary structure is carefully defined, and rules for evaluation of free energy are established, then work on prediction of secondary structure can start.

† Research performed under the auspices of the U.S. Department of Energy.

ISBN 0-12-599101-0

Much effort has been given to the prediction of secondary structure. One of the most important methods, due to Tinoco *et al.* [20], considers the base pairing matrix for a tRNA. This method has been modified and extended but appears to be the basis of the algorithms that have since been proposed or used [3, 7, 8, 13, 17, 19]. Below a connection is made between a Tinoco-like method and the adjacency matrix for the graph corresponding to the secondary structure. This connection allows a clear understanding of the method and its severe limitations.

This paper gives a careful and general definition of secondary structure and derives some properties that follow from the definition. A classification of secondary structure by complexity is given, and this classification is used as a basis for efficient algorithms to find the optimal secondary structure. The algorithms are new, and careful proofs are given that show they solve the general problem of optimal structure of single-stranded nucleic acids. Attention is also given to computational efficiency.

2. The Graph Theory of Secondary Structure

In this section, graph theoretic properties of secondary structures are studied. There will be no consideration of the specific pairing rules or properties of a specific single-stranded nucleic acid. Instead, the object is to give a precise definition of secondary structure and of the components of secondary structure for an arbitrary sequence of length n. The total number of secondary structures for a sequence of length n is considered. Also, the set of possible secondary structures will be decomposed into a disjoint union indexed by $k = 0, 1, \ldots$. The kth set in this union will be called the set of kth order secondary structures and will be given a simple definition based on the complexity of the secondary structure. This classification of secondary structures will allow a proof (Theorem 7.5) that the algorithms stated later in this paper do find the best secondary structure given a specific single stranded nucleic acid with pairing rules.

Figure 2.1 gives some examples of secondary structure. In the following discussion, a sequence of n points $(1\text{–}2\text{–}\cdots\text{–}n)$ will be assumed given in

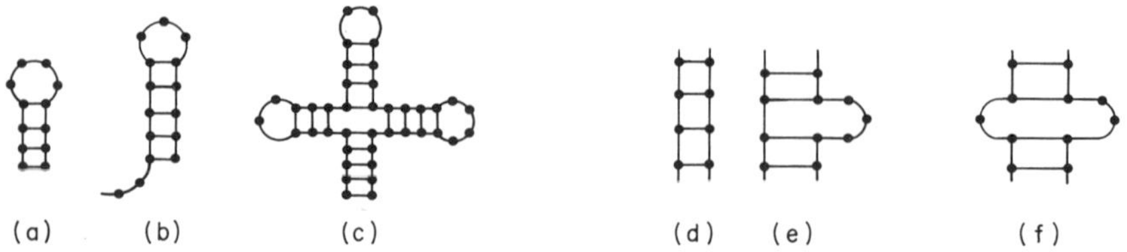

Fig. 2.1. Secondary structure.

which each point i, $1 < i < n$, is joined or bonded to $i - 1$ and $i + 1$. A point i unpaired will mean i is not joined to any points other than $i - 1$ and $i + 1$. Figure 2.1a is the configuration of the sequence (1–2–3· · ·–12) with, in addition to the bonding between adjacent members of the sequence, the bonds (1, 12), (2, 11), (3, 10), (4, 9). Thus, when labeled, Fig. 2.1a becomes:

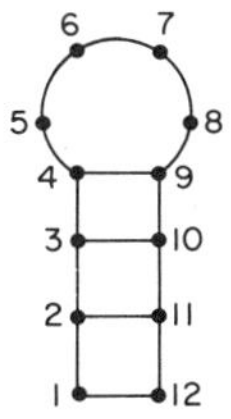

This structure is called a hairpin. The region (1–2–3–4) bonded with (9–10–11–12) is called a ladder (also called helical region in the literature) and appears in Fig. 2.1d. The unpaired region (5–6–7–8) forms what is called a hairpin loop in the literature. (In this paper hairpin loop and loop have identical meanings.) The unpaired region in the initial part of the sequence in Fig. 2.1b will be called a tail. The secondary structure in Fig. 2.1c is called a cloverleaf. The unpaired region in Fig. 2.1e is known as a bulge, while the unpaired members of Fig. 2.1f are known as an interior loop.

The problem is to construct a usable mathematical model of all possible secondary structures and to use this model to study secondary structure. It is important to realize that these molecules have a polarity (the 3′ end to the 5′ end), and that only bases of opposite polarity bond. The most useful approach to the model seems to be graph theoretic since secondary structures are clearly graphs in the plane. Also, a and b are said to be *adjacent* if (a, b) is a line of the graph. This leads to the definition of the *adjacency matrix* $A = (a_{ij})$ of a labeled graph G with P points. The element $a_{ij} = 1$ if i and j are adjacent, and $a_{ij} = 0$ otherwise. (Set the elements $a_{ii} = 0$.)

A study of the properties of these structures and the adjacency matrices has led to the following definition of secondary structure.

DEFINITION 2.1. A *secondary structure* is a graph on the set of n labeled points $\{1, 2, \ldots, n\}$ such that the adjacency matrix $A = (a_{ij})$ has the following three properties:

(i) $a_{i,\,i+1} = 1$ for $1 \leqslant i \leqslant n - 1$.

(ii) For each fixed i, $1 \leqslant i \leqslant n$, there is at most one $a_{ij} = 1$ where $j \neq i \pm 1$.

(iii) If $a_{ij} = a_{kl} = 1$, where $i < k < j$, then $i \leqslant l \leqslant j$.

If $a_{ij} = 1$, i and j are said to be *bonded*.

Part (i) of Definition 2.1 requires adjacent points to be bonded. Part (ii) states that each point can be bonded to at most one other point (besides the adjacent points). Finally, part (iii) is to assure that if i and j are bonded, then all bonding of points $i < k < j$ is with points l between i and j. Part (iii) is an essential part of the definition as it keeps the structure from "folding" and becoming a three-dimensional or tertiary structure.

It is interesting to note that Kleitman [16] and Hsieh [15] have studied irreducible diagrams which suggest a different representation of secondary structure. This representation, given in Fig. 2.2, is perhaps useful for combinatorial reasons but is not used here since the definitions and results here are motivated by the models of secondary structure in the biological literature.

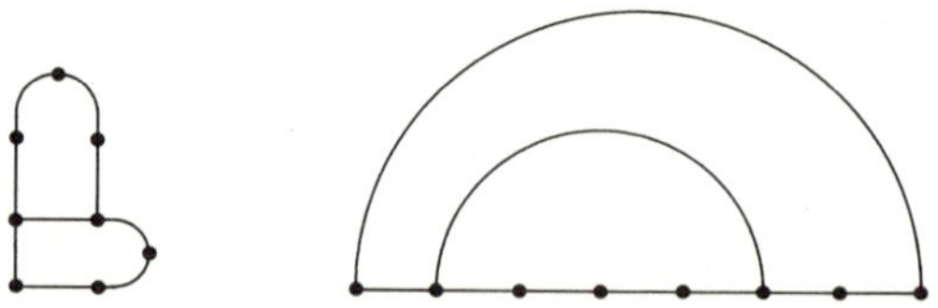

(a) (b)

FIG. 2.2. (a) Usual diagram of secondary structure. (b) "Loop" diagram of secondary structure.

Before proceeding to the enumeration of secondary structures, it is necessary to clearly indicate what structures are to be enumerated. Two secondary structures for $(1-2-\cdots-n)$ will be considered distinct if their adjacency matrices are not equal. Consider $n = 6$, where the seventeen secondary structures are enumerated in Fig. 2.3.

An examination of Fig. 2.3 shows that the enumeration depends heavily on the labeling of the sequence. In fact, there are eleven such structures if the sequence is unlabeled.

It would be an interesting problem to enumerate the unlabeled secondary structures, although of no direct use in this paper. Also, Paul Stein of Los Alamos Scientific Laboratory has suggested the problem of determining whether or not a given unlabeled graph has a labeling such that the graph is a secondary structure.

THEOREM 2.1. *Let $S(n)$ be the number of secondary structures for n points. Then $S(1) = S(2) = 1$, and for $n > 2$, $S(n)$ satisfies*

$$S(n+1) = S(n) + \sum_{k=0}^{n-2} S(k)S(n-k-1),$$

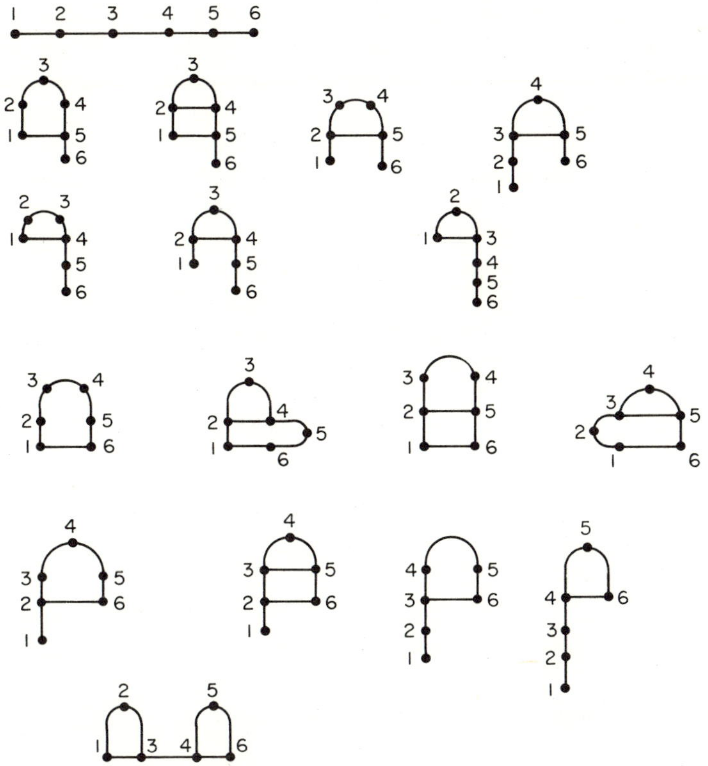

FIG. 2.3. The seventeen secondary structures for (1–2–3–4–5–6).

where $S(0) \equiv 1$. *Also,* $S(n) \geqslant 2^{n-2}$ *for* $n \geqslant 2$. (*The table below gives the values of* $S(n)$ *for* $n = 1, \ldots, 10$.)

n	0	1	2	3	4	5	6	7	8	9	10
$S(n)$	1	1	1	2	4	8	17	37	82	185	423

Proof. It is easy to see from Definition 2.1 that for $n = 1$ and $n = 2$ the only secondary structures are

$$\underset{(n=1)}{\overset{1}{\cdot}} \qquad \underset{(n=2)}{\overset{1\quad 2}{\cdot\!\!-\!\!-\!\!\cdot}}$$

and therefore $S(1) = S(2) = 1$.

Suppose $S(k)$ is known for $1 \leqslant k \leqslant n$. Now consider $(1\text{–}2\text{–}\cdots\text{–}n+1)$. Either $n+1$ is not paired or $n+1$ is paired with j for $1 \leqslant j \leqslant n-1$. If $n+1$ is not paired, then $(1\text{–}2\text{–}\cdots\text{–}n)$ can form any possible secondary structure.

If $n + 1$ is paired with j, then $(1\text{–}2\text{–}\cdots\text{–}(j-1))$ and $(j+1\text{–}\cdots\text{–}n)$ can each form any possible secondary structure. This technique forms secondary structures for $(1\text{–}2\text{–}\cdots\text{–}n+1)$ as a check of Definition 2.1 shows. That this technique enumerates all secondary structures for the sequence follows from (iii) of that definition.

From the preceeding paragraph, it follows that

$$S(n+1) = S(n) + S(n-1) + S(1)S(n-2) + \cdots + S(n-2)S(1).$$

Defining $S(0) = 1$, the equation takes the form $(n \geqslant 2)$

$$\begin{aligned} S(n+1) &= S(n) + \sum_{k=0}^{n-2} S(k)S(n-k-1) \\ &= S(n) + S(n-1) + \sum_{k=1}^{n-2} S(k)S(n-k-1) \\ &= S(n) + S(n-1) + \sum_{k=0}^{n-3} S(k+1)S(n-k-2). \end{aligned}$$

Since $S(k+1) \geqslant S(k)$,

$$S(n+1) \geqslant S(n) + S(n-1) + \sum_{k=0}^{n-3} S(k)S(n-k-2) = 2S(n).$$

Now $S(2) = 1$ so that the above inequality implies

$$S(n) \geqslant 2^{n-2}.$$

This completes the proof.

To show that 2^{n-2} is an unsatisfactory bound, assume $\lambda_2\alpha^n \leqslant S(n) < \lambda_1\alpha^n$, where $\lambda_2 > 0$. Then

$$\lambda_1\alpha^{n+1} \geqslant \lambda_2\left[\alpha^n + \lambda_2 \sum_{k=0}^{n-2} \alpha^{n-1}\right] = \lambda_2[\alpha^n + \lambda_2(n-2)\alpha^{n-1}],$$

and

$$(\lambda_1/\lambda_2)\alpha^2 \geqslant \alpha + \lambda_2(n-2),$$

which is a contradiction. Thus, the rate of growth of $S(n)$ is not geometric. However, it will be shown that $S(n)$ is bounded by a geometric growth rate. If the generating function $\phi(x)$ is defined by

$$\phi(x) = \sum_{n=0}^{\infty} S(n)x^n,$$

the recursion formula in Theorem 2.1 can be multiplied by x^{n+1} and summed to obtain

$$x^2\phi^2(x) + (x - 1 - x^2)\phi(x) + 1 = 0.$$

This equation can be solved to yield the next corollary.

COROLLARY 2.1. *If* $\phi(x) = \sum_{n=0}^{\infty} S(n)x^n$, *then*

$$\phi(x) = \frac{x^2 - x + 1 - [1 + x(x^3 - 2x^2 - x - 2)]^{1/2}}{2x^2}.$$

Next it is shown that $S(n)$ is bounded by a geometric series.

COROLLARY 2.2. *For* $n \geqslant 2$ *there is a fixed* $M > 0$ *such that*

$$2^{n-2} \leqslant S(n) \leqslant M4^n.$$

Proof. Now $S(n+1) \leqslant \sum_{k=0}^{n} S(k)S(n+1-k)$ so that if $g(0) = g(1) = 1$ and

$$g(n) = \sum_{k=0}^{n-1} g(k)g(n-k),$$

then $S(n) \leqslant g(n)$; and DeBruijn and Erdos [2] show

$$g(n) = \frac{1}{2^{n-1}}\binom{2n}{n} = O(4^n).$$

Now some additional definitions will be given so that several of the features in Fig. 2.1 can be identified from the adjacency matrix.

DEFINITION 2.2. Suppose A is the adjacency matrix for a secondary structure on $(1-2-\cdots-n)$.

(i) The point j is said to be *paired* if there is some point $i \neq j \pm 1$ such that $a_{ij} = 1$.

(ii) The region $(i+1-i+2-\cdots-(j-1))$ is a *loop* if $i+1, i+2, \ldots, j-1$ are all unpaired and $a_{ij} = 1$. The pair (i, j) is said to be the *foundation of the loop*.

(iii) The sequence $(i+1-i+2-\cdots-(j-1))$ is a *bulge* if $i+1, i+2, \ldots, j-1$ are all unpaired, i and j are both paired, and $a_{ij} \neq 1$.

(iv) An *interior loop* is two bulges $(i+1-i+2-\cdots-(j-1))$ and $(k+1-k+2-\cdots-(l-1))$ such that $a_{il} = 1$ and $a_{jk} = 1$. (Here $i < j < k < l$.)

(v) A *join* is a bulge $(i-i+1-\cdots-j)$ such that $a_{kl} = 1$ for $k < i$ implies $l \leqslant i$, and $a_{kl} = 1$ for $k > j$ implies $l \geqslant j$.

(vi) A *tail* is a sequence $(1–2–\cdots–j)$ where $1, 2, \ldots, j$ are unpaired and $j + 1$ is paired.

(vii) A *ladder* is two sequences $(i + 1–i + 2–\cdots–i + j)$ and $(k + 1–k + 2–\cdots–k + j)$ such that $i + j + 1 < k$, $a_{i+l, k+j-l+1} = 1$ for $1 \leqslant l \leqslant j$ and $a_{i, k+j+1} = a_{i+j+1, k} = 0$. If $i + j + 3 = k + 1$, this last requirement is dropped.

(viii) A *hairpin* is the longest sequence $(i + 1–i + 2–\cdots–(j - 1))$ containing exactly one loop such that $a_{i+1, j-1} = 1$ and $a_{i,j} = 0$. The paired points $i + 1$ and $j - 1$ will be called the *foundation* of the hairpin.

The above Definition 2.2 shows that the definition of secondary structure given here is rich enough to include the elementary structures in Fig. 2.1. These structures can be easily identified from the graph or from the adjacency matrix. For example, a ladder corresponds to a sequence of 1's on some negative diagonal. It is this observation that Tinoco *et al.* [20] utilize in their work on prediction of RNA secondary structure. It has been the basis of all previous algorithms to predict secondary structure (see, e.g., [19].) All these algorithms rely on the examination of all possible secondary structures. The combinatorial results shown later in this section indicate that the number of cases in such an examination is extremely large.

However, the question of whether the definition of secondary structure is too broad remains. The next theorem shows that any secondary structure is made up of loops, ladders, bulges, and tails. No secondary structure can be drawn with a point that is not a member of a loop, ladder, bulge, or tail.

THEOREM 2.2. *Any secondary structure can be uniquely decomposed into loops, ladders, bulges, and tails.*

Proof. If $a_{ij} = 1$, where $i \neq j \pm 1$, then i and j are members of sequences (possibly of length 1) which are a ladder. Thus, assume i is an unpaired point. Then let $(i - j)–\cdots–i–\cdots–i + k$ be the longest sequence of unpaired points that i is a member of. If $i - j = 1$ or $i + k = n$, then i belongs to a tail. Otherwise $i - j - 1$ and $i + k + 1$ are paired. If $a_{i-j-1, i+k+1} = 1$, then i belongs to a loop. If $a_{i-j-1, i+k+1} = 0$, then i belongs to a bulge.

The next theorem shows that any nontrivial secondary structure contains a loop.

THEOREM 2.3. *If a secondary structure has at least one pair, it has at least one loop.*

Proof. Clearly, there exists at least one ladder. Consider the ladders in the secondary structure: $L_1, L_2, \ldots, L_k$. By the definition of ladder, the union of the two sequences making up the ladder does not make a new sequence. Since there are a finite number of ladders and every paired point belongs to a ladder, there exists a ladder such that the nonempty sequence of points between the two sequences making up the ladder has the property that they are all unpaired. By definition this sequence is a loop.

The next result shows the secondary structure in more detail. This result is necessary to make a classification of secondary structures by complexity.

THEOREM 2.4. *Every secondary structure can be uniquely decomposed into* (i) *hairpins and* (ii) *ladders, bulges, and tails which are not members of a hairpin.*

Proof. Each loop is contained in a hairpin. Since by definition a hairpin has exactly one loop, there are as many (nonintersecting) hairpins as loops. The structures remaining must, by Theorem 2.2, be ladders, bulges, and tails.

Next, secondary structures are classified by a certain complexity criterion. A simple lemma is necessary to make certain this definition can be accomplished.

LEMMA 2.1. *If A is the adjacency matrix for some secondary structure and, if $A' = (a'_{ij})$ is formed from A by setting $a'_{ij} = a'_{ji} = 0$ for any set of choices of i and j ($i \neq j \pm 1$), then A' is the adjacency matrix for another secondary structure.*

Proof. It is easy to check Definition 2.1 for $A' = (a'_{ij})$.

DEFINITION 2.3. Let A be the adjacency matrix for a secondary structure. A sequence $A^{(i)}$ of adjacency matrices of secondary structure is formed as follows:

(i) $A^{(0)} = A$.
(ii) Form $A^{(i+1)}$ from $A^{(i)}$ by setting $a_{kl}^{(i+1)} = a_{lk}^{(i+1)} = 0$ whenever $a_{kl}^{(i)} = a_{lk}^{(i)} = 1$, k and l are members of some hairpin, and $k \neq l \pm 1$.

The secondary structure for A is said to be *kth order* if $A^{(k)}$ is the first matrix in the sequence $\{A^{(k)}\}_{k=0}^{\infty}$ such that the secondary structure for $A^{(k)}$ has no hairpins. (Of course, this means $A^{(k)}$ is a matrix such that $a_{ij}^{(k)} = 0$ if $i \neq j \pm 1$.)

It is clear from Lemma 2.1 that the algorithm of Definition 2.3 is well defined. The next theorem states some additional properties of order.

THEOREM 2.5. *Any secondary structure is a kth order secondary structure for some unique $k \geqslant 0$. If the sequence is of length n, then $k \leqslant [n/3]$.*

Proof. Since at least three points are required to make up a hairpin, there are a finite number of hairpins for any secondary structure. If there are no such hairpins, the secondary structure is 0th order and there is nothing to show. If there is exactly one hairpin, there can be no paired points not belonging to the hairpin; and therefore the secondary structure is first order. Otherwise there are at least two hairpins.

Assume $A^{(i)}$ has at least two hairpins. At least one pair of these hairpins has only unpaired points between them. Thus $A^{(i+1)}$ can have at least one less hairpin than $A^{(i)}$, and the algorithm must terminate in a finite number of steps.

In fact, each hairpin has at least one point in its loop and at least one pair of points on or above the foundation. Therefore, there are at most $[n/3]$ loops and the secondary structure cannot have order greater than $[n/3]$.

It is very convenient to determine the order of a secondary structure from a graph, and this is illustrated in Fig. 2.4. An arrow will denote passage from the secondary structure for $A^{(i)}$ to that for $A^{(i+1)}$ in Fig. 2.4.

It is clear that order is related to complexity and that, for sufficiently large n, secondary structures of arbitrarily large order exist. A proof of this can be based on Fig. 2.4f. If a secondary structure of order k can be constructed, add it to each side of a hairpin. The new structure is of order $k + 1$.

THEOREM 2.6. *For any* $k > 0$ *there is an* $n = n(k)$ *such that some secondary structure for* $(1\text{–}2\text{–}\cdots\text{–}n)$ *is of order* k.

Next, the secondary structures with exactly one loop are enumerated. These structures are the basis of the lower bound in Theorem 2.1.

THEOREM 2.7. *There are exactly* $2^{n-2} - 1$ *secondary structures of length* $n\,(n \geqslant 2)$ *which have exactly one loop.*

Proof. Let $L(n)$ be the number of secondary structures of length n which have *no more* than one loop. By inspection $L(1) = L(2) = 1$.

The proof follows that of Theorem 2.1. Assume $L(1), L(2), \ldots, L(n)$ are known and consider $(1\text{–}2\text{–}\cdots\text{–}n + 1)$. Either $n + 1$ is not paired or it is paired with j, where $1 \leqslant j \leqslant n - 1$. In the first case, $(1\text{–}2\text{–}\cdots\text{–}n + 1)$ can form $L(n)$ secondary structures of interest. In the second case, $(j + 1\text{–}\ldots\text{–}(n - 1))$ can form any possible secondary structure of interest. $(1\text{–}2\text{–}\cdots\text{–}(j - 1))$ cannot form a loop as this would make two loops.

Thus it is clear that

$$\begin{aligned} L(n + 1) &= L(n) + L(n - 1) + \cdots + L(1) \\ &= L(n) + L(n) = 2L(n) \qquad \text{for} \quad n \geqslant 2. \end{aligned}$$

Since $L(2) = 1$,

$$L(n) = 2^{n-2} \qquad \text{for} \quad n \geqslant 2.$$

The function $L(n)$ counts all secondary structures with one loop plus the unique case of no pairing. The result now follows.

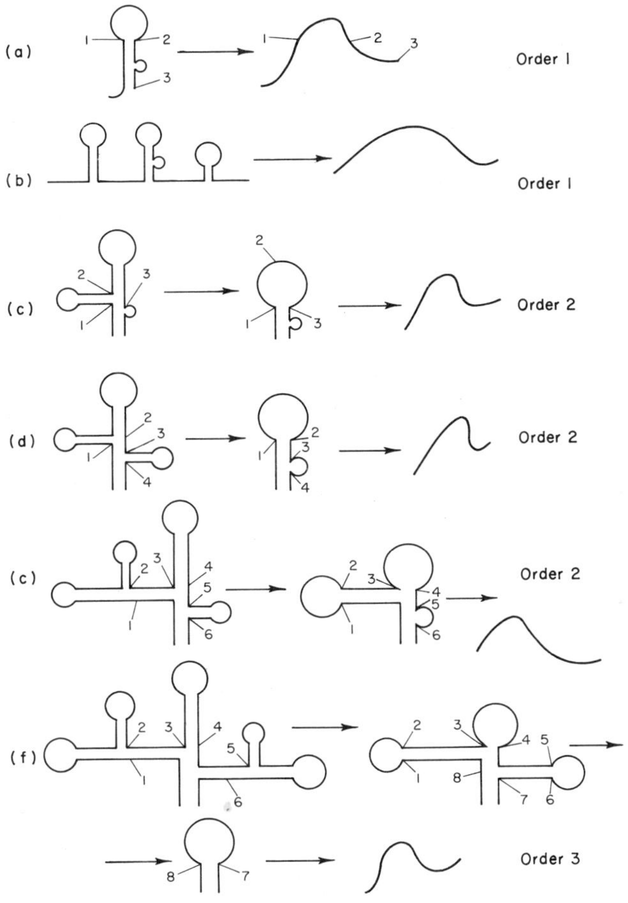

FIG. 2.4. Example of order.

Although Theorem 2.6 is a satisfactory answer to the question of how many (order 1) secondary structures have exactly one loop, the question of the number of order 1 secondary structures remains. This problem is dealt with in the next theorem, and the solution follows from an application of the discrete renewal theorem.

THEOREM 2.8. *Let $S_n^{(1)}$ be the number of secondary structures of order* 1 *for a sequence of length n. There is a $\lambda \in (2, 3)$ ($\lambda = 2.2055\ldots$) such that λ is a solution of $x^3 - 2x^2 - 1 = 0$ and*

$$\lim_{n\to\infty} \lambda^{-n} S_n^{(1)} = \mu^{-1},$$

where $\mu = \lambda^{-1} + 3\lambda^{-3} + 2^{-4}[2\lambda(\lambda-2)^{-2} - 2\lambda^{-1} - 8\lambda^{-2} - 24\lambda^{-3}]$.

Proof. Let s_n be the number of zero or first order secondary structures for a sequence of length n. (Clearly, $s_n = S_n^{(1)} - 1$.) Also, let $h_1 = 1$, $h_2 = 0$, and, for $k \geqslant 3$, let h_k be the number of secondary structures with no more than one loop for a sequence of $k - 2$ points.

It easily follows from the definition of first order that if i and j are paired there can be no more than one loop in the sequence $(i + 1-\cdots-(j-1))$. By an argument similar to that of Theorem 2.1, the system of equations below follows:

$$
\begin{aligned}
s_0 &= 1,\\
s_1 &= 1 = s_0 h_1,\\
s_2 &= 1 = s_0 h_2 + s_1 h_1,\\
s_3 &= s_2 + s_0 = s_0 h_3 + s_1 h_2 + s_2 h_1,\\
s_4 &= s_3 + s_0 1 + s_1 = s_0 h_4 + s_1 h_3 + s_2 h_2 + s_3 h_1\\
s_5 &= s_4 + s_0 h_5 + s_1 h_4 + s_2 = s_0 h_5 + s_1 h_4 + s_2 h_3 + s_3 h_2 + s_4 h_1,\\
&\vdots\\
s_n &= s_{n-1} + s_0 h_n + s_1 h_{n-1} + \cdots + s_{n-3} h_3,\\
&= s_0 h_n + s_1 h_{n-1} + \cdots + s_{n-3} h_3 + s_{n-2} h_2 + s_{n-1} h_1.
\end{aligned}
$$

Define $b_0 = 1$, and, for $\lambda > 2$,

$$v_n = s_n \lambda^{-n}, \qquad f_n = h_n \lambda^{-n}.$$

Then,

$$
\begin{aligned}
v_0 &= b_0,\\
v_n &= v_0 f_n + v_1 f_{n-1} + \cdots + v_{n-1} f_1, \qquad n \geqslant 1.
\end{aligned}
$$

Now,

$$
\begin{aligned}
f = \sum_{n=1}^{\infty} f_n &= \lambda^{-1} + \lambda^{-3} + \sum_{n=4}^{\infty} 2^{n-4} \lambda^{-n}\\
&= \lambda^{-1} + \lambda^{-3} + 2^{-4}(2/\lambda)^4 \sum_{m=0}^{\infty} (2/\lambda)^m\\
&= \lambda^{-1} + \lambda^{-3} + \lambda^{-3}(\lambda - 2)^{-1}.
\end{aligned}
$$

If $\lambda = 3$, $f = 3^{-1} + 3^{-3} + 3^{-3}(3-2)^{-1} < 1$. But $\lim_{\lambda \downarrow 2} f = +\infty$. Thus there is a $\lambda \in (2, 3)$ such that $f = 1$. The equation $f = 1$ can be rewritten as the equation in the theorem statement by factoring and deleting the factor $x - 1$.

Next apply the discrete renewal theorem as given in Feller [5, pp. 330–331]. The only additional computation to obtain the result is

$$\mu = \sum_{n=1}^{\infty} nf_n = \lambda^{-1} + 3\lambda^{-3} + \sum_{n=4}^{\infty} n2^{n-4}\lambda^{-n}$$
$$= \lambda^{-1} + 3\lambda^{-3} + 2^{-4}[2\lambda(\lambda-2)^{-2} - 2\lambda^{-1} - 8\lambda^{-2} - 24\lambda^{-3}].$$

Of course, there is only one secondary structure of order 0 so

$$\lim_{n\to\infty} \lambda^{-n} S_n^{(1)} = \lim_{n\to\infty} \lambda^{-n} s_n = \mu^{-1}.$$

The proof is now completed.

Since there are exactly $2^{n-2} - 1$ secondary structures with exactly one loop ($n \geqslant 2$), it follows that, if $S_n^{(1)*}$ is the number of first order secondary structures with more than one loop,

$$\lim_{n\to\infty} \lambda^{-n} S_n^{(1)*} = \mu^{-1}.$$

3. First Order Single Loop Secondary Structures

In this section results are derived which allow the determination of the minimum free energy first order single loop secondary structure for a single-stranded nucleic acid. The results are quite general and allow input of more precise information as it becomes available. Of course, restrictions on the possible secondary structure are imposed, and the set of possible secondary structures is not nearly as general as those described in Section 2. The algorithms which are presented in this section are motivated in form and proof by previous work [22] on sequence homology, but the results of [22] cannot be directly adapted to this problem. In fact, some quite distinct properties arise as a result of the assumptions and restrictions imposed here. Symmetries exist in the sequence homology problem that are absent here.

Some fundamental definitions are now made.

Definition 3.1. Let $\mathscr{A}$ be a finite set called the *alphabet*. Then the following functions are assumed to be given.

(i) A *pairing function* is a symmetric function $p(\cdot, \cdot)$ defined on $\mathscr{A} \times \mathscr{A}$ such that $p(a, b) \in \{0, 1\}$ for $(a, b) \in \mathscr{A} \times \mathscr{A}$.

(ii) A real valued symmetric function $\alpha(\cdot, \cdot)$ defined on $\mathscr{A} \times \mathscr{A}$ is called the free energy associated with the pair (a, b). Using $\alpha(\cdot, \cdot)$, the *ladder function* can be defined on $\mathscr{A}^k \times \mathscr{A}^k$ for $k \geqslant 1$ by

$$\alpha(a_1 \cdots a_k, b_1 \cdots b_k) = \sum_{i=1}^{k} \alpha(a_i, b_i).$$

(iii) A real valued function $\beta(\cdot)$ defined on $\mathscr{A}^k$ for $k \geqslant 1$ is called the *bulge function.* Its value $\beta(a_1 \cdots a_k)$ is said to be the free energy associated with a bulge $a_1 \cdots a_k$.

(iv) The real valued symmetric function $\gamma(\cdot, \cdot)$, defined on $\mathscr{A}^k \times \mathscr{A}^l$ for $\min\{k, l\} \geqslant 1$, is called the *interior loop function.* Its value $\gamma(a_1 \cdots a_k, b_1 \cdots b_l)$ is said to be the free energy associated with an interior loop $a_1 \cdots a_k$ and $b_1 \cdots b_l$.

(v) The real valued function $\zeta(\cdot; \cdot)$ defined on $\mathscr{A}^2 \times \mathscr{A}^k$ for $k \geqslant 1$ is called the *loop function.* The value $\zeta(cd; a_1 \cdots a_k)$ is called the free energy associated with a loop $a_1 \cdots a_k$ with foundation cd.

(vi) The real valued function $\xi(\cdot)$ defined on $\mathscr{A}^k$ for $k \geqslant 1$ is called the *join function.* Its value $\xi(a_1 \cdots a_k)$ is called the free energy associated with a join.

(vii) The real valued function $\tau(\cdot)$ defined on $\mathscr{A}^k$ for $k \geqslant 1$ is called the *tail function.* Its value $\tau(a_1 \cdots a_k)$ is called the free energy associated with a tail. In this work $\tau(a_1 \cdots a_k) \equiv 0$ for all elements of $\mathscr{A}^k$ for all k.

To establish an efficient algorithm, several assumptions are made about these five functions. While these assumptions might seem rather arbitrary, they appear natural from energy considerations; and all are satisfied by current estimates of the functions [6, 11, 12, 19, 20]. They are necessary for the proof of Theorem 3.1 below. For ease of expression below when, say, $\beta(k)$ is written, it is to be interpreted as the bulge function β evaluated at an arbitrary sequence of k letters. An inequality like

$$\beta(k + l) \leqslant \beta(k) + \beta(l)$$

means

$$\beta(a_1 \cdots a_k, b_1 \cdots b_l) \leqslant \beta(a_1 \cdots a_k) + \beta(b_1 \cdots b_l).$$

Also, since ζ is a function of $\mathscr{A}^2 \times \mathscr{A}^k$, ζ is written as $\zeta(k)$.

DEFINITION 3.2. The functions $(p, \alpha, \beta, \gamma, \zeta)$ with an alphabet are said to be *regular* if

(i) for all arguments, β, γ, ζ are positive,

$$\begin{aligned} &\alpha(a, b) \text{ is negative} && \text{if } p(a, b) = 1, \\ &\alpha(a, b) = +\infty && \text{if } p(a, b) = 0, \end{aligned}$$

and

(ii) the following inequalities hold:

$$\beta(k + l) \leqslant \beta(k) + \beta(l),$$
$$\gamma(k, l) \leqslant \beta(k) + \beta(l),$$
$$\max\{\gamma(k + q, l), \gamma(q + k, l), \gamma(k, l + q), \gamma(k, q + l)\} \leqslant \beta(q) + \gamma(k, l),$$
$$\max\{\zeta(k + l), \zeta(l + k)\} \leqslant \beta(k) + \zeta(l),$$
$$\zeta(l + q + n) \leqslant \gamma(l, n) + \zeta(q),$$
$$\gamma(k + q, l + n) \leqslant \gamma(k, l) + \gamma(q, n).$$

Let **a** be a finite sequence of elements of $\mathscr{A}$ with the functions $p, \alpha, \beta, \gamma, \zeta, \xi$ defined as in Definition 3.1. By Theorem 2.2, any secondary structure $S(\mathbf{a})$ for $(\mathbf{a})$ can be decomposed into loops (LO), bulges (B'), ladders (LA), and tails (T). If the interior loops are denoted by IL and joins by J, then let $B = (B' \sim IL) \sim J$.

The next definition concerns the free energy of a class of secondary structures. Free energy here refers to the difference in Helmholtz free energy between a given secondary structure $S(\mathbf{a})$ and the completely unpaired structure $S_0(\mathbf{a})$ (a tail). $S_0(\mathbf{a})$ is the random coil state and is a member of T. While the traditional notation for free energy is ΔF or ΔA [24], the delta has been dropped as its mathematical connotations might be misleading. The assumption of additivity for free energy is equivalent to independence of energy contributions of the various structural components. This results from the assumption of statistical mechanics that the partition function is a product for independent energy contributions [24].

DEFINITION 3.3. Assume **a** is a sequence as given above and that a first order secondary structure $S(\mathbf{a})$ has at least one pair. Then the *first order free energy associated with* $S(\mathbf{a})$ is defined by

$$F_1(S(a)) = \sum_{v \in J} \xi(v) + \sum_{w \in LO} \zeta(w) + \sum_{x \in B} \beta(x) + \sum_{y \in IL} \gamma(y) + \sum_{z \in LA} \alpha(z).$$

If $S(\mathbf{a})$ has no pair, $F_1(S(\mathbf{a})) = 0$ by definition. (This is consistent with the omission of T in the above equation.) The *first order free energy associated with* **a** is defined by

$$F_1(\mathbf{a}) = \min_{S(\mathbf{a}) \in \mathscr{S}} F_1(S(\mathbf{a})),$$

where

$$\mathscr{S} = \{S(a): S(a) \text{ is zero or first order},\ w \in LO \text{ implies } w \text{ is a sequence of at least } m \text{ elements, and } z \in LA \text{ implies } \alpha(z) \leqslant \delta < 0\}.$$

If $\mathscr{S}^* = \mathscr{S} \cap \{S(\mathbf{a}): S(a)$ has no more than one loop$\}$, then define

$$F_1^*(\mathbf{a}) = \min_{S(\mathbf{a}) \in \mathscr{S}^*} F(S(\mathbf{a})).$$

Since any loop must have at least one element, the requirement of at least m elements is no loss of generality and can be used to eliminate computation. The requirement of $\alpha(z) \leqslant \delta < 0$ is to assure that ladders have enough free energy to be stable. Previously, at least three (consecutive) pairs in a ladder had been required; the present requirement is more realistic but can include the previous requirement.

Definition 3.3 is a mathematical statement of the problem of "best" secondary structure. It will require some slight alterations in succeeding sections but is entirely adequate for first order secondary structures.

A few more comments are in order before the next theorem. The algorithm given by the theorem essentially considers, for $a_1 \cdots a_i$ and $a_n \cdots a_{n-j+1}$, the free energies associated with pairs, bulges, and interior loops between the two sequences. The algorithm chooses the "structure" which has the smallest free energy. As the "structure" formed has no loop and is not a secondary structure, the word *alignment* is used. This usage is familiar from the literature on homology of biological sequences [22]. Another convention is that, if an alignment ends in a_i and a_{n-j+1} paired, it is said that that alignment ends in a pair. Of course, the final step of the theorem obtains the value of $F_1^*(\mathbf{a})$.

THEOREM 3.1. *Let* $\mathbf{a} = a_1 \cdots a_n$ *be a sequence from* $\mathscr{A}^n$ *and assume the functions* $(p, \alpha, \beta, \gamma, \zeta)$ *are regular. Let* $\delta < 0$ *and* m*, a positive integer, be used in the definition of* F_1^**. Define*

$$b_j = a_{n-j+1}, \qquad 1 \leqslant j \leqslant n$$

and $\rho(0, j) = \rho(i, 0) = 0$ *for* $0 \leqslant i, j \leqslant n$*. Then inductively define* $\rho(i, j)$*, for* $0 < ij; i + j \leqslant n - m$*, to be the minimum of zero and:*

(i)

$$\rho(i-k, j-k) + \alpha(a_{i-k+1} \cdots a_i, b_{j-k+1} \cdots b_j),$$

where $1 \leqslant k \leqslant \min\{i, j\}$ *and* $\alpha(a_{i-k+1} \cdots a_i, b_{j-k+1} \cdots b_j) \leqslant \delta$;

(ii)

$$\rho(i-k, j) + \beta(a_{i-k+1} \cdots a_i),$$

where $1 \leqslant k \leqslant i - 1$ *and some alignment for* $\rho(i-k, j)$ *ends in a pair;*

(iii)

$$\rho(i, j-k) + \beta(b_{j-k+1} \cdots b_j),$$

where $1 \leqslant k \leqslant j - 1$ *and some alignment for* $\rho(j, j-k)$ *ends in a pair; and*

(iv)

$$\rho(i-k, j-l) + \gamma(a_{i-k+1} \cdots a_i, b_{j-l+1} \cdots b_j),$$

where $1 \leqslant k \leqslant i-1$, $1 \leqslant l \leqslant j-1$, *and some alignment for* $\rho(i-k, j-l)$ *ends in a pair.*

Then

$$F_1^*(\mathbf{a}) = \min\{\rho(i, j) + \zeta(a_i b_j; a_{i+1} \cdots a_{n-j}) : i + j \leqslant n - m \text{ and some alignment for } \rho(i, j) \text{ ends in a pair}\},$$

where $F_1^*(\mathbf{a}) = 0$ *if the above set is empty.*

Proof. The proof that $\rho(i, j)$ is the minimum free energy associated with $a_1 \cdots a_i$ and $b_1 \cdots b_j = a_n \cdots a_{n-j+1}$ is postponed at present, and the statement is assumed true.

If a best first order secondary structure with no more than one loop has at least one pair, then it has exactly one loop. Let $a_i b_j$ be the foundation of that loop. Then the free energy e for $a_1 \cdots a_i$ and $b_1 \cdots b_j$ in this structure satisfies $\rho(i, j) \leqslant e < 0$. If $\rho(i, j)$ has an alignment with i and j paired, then $\rho(i, j) < e$ is a contradiction and $\rho(i, j) = e$. Otherwise the assumption of the paragraph above requires $\rho(i, j) < e$. There are two situations for $\rho(i, j)$. The first situation is

$$\rho(i, j) = \rho(i-k, j-l) + \gamma(a_{i-k+1} \cdots a_i, b_{j-l+1} \cdots b_j),$$

which implies

$$\begin{aligned}\rho(i-k, j-l) &+ \zeta(a_{i-k} b_{j-l}; a_{i-k+1} \cdots a_{n-j+l}) \\ &\leqslant \rho(i-k, j-l) + \gamma(a_{i-k+1} \cdots a_i, b_{j-l+1} \cdots b_j) + \zeta(a_i b_j; a_{i+1} \cdots a_{n-j}) \\ &< e + \zeta(a_i b_j; a_i \cdots a_{n-j}) = f_1^*(\mathbf{a}).\end{aligned}$$

The other situation is

$$\rho(i, j) = \rho(i-k, j) + \beta(a_{i-k+1} \cdots a_i)$$

so that

$$\begin{aligned}\rho(i-k, j) &+ \zeta(a_{i-k} b_j; a_{i-k+1} \cdots a_{n-j}) \\ &\leqslant \rho(i-k, j) + \beta(a_{i-k+1} \cdots a_i) + \zeta(a_i b_j; a_{i+1} \cdots a_{n-j}) \\ &< e + \zeta(a_i b_j; a_{i+1} \cdots a_{n-j}) = f_1^*(\mathbf{a}).\end{aligned}$$

Each of these situations (the second covers two cases, one of which is omitted) results in a contradiction since a member of $\mathcal{S}^*$ has been exhibited with free energy smaller than F_1^*. Therefore, subject to the assumption on $\rho(i, j)$ made above, the theorem holds.

To show $\rho(i, j)$ has the required property, assume that the property holds for $\rho(k, l)$, where $0 \leqslant k \leqslant i$, $0 \leqslant l \leqslant j$, $k + l < i + j$. Consider an optimal (minimum free energy) alignment for $a_1 \cdots a_i$ and $b_1 \cdots b_j$. Either (i) a_i and b_j are paired, (ii) exactly one of a_i and b_j is paired, or (iii) neither a_i or b_j is paired. Each case will be handled below. It must be shown that, in each case, the optimal alignment is one that is obtained by the theorem.

In case (i), the pair $a_i b_j$ belongs to a ladder z satisfying $\alpha(z) \leqslant \delta$. The free energy associated with the optimal alignment for $a_1 \cdots a_i$ and $b_1 \cdots b_j$ has, by the equation of Definition 3.3, the form

$$r(i - k, j - k) + \alpha(a_{i-k+1} \cdots a_i, b_{j-k+1} \cdots b_j),$$

where an alignment for $r(i - k, j - k)$ does not end with a_{i-k} and b_{j-k} paired. If $\rho(i - k, j - k) < r(i - k, j - k)$, an easy contradiction results so that, in case (i), $\rho(i, j)$ has the required property.

For case (ii), suppose, without loss of generality, that a_i is paired. Then the minimum free energy associated with $a_1 \cdots a_i$ and $b_1 \cdots b_j$ is of the form

$$r(i, j - k) + \beta(b_{j-k+1} \cdots b_j)$$

since b_j must belong to a bulge. In the alignment for $r(i, j - k)$, a_i and b_{j-k} are paired. If $\rho(i, j - k) = r(i, j - k)$, then the result holds. Therefore, assume $\rho(i, j - k) < r(i, j - k)$. If a_i and b_{j-k} are paired in some alignment for $\rho(i, j - k)$, then an easy contradiction results. There are three more possibilities to be considered.

First, suppose an alignment for $\rho(i, j - k)$ is of the form

$$\rho(i - l, j - k) + \beta(a_{i-l+1} \cdots a_i) = \rho(i, j - k).$$

Then,

$$\begin{aligned} &\rho(i - l, j - k) + \gamma(a_{i-l+1} \cdots a_i, b_{j-k+1} \cdots b_j) \\ &\quad \leqslant \rho(i - l, j - k) + \beta(a_{i-l+1} \cdots a_i) + \beta(b_{j-k+1} \cdots b_j) \\ &\quad < r(i, j - k) + \beta(b_{j-k+1} \cdots b_j), \end{aligned}$$

and a contradiction has been obtained.

Next, assume an alignment for $\rho(i, j - k)$ has the form

$$\rho(i, j - k - l) + \beta(b_{j-k-l+1} \cdots b_{j-k}) = \rho(i, j - k).$$

Then,

$$\begin{aligned} &\rho(i, j - k - l) + \beta(b_{j-k-l+1} \cdots b_j) \\ &\quad \leqslant \rho(i, j - k - l) + \beta(b_{j-k-l+1} \cdots b_{j-k}) + \beta(b_{j-k+1} \cdots b_j) \\ &\quad < r(i, j - k) + \beta(b_{j-k+1} \cdots b_j) \end{aligned}$$

and, again, a contradiction is shown.

To conclude case (ii), suppose an alignment for $\rho(i, j-k)$ is of the form

$$\rho(i-l, j-k-q) + \gamma(a_{i-l+1} \cdots a_i, b_{j-k-q+1} \cdots b_{j-k}) = \rho(i, j-k).$$

Then,

$$\begin{aligned}
&\rho(i-l, j-k-q) + \gamma(a_{i-l+1} \cdots a_i, b_{j-k-q+1} \cdots b_j) \\
&\quad \leqslant \rho(i-l, j-k-q) + \gamma(a_{i-l+1} \cdots a_i, b_{j-k-q+1} \cdots b_{j-k}) \\
&\qquad + (b_{j-k+1} \cdots b_j) \\
&\quad < r(i, j-k) + \beta(b_{j-k+1} \cdots b_j)
\end{aligned}$$

and, with this contradiction, case (ii) is concluded.

In case (iii), neither a_i nor b_j is paired. Then the minimum free energy associated with $a_1 \cdots a_i$ and $b_1 \cdots b_j$ has the form

$$r(i-l, j-k) + \gamma(a_{i-l+1} \cdots a_i, b_{j-k+1} \cdots b_j).$$

If $\rho(i-l, j-k) = r(i-l, j-k)$, the proof is complete. Therefore, assume $\rho(i-l, j-k) < r(i-l, j-k)$. If a_{i-l} and b_{j-k} are paired in some alignment for $\rho(i-l, j-k)$, an immediate contradiction results. There are essentially two distinct situations to consider.

First, suppose an alignment for $\rho(i-l, j-k)$ has the form

$$\rho(i-l-q, j-k) + \beta(a_{i-l-q+1} \cdots a_{i-l}) = \rho(i-l, j-k).$$

Then,

$$\begin{aligned}
&\rho(i-l-q, j-k) + \gamma(a_{i-l-q+1} \cdots a_i, b_{j-k+1} \cdots b_j) \\
&\quad \leqslant \rho(i-l-q, j-k) + \beta(a_{i-l-q+1} \cdots a_{i-l}) + \gamma(a_{i-l+1} \cdots a_i, b_{j-k+1} \cdots b_j) \\
&\quad < r(i-l, j-k) + \gamma(a_{i-l+1} \cdots a_i, b_{j-k+1} \cdots b_j),
\end{aligned}$$

which is a contradiction.

The last (nonredundant) situation to consider is that an alignment for $\rho(i-l, j-k)$ has the form

$$\rho(i-l-q, j-k-p) + \gamma(a_{i-l-q+1} \cdots a_{i-l}, b_{j-k-p+1} \cdots b_{j-k}).$$

Then,

$$\begin{aligned}
&\rho(i-l-q, j-k-p) + \gamma(a_{i-l-q+1} \cdots a_i, b_{j-k-p+1} \cdots b_j) \\
&\quad \leqslant \rho(i-l-q, j-k-p) + \gamma(a_{i-l-q+1} \cdots a_{i-l}, b_{j-k-p+1} \cdots b_{j-k}) \\
&\qquad + \gamma(a_{i-l+1} \cdots a_i, b_{j-k+1} \cdots b_j) \\
&\quad < r(i-l, j-k) + \gamma(a_{i-l+1} \cdots a_i, b_{j-k+1} \cdots b_j).
\end{aligned}$$

With this last contradiction, case (iii) and therefore the proof of the theorem is complete.

Hopefully, it is evident that the algorithm of Theorem 3.1 can be implemented on a computer and that a matrix formulation is the most convenient. The only points (i, j), except for ladders, to search over are those whose optimal alignments ended in a ladder. It is possible to make other observations to reduce the search even more, and this is handled in the next corollaries and theorem. For the theorem it is necessary to make further assumptions about β and γ. These assumptions are quite reasonable and satisfied by current estimates of β and γ [19]. The set I denotes the set of integers.

COROLLARY 3.1. *For Theorem 3.1, let* $\lambda = \min_{\mathscr{A}^2} \alpha(a, b) < 0$. *Then*

(i) *if* $\delta/\lambda \in I$, *then* $\rho(i, j) = 0$ *for* $\min\{i, j\} < \delta/\lambda$,
(ii) *if* $\delta/\lambda \notin I$, *then* $\rho(i, j) = 0$ *for* $\min\{i, j\} \leqslant \delta/\lambda$.

Proof. An examination of Theorem 3.1 shows that $\rho(i, j) \neq 0$ must first be achieved by part (i) of that theorem; but part (i) does not have a contribution unless $\alpha(a_{i-k+1} \cdots a_i, b_{j-k+1} \cdots b_j) \leqslant \delta$. The first occasion that this can occur is for the smallest integer k such that $k\lambda \leqslant \delta$ or $k \leqslant \delta/\lambda < 0$. The corollary follows.

COROLLARY 3.2. *In Theorem* 3.1, (i) *can be replaced by*

(i)′

$$\rho(i - k, j - k) + \alpha(a_{i-k+1} \cdots a_i, b_{j-k+1} \cdots b_j),$$

where $1 \leqslant k \leqslant \min\{i, j\}$, $\alpha(a_{i-k+1} \cdots a_i, b_{j-k+1} \cdots b_j)$, *and no alignments for* $\rho(i - k, j - k), \rho(i - k + 1, j - k + 1), \ldots, \rho(i - 1, j - 1)$ *end in a pair, and* (ii) *can be replaced by*

(ii)′

$$\rho(i - k, j - k) + \alpha(a_{i-k+1} \cdots a_i, b_{j-k+1} \cdots b_j),$$

where $1 \leqslant k \leqslant \min\{i, j\}$, $\alpha(a_{i-k+1} \cdots a_i, b_{j-k+1} \cdots b_j) \leqslant 0$, *no alignments for* $\rho(i - k + 1, j - k + 1), \ldots, \rho(i - 1, j - 1)$ *end in a pair, and some alignment for* $\rho(i - k, j - k)$ *ends in a pair.*

Proof. The proof follows easily from the additivity of α.

THEOREM 3.2. *Assume the situation of Theorem* 3.1 *and, in addition, that*

$$\beta(k) \leqslant \min\{\beta(k + l), \beta(l + k)\},$$
$$\beta(k) \leqslant \min\{\gamma(k + l), \gamma(l, k)\},$$

and

$$\gamma(k, l) \leqslant \min\{\gamma(k + i, l + j), \gamma(i + k, j + l)\},$$

where i, j, k, *and* l *are arbitrary nonnegative integers. Suppose that* $\rho(i - k, j - l)$ *and* $\rho(i, j)$ *each possess an alignment that ends in a pair. Then, if* $\rho(i, j) \leqslant \rho(i - k, j - l)$, $\rho(i - k, j - l)$ *need not be used in steps* (ii), (iii), *or* (iv) *to determine* $\rho(i + p, j + q)$ (*where* $0 < p + q$). *Also,* $\rho(i - k, j - l)$ *need not be used to determine* $F_1^*(\mathbf{a})$.

Proof. Suppose $q = 0$ and $p \neq 0$. Then three cases need to be considered:

(i) $l = 0, k \neq 0$,
(ii) $l \neq 0, k \neq 0$, and
(iii) $l \neq 0, k = 0$.

For case (i),

$$\begin{aligned}\rho(i, j) + \beta(a_{i+1} \cdots a_{i+p}) &\leqslant \rho(i - k, j) + \beta(a_{i+1} \cdots a_{i+p}) \\ &\leqslant \rho(i - k, j) + \beta(a_{i-k+1} \cdots a_{i+p}).\end{aligned}$$

For case (ii),

$$\begin{aligned}\rho(i, j) + \beta(a_{i+1} \cdots a_{i+p}) &\leqslant \rho(i - k, j - l) + \beta(a_{i+1} \cdots a_{i+p}) \\ &\leqslant \rho(i - k, j - l) + \beta(a_{i-k+1} \cdots a_{i+p}) \\ &\leqslant \rho(i - k, j - l) + \gamma(a_{i-k+1} \cdots a_{i+p}, b_{j-l+1} \cdots b_j).\end{aligned}$$

In case (iii),

$$\begin{aligned}\rho(i, j) + \beta(a_{i+1} \cdots a_{i+p}) &\leqslant \rho(i, j - l) + \beta(a_{i+1} \cdots a_{i+p}) \\ &\leqslant \rho(i, j - l) + \gamma(a_{i+1} \cdots a_{i+p}, b_{j-l+1} \cdots b_j).\end{aligned}$$

In each of the three cases above, the quantity for $\rho(i, j)$ was smaller than the corresponding quantity for $\rho(i - k, j - l)$. The only (nonredundant) situation remaining is $p \neq 0$, $q \neq 0$. Then,

$$\begin{aligned}&\rho(i - j) + \gamma(a_{i+1} \cdots a_{i+p}, b_{j+1} \cdots b_{j+q}) \\ &\quad\leqslant \rho(i - k, j - l) + \gamma(a_{i+1} \cdots a_{i+p}, b_{j+1} \cdots b_{j+q}) \\ &\quad\leqslant \rho(i - k, j - l) + \gamma(a_{i-k+1} \cdots a_{i+p}, b_{j-l+1} \cdots b_{j+q}),\end{aligned}$$

and the proof is concluded.

EXAMPLE. For a simple example, let $A = \{a, u, g, c, n\}$ with $p(g, c) = p(a, u) = 1$ and $p(x, y) = 0$ elsewhere. (The base n is to be thought of as a neutral element.) Define

$$\alpha(g, c) = \alpha(a, u) = -2,$$
$$\beta(k) = k,$$
$$\gamma(k, l) = k + l,$$

and

$$\zeta(k) = k.$$

The functions $(p, q, \beta, \gamma, \zeta)$ are easily seen to be regular. With $\delta = -2 < 0$ and $m = 2$, Theorems 3.1, 3.2, and Corollary 3.1 all hold.

Now consider

$$\mathbf{a} = ggguaunnnauagggnnncccauannnuauccc.$$

The elements of this sequence of length will be numbered from 2 to 34. The structure for $F_1^*(\mathbf{a}) = -7$ has four ladder regions and appears in Fig. 4.1.

Base number	14 pairs with base number 22	: *gc*
Base number	15 pairs with base number 21	: *gc*
Base number	16 pairs with base number 20	: *gc*
Base number	11 pairs with base number 24	: *au*
Base number	12 pairs with base number 23	: *ua*
Base number	5 pairs with base number 30	: *ua*
Base number	6 pairs with base number 29	: *au*

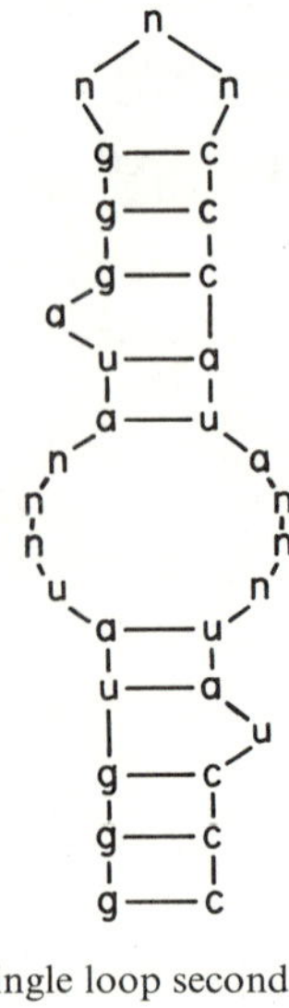

FIG. 4.1. Best single loop secondary structure for **a**.

Base number	2 pairs with base number 34	:	*gc*
Base number	3 pairs with base number 33	:	*gc*
Base number	4 pairs with base number 32	:	*gc*

4. The Hairpin Matrix

The purpose of this section is to define a matrix which will allow efficient calculation of the best first order secondary structure $F_1(\mathbf{a})$ and the best second order secondary structure. In addition, the concepts of the last section and of this section motivate the solution of the problem of finding the best Nth order secondary structure.

Motivation for the matrix defined below was the computation of the best single loop secondary structures for $a_i a_{i+1} \cdots a_j$ for $1 \leqslant i < j \leqslant n$. There are, of course, $n(n-1)/2$ such sequences. The use for such a matrix is to use the free energies

$$F_1{}^*(a_i a_{i+1} \cdots a_j)$$

in much the same way $\beta(a_i \cdots a_j)$ was used in Section 3. Certain problems immediately arise. A major difficulty is that $F_1{}^*(a_i \cdots a_j)$ gives tails weight zero when a tail of such a structure could become a bulge or a join when the entire sequence $a_1 a_2 \cdots a_n$ is considered. This difficulty is overcome by the next definition.

Definition 4.1. For a given sequence $\mathbf{a} = a_1 \cdots a_n$, regular functions $(p, \alpha, \beta, \gamma, \zeta)$ and parameters $\delta < 0$ and m, define the *hairpin matrix* H to be the symmetric $n \times n$ matrix with $h_{ii} = 0$ for $1 \leqslant i \leqslant n$ and, for $1 \leqslant i < j \leqslant n$,

$$h_{ij} = \min_{S \in \mathscr{S}_{ij}} F_1(S),$$

where $\mathscr{S}_{ij} = \{S(a_i \cdots a_j) : S(a_i \cdots a_j)$ is first order with exactly one loop, $w \in LO$ implies w has at least m elements, $z \in LA$ implies $\alpha(z) \leqslant \delta < 0$, and $p(a_i, a_j) = 1\}$. If $\mathscr{S}_{ij} = \varnothing$, let $h_{ij} = K$, where K is large enough so that the structure corresponding to h_{ij} is not used.

Since this definition is fundamental in the next two sections, it must be carefully examined. The value h_{ij} is the minimum free energy of single loop secondary structures for $a_i \cdots a_j$ with the restriction that a_i and a_j are paired. This avoids the problem of tails with weight zero and also allows the omission of calculation of

$$F_1{}^*(a_i \cdots a_j)$$

for many values of i and j.

The calculation of h_{ij} cannot be handled directly by Theorem 3.1. The next theorem modifies Theorem 3.1 to the present problem for the sequence $a_1 \cdots a_n$. Of course, $a_i \cdots a_j$ can be similarly handled. Notice that $h_{1,n} > 0$ is allowed.

THEOREM 4.1. *Let $a_1 \cdots a_n$ be a sequence from $\mathscr{A}^n$ with regular functions $(p, \alpha, \beta, \gamma, \zeta)$, $\delta < 0$, and m a positive integer. Define $b_j = a_{n-j+1}$, $1 \leqslant j \leqslant n$. Let*

$$v = \min\{k \geqslant 1 : \alpha(a_1 \cdots a_k, b_1 \cdots b_k) \leqslant \delta, \text{ and } 2k \leqslant n - m\}.$$

If v does not exist, then $h_{1n} = K$. Otherwise, let

$$\rho(v, v) = \alpha(a_1 \cdots a_v, b_1 \cdots b_v),$$

where the alignment for $\rho(v, v)$ is said to end in a pair, and inductively define $\rho(i, j)$, $v \leqslant i$, $v \leqslant j$, $2v < i + j \leqslant n - m$, to be the minimum of zero and:

(i)

$$\rho(i-k, j-k) + \alpha(a_{i-k+1} \cdots a_i, b_{j-k+1} \cdots b_j),$$

where $1 \leqslant k \leqslant \min\{i - v, j - v\}$ and either (A) *$\alpha(a_{i-k+1} \cdots a_i, b_{j-k+1} \cdots b_j) \leqslant \delta$ and no alignments for $\rho(i-k, j-k), \cdots, \rho(i-1, j-1)$ end in a pair, or* (B) *$\alpha(a_{i-k+1} \cdots a_i, b_{j-k+1} \cdots b_j) \leqslant 0$ and no alignments for $\rho(i-k+1, j-k+1), \ldots, \rho(i-1, j-1)$ end in a pair, and some alignment for $\rho(i-k, j-k)$ ends in a pair;*

(ii)

$$\rho(i-k, j) + \beta(a_{i-k+1} \cdots a_i),$$

where $1 \leqslant k \leqslant i - v$ and some alignment for $\rho(i-k, j)$ ends in a pair;

(iii)

$$\rho(i, j-k) + \beta(b_{j-k+1} \cdots b_j),$$

where $1 \leqslant k \leqslant j - v$ and some alignment for $\rho(i, j-k)$ ends in a pair; and

(iv)

$$\rho(i-k, j-l) + \gamma(a_{i-k+1} \cdots a_i, b_{j-l+1} \cdots b_j)$$

where $1 \leqslant k \leqslant i - v$, $1 \leqslant l \leqslant j - v$ and some alignment for $\rho(i-k, j-l)$ ends in a pair.

Then

$$h_{1,n} = \min\{\rho(i, j) + \zeta(a_i b_j; a_{i+1} \cdots a_{n-j}) : v \leqslant i, v \leqslant j, \\ i + j \leqslant n - m, \text{ and some alignment for } \rho(i, j) \text{ ends in a pair}\}.$$

Proof. If $\{k \leqslant 1 : \alpha(a_1 \cdots a_k, b_1 \cdots b_k) \leqslant \delta$ and $2k \leqslant n - m\} = \varnothing$, then $\mathscr{S}_{1,n} = \varnothing$ and $h_{1,n} = K$.

Otherwise, v is well defined. The main thing to be proved is that $\rho(i, j)$ is the minimum free energy of alignments of $a_1 \cdots a_i$ and $b_1 \cdots b_j$ subject to $a_1 \cdots a_v$ and $b_1 \cdots b_v$ paired. The only step that can be taken, where the previous $\rho(l, m)$ does not end in a pair, is part (A) of step (i), where $\rho(i - k, j - l)$ cannot have an alignment which ends in a pair. But, $\rho(i - k, j - l)$ *must* have been obtained in steps (ii), (iii), or (iv), where its previous $\rho(l, m)$ did end in a pair. Since the induction began with $\rho(v, v)$, which ends in a pair, all alignments trace back to $\rho(v, v)$ and therefore have $a_1 \cdots a_v$ and $b_1 \cdots b_v$ paired.

That $h_{1,n}$ is the minimum value follows as in Theorem 3.1.

A few cases can be eliminated by the next corollary.

COROLLARY 4.1. *For Theorem* 4.1, *let* $\lambda = \min_{\mathscr{A}^2} \alpha(a, b) < 0$. *Then* $h_{i,j} = 0$ *if*

(i) $i - j + 1 < m + 2(\delta/\lambda)$ *when* $\delta/\lambda \in I$, *or*
(ii) $i - j + 1 < m + 2(\delta/\lambda) + 2$ *when* $\delta/\lambda \notin I$.

Proof. Obviously, the sequence $a_i \cdots a_j$ must be long enough to have a loop of length m and to have pairing of weight $\leqslant \delta$. To have the pairing in the shortest possible time, k pairs are needed, where k is the smallest integer k such that $0 < \delta/\lambda \leqslant k$ (see Corollary 3.1). The result follows as two elements are needed to form each pair.

EXAMPLE. The simple example of Section 3, **a** = *ggguaunnnauaggg-nnncccauannnuauccc*, has the 33×33 hairpin matrix shown in Table 4.1. Here there is no harm in letting $K = 0$.

5. FIRST ORDER SECONDARY STRUCTURES

In this section a simple algorithm is given to find the best first order secondary structure from the hairpin matrix H. To review first order secondary structures, a simple theorem is stated. This theorem follows easily from Definition 2.3 and was used in the proof of Theorem 2.8.

THEOREM 5.1. *A secondary structure is first order if and only if* (i) *there is at least one pair and* (ii) *if i is paired with j then there is no more than one loop in* $a_i \cdots a_j$.

TABLE 4.1
Hairpin matrix for a ($K = 0$)

	g	g	g	u	a	u	n	n	n	a	u	a	g	g	g	n	n	n	c	c	c	a	u	a	n	n	n	u	a	u	c	c	c
g	0	0	0	0	0	0	0	0	0	0	0	0	0	0	0	0	0	0	0	0	-3	0	0	0	0	0	0	0	0	0	0	-4	-7
g	0	0	0	0	0	0	0	0	0	0	0	0	0	0	0	0	0	0	0	-1	0	0	0	0	0	0	0	0	0	0	0	-5	-4
g	0	0	0	0	0	0	0	0	0	0	0	0	0	0	0	0	0	0	0	0	0	0	0	0	0	0	0	0	0	0	0	0	0
u	0	0	0	0	0	0	0	0	0	0	0	-3	0	0	0	0	0	0	0	0	0	0	0	-3	0	0	0	0	-2	0	0	0	0
a	0	0	0	0	0	0	0	0	0	0	-1	0	0	0	0	0	0	0	0	0	0	0	-1	0	0	0	0	0	0	-2	0	0	0
u	0	0	0	0	0	0	0	0	0	0	0	0	0	0	0	0	0	0	0	0	0	0	0	0	0	0	0	0	0	0	0	0	0
n	0	0	0	0	0	0	0	0	0	0	0	0	0	0	0	0	0	0	0	0	0	0	0	0	0	0	0	0	0	0	0	0	0
n	0	0	0	0	0	0	0	0	0	0	0	0	0	0	0	0	0	0	0	0	0	0	0	0	0	0	0	0	0	0	0	0	0
n	0	0	0	0	0	0	0	0	0	0	0	0	0	0	0	0	0	0	0	0	0	0	0	0	0	0	0	0	0	0	0	0	0
a	0	0	0	0	0	0	0	0	0	0	0	0	0	0	0	0	0	0	0	0	0	0	-6	0	0	0	0	0	0	-3	0	0	0
u	0	0	0	0	-1	0	0	0	0	0	0	0	0	0	0	0	0	0	0	0	0	0	0	-6	0	0	0	0	-1	0	0	0	0
a	0	0	0	-3	0	0	0	0	0	0	0	0	0	0	0	0	0	0	0	0	0	0	0	0	0	0	0	0	0	0	0	0	0
g	0	0	0	0	0	0	0	0	0	0	0	0	0	0	0	0	0	0	0	0	-3	0	0	0	0	0	0	0	0	0	0	0	-3
g	0	0	0	0	0	0	0	0	0	0	0	0	0	0	0	0	0	0	0	-1	0	0	0	0	0	0	0	0	0	0	0	-1	0
g	0	0	0	0	0	0	0	0	0	0	0	0	0	0	0	0	0	0	0	0	0	0	0	0	0	0	0	0	0	0	0	0	0
n	0	0	0	0	0	0	0	0	0	0	0	0	0	0	0	0	0	0	0	0	0	0	0	0	0	0	0	0	0	0	0	0	0
n	0	0	0	0	0	0	0	0	0	0	0	0	0	0	0	0	0	0	0	0	0	0	0	0	0	0	0	0	0	0	0	0	0
n	0	0	0	0	0	0	0	0	0	0	0	0	0	0	0	0	0	0	0	0	0	0	0	0	0	0	0	0	0	0	0	0	0
c	0	0	0	0	0	0	0	0	0	0	0	0	0	0	0	0	0	0	0	0	0	0	0	0	0	0	0	0	0	0	0	0	0
c	0	-1	0	0	0	0	0	0	0	0	0	0	0	-1	0	0	0	0	0	0	0	0	0	0	0	0	0	0	0	0	0	0	0
c	-3	0	0	0	0	0	0	0	0	0	0	0	-3	0	0	0	0	0	0	0	0	0	0	0	0	0	0	0	0	0	0	0	0
a	0	0	0	0	0	0	0	0	0	0	0	0	0	0	0	0	0	0	0	0	0	0	0	0	0	0	0	0	0	-3	0	0	0
u	0	0	0	0	-1	0	0	0	0	-6	0	0	0	0	0	0	0	0	0	0	0	0	0	0	0	0	0	0	-1	0	0	0	0
a	0	0	0	-3	0	0	0	0	0	0	-6	0	0	0	0	0	0	0	0	0	0	0	0	0	0	0	0	0	0	0	0	0	0
n	0	0	0	0	0	0	0	0	0	0	0	0	0	0	0	0	0	0	0	0	0	0	0	0	0	0	0	0	0	0	0	0	0
n	0	0	0	0	0	0	0	0	0	0	0	0	0	0	0	0	0	0	0	0	0	0	0	0	0	0	0	0	0	0	0	0	0
n	0	0	0	0	0	0	0	0	0	0	0	0	0	0	0	0	0	0	0	0	0	0	0	0	0	0	0	0	0	0	0	0	0
u	0	0	0	0	0	0	0	0	0	0	0	0	0	0	0	0	0	0	0	0	0	0	0	0	0	0	0	0	0	0	0	0	0
a	0	0	0	-2	0	0	0	0	0	0	-1	0	0	0	0	0	0	0	0	0	0	0	-1	0	0	0	0	0	0	0	0	0	0
u	0	0	0	0	-2	0	0	0	0	-3	0	0	0	0	0	0	0	0	0	0	0	-3	0	0	0	0	0	0	0	0	0	0	0
c	0	0	0	0	0	0	0	0	0	0	0	0	0	0	0	0	0	0	0	0	0	0	0	0	0	0	0	0	0	0	0	0	0
c	-4	-5	0	0	0	0	0	0	0	0	0	0	0	-1	0	0	0	0	0	0	0	0	0	0	0	0	0	0	0	0	0	0	0
c	-7	-4	0	0	0	0	0	0	0	0	0	0	-3	0	0	0	0	0	0	0	0	0	0	0	0	0	0	0	0	0	0	0	0

It was not necessary to consider joins in first order structures with exactly one loop; but if first order structures have more than one loop, they must have joins. A problem is created by two hairpins separated by no elements of $\mathscr{A}$. This is said to be a join of length 0 and is assigned a free energy of $0 \leqslant \xi(0)$. In general, all joins have been assigned free energy 0 in the literature; but as the addition is not too difficult, it is included here. These considerations will be important in second and higher order secondary structures.

THEOREM 5.2. *For* $\mathbf{a} = a_1 \cdots a_n$, *a sequence from* $\mathscr{A}^n$, *with regular functions* $(p, \alpha, \beta, \gamma, \zeta)$, $\delta > 0$, *and* m *a positive integer, let* $H = (h_{ij})$ *be the hairpin matrix. Assume* $\xi \geqslant 0$, $\xi(k + l) \leqslant \xi(k) + \xi(l)$, $F_1(0) = 0$, *and inductively define* $F_1(i)$ *to be the minimum of zero and the following quantities for* $1 \leqslant j \leqslant i < n$:

(i)

$$F_1(i - j) + h_{i-j+1,\,i},$$

if $h_{i-j+1,\,i} \neq 0$ *and* $F_1(i - j)$ *ends in a (nonempty) join;*

(ii)

$$F_1(i - j) + \xi(0) + h_{i-j+1,\,i},$$

if $h_{i-j+1,\,i} \neq 0$ *and* $F_1(i - j)$ *ends in a hairpin;*

(iii)

$$F_1(i - j) + \xi(a_{i-j+1} \cdots a_i),$$

if $h_{i-j+1,\,i} = 0$ *and* $F_1(i - j)$ *ends in a hairpin.*

Then $F_1(\mathbf{a})$ *is the minimum of the following quantities for* $1 \leqslant j \leqslant n$:

(i)′

$$F_1(n - j) + h_{n-j+1,\,n},$$

if $h_{n-j+1,\,n} \neq 0$ *and* $F_1(i - j)$ *ends in a join;*

(ii)′

$$F_1(n - j) + \xi(0) + h_{n-j+1,\,n},$$

if $h_{n-j+1,\,n} \neq 0$ *and* $F_1(n - j)$ *ends in a hairpin;*

(iii)′

$$F_1(n - j),$$

if $h_{n-j+1,\,n} = 0$ *and* $F_1(n - j)$ *ends in a hairpin.*

Proof. If $S(\mathbf{a})$ is a first order secondary structure whose l hairpins have foundations

$$i_{j_1} i_{k_1}; \qquad i_{j_2} i_{k_2}; \qquad \ldots; \qquad i_{j_l} i_{k_l};$$

then

$$F(S(a)) = \sum_{m=1}^{l-1} \left(h_{i_{j_m} i_{k_m}} + \xi(a_{i_{k_m}+1} \cdots a_{i_{j_{m+1}}-1} \right) + h_{i_{j_l} i_{k_l}}.$$

The proof is by induction. Assume, for $1 \leqslant j \leqslant i-1 < n-1$, that $f(j)$ is the minimum free energy of $a_1 \cdots a_j$. Then the best such structure for $a_1 \cdots a_i$ either is a tail or ends in a join or a hairpin. The number of ways $a_1 \cdots a_i$ can end in a hairpin are those hairpins whose numbers $h_{i-j+1,i} \neq 0$ $(1 \leqslant j \leqslant i-1)$. This handles (i) and (ii) of the theorem. The only other possible cases are where $a_1 \cdots a_i$ ends in a join $a_{i-j+1} \cdots a_i$. If the optimal structure for $a_1 \cdots a_{i-j}$ also ended in a join, say $a_{k+1} \cdots a_{i-j}$, then

$$\begin{aligned} F(i) &= F(k) + \xi(a_{k+1} \cdots a_{i-j}) + \xi(a_{i-j+1} \cdots a_i) \\ &\geqslant F(k) + \xi(a_{k+1} \cdots a_i). \end{aligned}$$

Therefore, all necessary situations are covered by (iii).

To complete the proof, simply note that for $i = n$ ending in a "join" is actually ending in a tail and (iii) becomes (iii)$'$.

As the join function complicates Theorem 5.2, the theorem is restated with $\xi \equiv 0$. This restatement as a corollary is the practical version of the theorem for first order secondary structures.

COROLLARY 5.1. *Assume $\xi \equiv 0$ in Theorem 5.2. Then*

$$F_1(a_1 \cdots a_i) = F_1(i) = \min\{F_1(i-j) + h_{i-j+1,i} : 1 \leqslant j \leqslant i\}.$$

EXAMPLE. The example $\mathbf{a}$ = *ggguaunnnauagggnnncccauannnuauccc* given at the end of Section 3 has $F_1(\mathbf{a}) = -9$ and the structure is shown in Fig. 5.1.

1st hairpin from 23 to 31; hairpin free energy = −3.0:
Base number 23 pairs with base number 31 : *au*
Base number 24 pairs with base number 30 : *ua*
Base number 25 pairs with base number 29 : *au*

2nd hairpin from 14 to 22; hairpin free energy = −3.0:
Base number 14 pairs with base number 22 : *gc*
Base number 15 pairs with base number 21 : *gc*
Base number 16 pairs with base number 20 : *gc*

3rd hairpin from 5 to 13; hairpin free energy = −3.0:
Base number 5 pairs with base number 13 : *ua*
Base number 6 pairs with base number 12 : *au*
Base number 7 pairs with base number 11 : *ua*

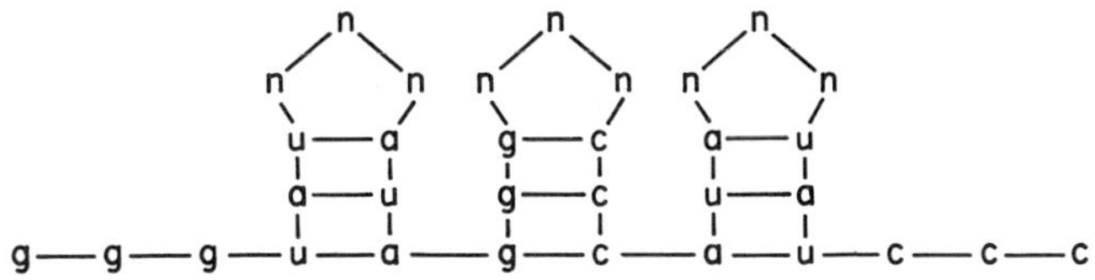

FIG. 5.1. Best first order secondary structure for **a**.

6. SECOND ORDER SECONDARY STRUCTURES

Next, second order secondary structures are considered. The definitions and algorithms of this section are a crucial step in solving the problem of the best Nth order secondary structure. The hairpin matrix of Section 4 and the algorithm of Section 5 for the best first order secondary structure are the tools for the solution obtained in this section.

The first task is to study some graph theory of second order secondary structures. Assume the adjacency matrix of a secondary structure is A. If

$$A = A^{(0)} \to A^{(1)} \to A^{(2)} = (a_{ij}^{(2)}),$$

where $A^{(2)}$ is the first $A^{(i)}$ such that $a_{ij}^{(2)} = 0$ if $i \neq j \pm 1$, then A is second order by Definition 2.3. Consider the matrix $A^{(1)}$ in this sequence. By definition, $A^{(1)}$ is the matrix for some first order secondary structure. Therefore, this first order secondary structure has exactly one loop or more than one loop. The problem of best second order structures will first be solved for the one loop case and then extended to the multiple loop case.

Next, a definition is given to classify the parts of a second order secondary structure in such a way to make the computation of free energy possible. Theorems 2.2 and 2.4 are not adequate for this purpose. A lemma is necessary to justify the definition.

LEMMA 6.1. *Let A be a second order secondary structure and suppose $A = A^{(0)} \to A^{(1)} \to A^{(2)}$. Assume $a_i a_{i+1} \cdots a_j$ is a sequence of unpaired elements in $A^{(1)}$ and that a_{i-1} and a_{j+1} are either paired or not in the sequence. Then $a_i a_{i+1} \cdots a_j$ is zero or first order in $A^{(2)}$.*

Proof. If $a_i a_{i+1} \cdots a_j$ were (at least) second order in $A^{(0)}$, then $a_i a_{i+1} \cdots a_j$ could not be unpaired in $A^{(1)}$.

DEFINITION 6.1. Let A be a second order secondary structure and suppose $A = A^{(0)} \to A^{(1)} \to A^{(2)}$ as in Definition 2.3. For $A^{(1)}$, let LO be the set of loops, LA_0 the set of ladders, IL the set of interior loops, B' the set of bulges, J the set of joins, T the set of tails, and let $B = B' \sim IL \sim J$.

(i) Assume $a_1 \cdots a_k \in LO$. If $a_1 \cdots a_k$ is a loop in $A^{(0)}$, $a_1 \cdots a_k$ is said to be a *zero order loop in* $A^{(0)}$. Otherwise, $a_1 \cdots a_k$ is said to be a *first order loop in* $A^{(0)}$. These relations are written as $a_1 \cdots a_k \in LO_0$ and $a_1 \cdots a_k \in LO_1$, respectively.

(ii) If $a_1 \cdots a_k \in B$, then, if $a_1 \cdots a_k$ is a member of B in $A^{(0)}$, $a_1 \cdots a_k$ is a *zero order bulge in* $A^{(0)}$ or $a_1 \cdots a_k \in B_0$. Otherwise, $a_1 \cdots a_k$ is called a *first order bulge in* $A^{(0)}$, which is written as $a_1 \cdots a_k \in B_1$.

(iii) Assume $(a_1 \cdots a_k, b_1 \cdots b_l) \in IL$. If this pair of sequences is an interior loop in $A^{(0)}$, then the pair is said to be a *zero order interior loop in* $A^{(0)}$. Otherwise, the pair is said to be a *first order interior loop in* $A^{(0)}$. These relations are written as $(a_1 \cdots a_k, b_1 \cdots b_l) \in IL_0$ or $(a_1 \cdots a_k, b_1 \cdots b_l) \in IL_1$, respectively.

(iv) Assume $a_1 \cdots a_k \in J$. If this sequence is a join in $A^{(0)}$, then it is said to be a *zero order join in* $A^{(0)}$. Otherwise, $a_1 \cdots a_k$ is said to be a *first order join in* $A^{(0)}$. These relations are written as $a_1 \cdots a_k \in J_0$ or $a_1 \cdots a_k \in J_1$, respectively.

(v) If $a_1 \cdots a_k \in T$, then if $a_1 \cdots a_k$ is also a tail in $A^{(0)}$, $a_1 \cdots a_k$ is said to be a *zero order tail in* $A^{(0)}$. Otherwise, $a_1 \cdots a_k$ is said to be a *first order tail in* A^0. These relations are written as $a_1 \cdots a_k \in T_0$ and $a_1 \cdots a_k \in T_1$, respectively.

(vi) LA_0 will be called the set of *zero order ladders in* $A^{(0)}$.

(vii) A sequence in $A^{(0)}$ is called a *cloverleaf* if the sequence is a hairpin in $A^{(1)}$ and the loop sequence in $A^{(1)}$ has exactly three loops in $A^{(0)}$.

The following theorem shows that Definition 6.1 adequately describes second order secondary structures.

THEOREM 6.1. *Any second order secondary structure can be uniquely decomposed into the sets* $LO_0, LO_1, B_0, B_1, IL_0, IL_1, J_0, J_1, T_0, T_1$, *and* LA_0.

Proof. Theorem 2.2 asserts that $A^{(1)}$ can be uniquely decomposed into loops, ladders, bulges, joins, and tails. These ladders are LA_0. Lemma 6.1 and Definition 6.1 allow decomposition of the remaining sets, say Q, into Q_0 and Q_1.

The free energy functions of Definition 3.1 are not adequate to directly define the free energy of a second order secondary structure. The main difficulty is associated with IL_1 and B_1, where it is not entirely clear from the literature how to assign the free energy. The approach taken here is motivated by DeLisi [4].

DEFINITION 6.2. Let $\mathscr{A}$ be an alphabet, ξ a join function, and $(p, \alpha, \beta, \gamma, \zeta)$ regular functions.

(i) A function r is called a replacement function if, for each $(a, b) \in \mathscr{A}^2$ such that $p(a, b) = 1$, $r(a, b)$ is a finite sequence of elements of $\mathscr{A}$.

(ii) It will be useful to let $\xi_0 = \xi$, $\alpha_0 = \alpha$, $\beta_0 = \beta$, $\gamma_0 = \gamma$, and $\zeta_0 = \zeta$.

(iii) Let $a_1 \cdots a_k$ be a sequence in B_1, J_1, LO_1, T_1, or one of the arguments in IL_1. Then, as a sequence in $A^{(0)}$, $a_1 \cdots a_k$ is composed of a tail $a_1 \cdots a_{t_1}$, a hairpin, a join $a_{i_1} \cdots a_{j_1}, \ldots,$ a hairpin, a join $a_{i_n} \cdots a_{j_n}$, a hairpin, and a tail $a_{t_2} \cdots a_k$. Notice that there must be at least one hairpin in this sequence, unless, perhaps, $a_1 \cdots a_k$ is an argument of IL_1. Now define

$$R(a_1 \cdots a_k)$$
$$= a_1 \cdots a_{t_1} r(a_{t_1+1}, a_{i_1-1}) \cdots a_{i_n} \cdots a_{j_n} r(a_{j_n+1}, a_{t_2-1}) a_{t_2} \cdots a_k,$$
$$\zeta_1(a, b; a_1 \cdots a_k) = \zeta_0(a, b; R(a_1 \cdots a_k)),$$
$$\beta_1(a_1 \cdots a_k) = \beta_0(R(a_1 \cdots a_k)),$$
$$\gamma_1(a_1 \cdots a_k, b_1 \cdots b_l) = \gamma_0(R(a_1 \cdots a_k), R(b_1 \cdots b_l)),$$

where, if $a_1 \cdots a_k$ is unpaired in $A^{(0)}$, $R(a_1 \cdots a_k) = a_1 \cdots a_k$.

Next, the above concepts are used to give a definition of the free energy of a second order secondary structure.

DEFINITION 6.3. Assume $\mathbf{a}$ is a sequence in $\mathscr{A}^n$, where ξ is a join function, $(p, \alpha, \beta, \gamma, \zeta)$ are regular functions, and r is a replacement function. Let $S(\mathbf{a})$ be a secondary structure of no more than second order. Then the *second order free energy associated with* $S(\mathbf{a})$ is defined by

$$F_2(S(a)) = \sum_{w_0 \in LO_0} \zeta_0(w_0) + \sum_{w_1 \in LO_1} \zeta_1(w_1) + \sum_{t_1 \in T_1} F_1(t_1)$$
$$+ \sum_{v_0 \in J_0} \xi_0(v_0) + \sum_{v_1 \in J_1} \xi_1(v_1) + \sum_{x_0 \in B_0} \beta_0(x_0) + \sum_{x_1 \in B_1} \beta_1(x_1)$$
$$+ \sum_{x_0 \in IL_0} \gamma_0(y_0) + \sum_{y_1 \in IL_1} \gamma_1(y_1) + \sum_{z_0 \in LA_0} \alpha_0(z_0).$$

Of course, $F_2(S(\mathbf{a})) = 0$ if $S(\mathbf{a})$ has no pair. The *second order free energy for* $\mathbf{a}$ is given by

$$F_2(\mathbf{a}) = \min_{S(\mathbf{a}) \in \mathscr{S}} F_2(S(\mathbf{a})),$$

where $\mathscr{S} = \{S(\mathbf{a}) : S(\mathbf{a})$ is of no more than second order, w a loop in $S(\mathbf{a})$ implies w has no more than m elements, and z a ladder in $S(\mathbf{a})$ implies $\alpha(z) \leqslant \delta\}$. If $\mathscr{S}^* = \mathscr{S} \cap \{S(\mathbf{a}) : A^{(1)}$ has no more than one loop$\}$, then let

$$F_2^*(\mathbf{a}) = \min_{S(\mathbf{a}) \in \mathscr{S}^*} F_2(S(\mathbf{a})).$$

Remark. It is easy to see from Definitions 3.2 and 3.3 that $F_2(\mathbf{a})$ and $F_2^*(\mathbf{a})$ can be found from the functions r, ξ, p, α, β, γ, ζ, and a procedure for calculating $F_1(S(\mathbf{b}))$ for any sequence b.

The next theorem generalizes the algorithm for $F_1^*(\mathbf{a})$ to $F_2^*(\mathbf{a})$.

THEOREM 6.2. *Let $a_1 \cdots a_n$ be a sequence from $\mathscr{A}^n$ and assume the functions $(p, \alpha, \beta, \gamma, \zeta)$ are regular, ξ is a join function, r is a replacement function, and δ and m satisfy the conditions of the definition of F_2^*. Define*

$$b_j = a_{n-j+1}, \qquad 1 \leqslant j \leqslant n,$$

and $\rho(0, j) = \rho(i, 0) = 0$ for $0 \leqslant i, j \leqslant n$. Then, for $0 < ij$ and $i + j \leqslant n - m$, inductively define $\rho(i, j)$ to be the minimum of zero and:

(i)

$$\rho(i-k, j-k) + \alpha(a_{i-k+1} \cdots a_i, b_{j-k+1} \cdots b_j),$$

where $1 \leqslant k \leqslant \min\{i, j\}$ and $\alpha(a_{i-k+1} \cdots a_i, b_{j-k+1} \cdots b_j) \leqslant \delta$;

(ii)

$$\rho(i-k, j) + \min_{v=0,1} \beta_v(a_{i-k+1} \cdots a_i);$$

where $1 \leqslant k \leqslant i-1$ and some alignment for $\rho(i-k, j)$ ends in a pair;

(iii)

$$\rho(i, j-k) + \min_{v=0,1} \beta_v(b_{j-k+1} \cdots b_j),$$

where $1 \leqslant k \leqslant j-1$ and some alignment for $\rho(i, j-k)$ ends in a pair, and

(iv)

$$\rho(i-k, j-l) + \min_{v=0,1} \gamma_v(a_{i-k+1} \cdots a_i, b_{j-l+1} \cdots b_j),$$

where $1 \leqslant k \leqslant i-1$, $1 \leqslant l \leqslant j-1$, and some alignment for $\rho(i-k, j-l)$ ends in a pair.

Then

$$F_2^*(\mathbf{a}) = \min\{\rho(i, j) + \min_{v=0,1} \zeta_v(a_i b_j; a_{i+1} \cdots a_{n-j})$$

$$: i + j \leqslant n - m \text{ and some alignment for } \rho(i, j) \text{ ends in a pair}\},$$

where $F_2^*(a) = 0$ if the above set is empty.

Proof. The proof proceeds exactly as in Theorem 3.1 as soon as the inequalities of $(p, \alpha, \beta, \gamma, \zeta)$ regular are shown to hold with subscripts of 0 and 1 inserted in a consistent manner. For example, to show the inequality $\beta(k+l) \leqslant \beta(k) + \beta(l)$, it may be necessary to show that

$$\beta_1(k+l) \leqslant \beta_1(k) + \beta_0(l).$$

The difficulty is in inserting the proper choice of 0 or 1 as subscripts. This is done below in a case-by-case fashion for the steps of Theorem 3.1.

The first situation is to assume $\rho(i,j)$ is the minimum free energy for alignments of $a_1 \cdots a_i$ and $b_1 \cdots b_j$. If $A = A^{(0)}$ is the adjacency matrix for the best second order structure with the structure for $A^{(1)}$ having exactly one loop, let $a_i b_j$ be the foundation of that loop. The free energy for $a_1 \cdots a_i$ and $b_1 \cdots b_j$ in this structure satisfies $\rho(i,j) \leqslant e < 0$. As before, assume $\rho(i,j) < e$. Suppose first that

$$F_2{}^*(a) = e + \zeta_{v_3}(a_i b_j;\, a_{i+1} \cdots a_{n-j})$$

and

$$\rho(i,j) = \rho(i-k, j-l) + \gamma_{v_1}(a_{i-k+1} \cdots a_i,\, b_{j-l+1} \cdots b_j).$$

Then

$$\begin{aligned}\rho(i-k, j-l) &+ \zeta_{v_2}(a_{i-k} b_{j-l};\, a_{i-k+1} \cdots a_{n-j+1}) \\ &\leqslant \rho(i,j) + \zeta_{v_3}(a_i b_j;\, a_{i+1} \cdots a_{n-j}) < F_2{}^*(\mathbf{a}),\end{aligned}$$

where the structure for ζ_{v_2} is formed by "addition" of the structures for ζ_{v_1} and ζ_{v_3}. Therefore, $v_2 = 1$ if and only if $\max\{v_1, v_3\} = 1$. The first inequality holds since $\zeta_0(l+q+n) \leqslant \gamma_0(l,n) + \zeta_0(q)$ holds for the replacement bulges and loops.

To complete the first situation, assume $F_2{}^*(\mathbf{a})$ is as above and suppose

$$\rho(i,j) = \rho(i-k, j) + \beta_{v_1}(a_{i-k+1} \cdots a_i).$$

Then,

$$\begin{aligned}\rho(i-k, j) &+ \zeta_{v_2}(a_{i-k} b_j;\, a_{i-k+1} \cdots a_{n-j}) \\ &\leqslant \rho(i,j) + \zeta_{v_3}(a_i b_j;\, a_{i+1} \cdots a_{n-j}) < F_2{}^*(a).\end{aligned}$$

Again, $v_2 = \max\{v_1, v_3\}$.

To show $\rho(i,j)$ had the optimal property desired, three cases were considered in Theorem 3.1. Case (i) proceeds without change. For Case (ii), assume the minimum free energy for $a_1 \cdots a_i$ and $b_1 \cdots b_j$ has the form

$$r^*_{(i-j)} = r(i, j-k) + \beta_{v_3}(b_{j-k+1} \cdots b_j),$$

and $\rho(i, j-k) < r(i, j-k)$. First, suppose an alignment for $\rho(i, j-k)$ is of the form

$$\rho(i, j-k) = \rho(i-l, j-k) + \beta_{v_1}(a_{i-l+1} \cdots a_i).$$

Then ($v_2 = \max\{v_1, v_3\}$)

$$\begin{aligned}\rho(i-l, j-l) &+ \gamma_{v_2}(a_{i-l+1} \cdots a_i, b_{j-k+1} \cdots b_j) \\ &\leqslant \rho(i, j-k) + \beta_{v_3}(b_{j-k+1} \cdots b_j) < r_{ij}^*.\end{aligned}$$

Next, suppose $\rho(i, j-k)$ is of the form

$$\rho(i, j-k-l) + \beta_{v_1}(b_{j-k-l+1} \cdots b_{j-k}).$$

Then, with $v_2 = \max\{v_1, v_3\}$,

$$\begin{aligned}\rho(i, j-k-l) &+ \beta_{v_2}(b_{j-k-l+1} \cdots b_j) \\ &\leqslant \rho(i, j-k) + \beta_{v_1}(b_{j-k+1} \cdots b_j) < r^*(i, j).\end{aligned}$$

The last situation for case (ii) is

$$\rho(i, j-k) = \rho(i-l, j-k-q) + \gamma_{v_1}(a_{i-l+1} \cdots a_i, b_{j-k-q+1} \cdots b_{j-k}).$$

Then with v_2 as usual,

$$\begin{aligned}\rho(i-l, j-k-q) &+ \gamma_{v_2}(a_{i-l+1} \cdots a_i, b_{j-k-q+1} \cdots b_j) \\ &\leqslant \rho(i, j-k) + \beta_{v_3}(b_{j-k+1} \cdots b_j) < r^*(i, j).\end{aligned}$$

Now, proceed to case (iii), where

$$r^*(i, j) = r(i-l, j-k) + \gamma_{v_3}(a_{i-l+1} \cdots a_i, b_{j-k+1} \cdots b_j),$$

and assume $\rho(i-l, j-k) < r(i-l, j-k)$. First, assume an alignment for $\rho(i-l, j-k)$ such that

$$\rho(i-l, j-k) = \rho(i-l-q, j-k) + \beta_{v_1}(a_{i-l-q+1} \cdots a_{i-l}).$$

Then

$$\begin{aligned}\rho(i-l-q, j-k) &+ \gamma_{v_2}(a_{i-l-q+1} \cdots a_i, b_{j-k+1} \cdots b_j) \\ &\leqslant \rho(i-l, j-k) + \gamma_{v_3}(a_{i-l+1} \cdots a_i, b_{j-k+1} \cdots b_j) < r^*(i, j).\end{aligned}$$

The last situation is for

$$\begin{aligned}\rho(i-l, j-k) = \rho(i-l-q, j-k-p)& \\ + \gamma_{v_1}(a_{i-l-q+1} &\cdots a_{i-l}, b_{j-k-p+1} \cdots b_{j-k}).\end{aligned}$$

Then

$$\begin{aligned}\rho(i-l-q, j-k-p) &+ \gamma_{v_2}(a_{i-l-q+1} \cdots a_i, b_{j-k-p+1} \cdots b_j) \\ &\leqslant \rho(i-l, j-k) + \gamma_{v_3}(a_{i-l+1} \cdots a_i, b_{j-k+1} \cdots b_j) < r^*(i, j).\end{aligned}$$

This completes the proof.

It is possible to obtain results analogous to Corollaries 3.1 and 3.2 and to Theorem 3.2. Those results were to reduce computation in $F_1^*(\mathbf{a})$. But for $F_2^*(\mathbf{a})$ much more serious difficulties are encountered, and a discussion to make computation feasible will be given later in this section. Presently, the discussion will be aimed at algorithms for $F_2(\mathbf{a})$.

The first task is to generalize the hairpin matrix H.

DEFINITION 6.4. Let the sequence $\mathbf{a} = a_1 \cdots a_n$ be given along with regular functions $(p, \alpha, \beta, \gamma, \zeta)$, a join function ξ, and a replacement function r. Define *the second order hairpin matrix* $H^{(2)}$ to be the symmetric $n \times n$ matrix with $h_{ii}^{(2)} = 0$ for $1 \leqslant i \leqslant n$ and, for $1 \leqslant i < j \leqslant n$,

$$h_{ij}^{(2)} = \min_{S \in \mathscr{S}_{ij}} F_2(S),$$

where

$\mathscr{S}_{ij} = \{S(a_i \cdots a_j): S = S(a_i \cdots a_j)$ is no more than second order, w a loop implies w has at least m elements, z a ladder implies $\alpha(z) \leqslant \delta < 0$. If S is first order, S has exactly one loop and $p(a_i, a_j) = 1$. If S is second order, the structure corresponding to $A^{(1)}$ has exactly one loop and $p(a_i, a_j) = 1\}$.

If $\mathscr{S}_{ij} = \varnothing$, $h_{ij}^{(2)} = K$.

Theorem 4.1 handled the algorithm for computing H. The modification of Theorem 4.1 is easily made for $H^{(2)}$.

THEOREM 6.3. *Theorem* 4.1 *holds for* $H^{(2)}$ *if* $\tau \in \{\beta, \gamma, \zeta\}$ *is replaced by* $\min\{\tau_0, \tau_1\}$ *and* $h_{1,n}$ *is replaced by* $h_{1,n}^{(2)}$.

It is also easy to generalize Theorem 5.2 and Corollary 5.1. Only the generalization of the corollary is given here.

THEOREM 6.4. *Assume* $\xi \equiv 0$ *in Theorem* 6.2. *Then*

$$F_2(a_1 \cdots a_i) = F_2(i) = \min\{F_2(i-j) + h_{i-j+1,i}^{(2)}: 1 \leqslant j \leqslant i\}.$$

Theorem 6.4 completes the general discussion of the best second order structure. The next section will handle the problem of best Nth order secondary structures.

Now the computational problems associated with Theorem 6.2 are taken up. Results corresponding to Corollary 3.2 and Theorem 3.2 no longer hold for second order structures. Even more important is the computation of β_1 and γ_1. Attention is now restricted to a smaller class of first order bulges and interior loops.

DEFINITION 6.5. Let $\mathscr{S}^{**} = \mathscr{S}^{*} \cap \{S(a): b_1 \cdots b_k$ a bulge, join, or tail in $A^{(1)}$ implies $b_1 \cdots b_k$ has no more than one loop in $A^{(2)}\}$. Then define F_2^{**} by

$$F_2^{**}(a) = \min_{S(a) \in \mathscr{S}^{**}} F_2(S(a)).$$

An $n \times n$ symmetric matrix H^β, called the *hairpin bulge matrix*, is defined by

$$h_{ij}^\beta = \min\{h_{kl} + \beta_1(a_i \cdots a_j): i \leqslant k \leqslant l \leqslant j\}.$$

Similarly, an $(n \times n)^2$ symmetric matrix H^γ called the *hairpin interior loop matrix* is defined by

$$h^\gamma(i, j; k, l) = \min\{h_{i'j'} + h_{k'l'} + \gamma_1(a_i \cdots a_j, a_k \cdots a_l): \\ i \leqslant i' \leqslant j' \leqslant j \text{ and } k \leqslant k' \leqslant l' \leqslant j\}.$$

Computation of H^β is now considered. The assumptions on β are r are consistent with the literature.

THEOREM 6.5. *Let the assumptions of Theorem 6.2 hold. Assume*

$$\beta(c_1 \cdots c_k) = \beta(k) \qquad \textit{independent of} \quad c_1 \cdots c_k,$$

$$\beta \qquad \text{is a strictly increasing function,}$$

and

$$r(a, b) = r \qquad \textit{for all} \quad (a, b) \quad \textit{such that} \quad p(a, b) = 1.$$

Then, if $\min_{i \leqslant k \leqslant l \leqslant j} h_{kl} = h_{i_0 j_0}$, (i) *and* (ii) *below hold*:

(i) If

$$h_{i_0 j_0} + \beta(r + j + i_0 - i - j_0) - \min_{\substack{i \leqslant k \leqslant l \leqslant j \\ (k,l) \neq (i_0, j_0)}} h_{kl} - \beta(r) \leqslant 0,$$

then

$$h_{ij}^\beta = h_{i_0 j_0} + \beta(r + j + i_0 - i - j_0).$$

(ii) *If*

$$h_{ij}^\beta = h_{i'j'} + \beta(r + j + i' - i - j'),$$

then

$$j_0 - i_0 \leqslant j' - i'.$$

Proof. (i) The hairpin weight for $i_0 \cdots j_0$ is

$$h_{i_0 j_0} + \beta(r + j + i_0 - i - j_0),$$

and the best possible weight for any remaining structure is

$$\min_{(k,l) \neq (i_0, j_0)} h_{kl} + \beta(r) = h' + \beta(r).$$

The conclusion of (i) holds if

$$h_{i_0 j_0} + \beta(r + j + i_0 - i - j_0) \leqslant h' + \beta(r).$$

(ii) Assume the conclusion is false. Then

$$j' - i' < j_0 - i_0,$$

or

$$i_0 - j_0 < i' - j',$$

and

$$\beta(r + j - i + i_0 - j_0) < \beta(r + j - i + i' - j').$$

Then

$$h_{i_0 j_0} + \beta(r + j - i + i_0 - j_0) < h_{i'j'} + \beta(r + j - i + i' - j'),$$

which is a contradiction.

Next it is seen that if β is linear for arguments greater than or equal to r then H^β can easily be computed.

THEOREM 6.6. *Let the assumptions of Theorem 6.5 hold and suppose*

$$\beta(r + k) = \beta(r) + s_\beta k \qquad \text{where} \quad s_\beta > 0.$$

Then

$$h^\beta_{i,\, j+1} = \min\{h^\beta_{ij} + s_\beta;\, h_{k,\, j+1} + \beta(r + k - i) : h_{k,\, j+1} < 0,\, i \leqslant k \leqslant j\}.$$

Proof. Let

$$\begin{aligned} h^\beta_{i,j} &= h_{i_0 j_0} + \beta(r + j - i + i_0 - j_0) \\ &= h_{i_0 j_0} + \beta(r) + s_\beta(j - i + i_0 - j_0). \end{aligned}$$

Now, if

$$\begin{aligned} h^\beta_{i,\, j+1} &= h_{i'j'} + \beta(r + 1 + j - i + i' - j') \\ &= h_{i'j'} + \beta(r) + s_\beta(j - i + i_0 - j_0) + s_\beta, \end{aligned}$$

where $i \leqslant i' \leqslant j' \leqslant j$, then $i' = i_0$ and $j' = j_0$. The other possibilities are covered by $j' = j + 1$ and are included in the above minimization.

It is now clear that H^β can easily and efficiently be computed. However, H^γ is a very large computational job. By making another linearity assumption, H^γ can be computed from the algorithm for H^β.

THEOREM 6.7. *Let the assumptions of Theorem 6.2 hold and assume*

$\gamma(c_1 \cdots c_k, d_1 \cdots d_l) = \gamma(k + l)$ independent of $c_1 \cdots c_k$ and $d_1 \cdots d_l$;
γ is a strictly increasing function in each argument;
$r(a, b) = r$ for (a, b) such that $p(a, b) = 1$; and
$\gamma(r + k) = \gamma(r) + s_\gamma k$ where $s_\gamma > 0$.

Then, if $h_{ij}^{\beta *} = \min\{h_{i'j'} + s_\gamma(j - i + i' - j') : h_{i'j'} < 0\}$,

$$h^\gamma(i, j; k, l) = \min\{\gamma(r) + s_\gamma r + h_{ij}^{\beta *} + h_{kl}^{\beta *}, \gamma(r) + s_\gamma(l - k + 1) + h_{ij}^{\beta *}, \gamma(r) + s_\gamma(i - j + 1) + h_{kl}^{\beta *}, \gamma(j - i + 1 + l - k + 1)\}.$$

Proof. Assume $h^\gamma(i, j; k, l)$ has a hairpin in $a_i \cdots a_j$ or $a_k \cdots a_l$. Then, in the case both have hairpins,

$$\begin{aligned} h^\gamma(i, j; k, l) &= h_{i_0 j_0} + h_{k_0 l_0} + \gamma_1(a_i \cdots a_j, a_k \cdots a_l) \\ &= h_{i_0 j_0} + h_{k_0 l_0} + \gamma(2r + j - i + i_0 - j_0 + l - k + k_0 - l_0) \\ &= h_{i_0 j_0} + h_{k_0 l_0} + \gamma(r) + s_\gamma(r + j - i + i_0 - j_0 + l - k + k_0 - l_0) \\ &= \gamma(r) + s_\gamma r + (h_{i_0 j_0} + s_\gamma(j - i + i_0 - j_0)) \\ &\quad + (h_{k_0 l_0} + s_\gamma(l - k + k_0 - l_0)). \end{aligned}$$

In case exactly one has a hairpin, say $a_i \cdots a_j$,

$$\begin{aligned} h^\gamma(i, j; k, l) &= h_{i_0 j_0} + \gamma_1(a_i \cdots a_j, a_k \cdots a_l) \\ &= h_{i_0 j_0} + \gamma((r + j - i + i_0 - j_0) + (l - k + 1)) \\ &= \gamma(r) + (h_{i_0 j_0} + s_\gamma(j - i + i_0 - j_0)) + (0 + s_\gamma(l - k + 1)). \end{aligned}$$

In case neither have hairpins,

$$h^\gamma(i, j; k, l) = \gamma_1(a_i \cdots a_j, a_k \cdots a_l) = \gamma(j - i + 1 + l - k + 1).$$

Now H^β can be used in steps (ii) and (iii) of Theorem 6.2 and H^γ in step (iv) to compute $F_2^{**}(\mathbf{a})$.

EXAMPLE. The example $\mathbf{a}$ = *ggguaunnnauagggnnncccauannnuauccc* given at the end of Section 3 and 5 has $F_2{}^*(\mathbf{a}) = -12$ if $R(a_1 \cdots a_k) = 1$. (That is, any hairpin is replaced by a sequence of length 1.) The structure for $F_2{}^*(\mathbf{a})$ is a cloverleaf and is shown in Fig. 6.1.

Ladder regions for the second order hairpin:

Base number	2 pairs with base number 34	:	*gc*
Base number	3 pairs with base number 33	:	*gc*
Base number	4 pairs with base number 32	:	*gc*

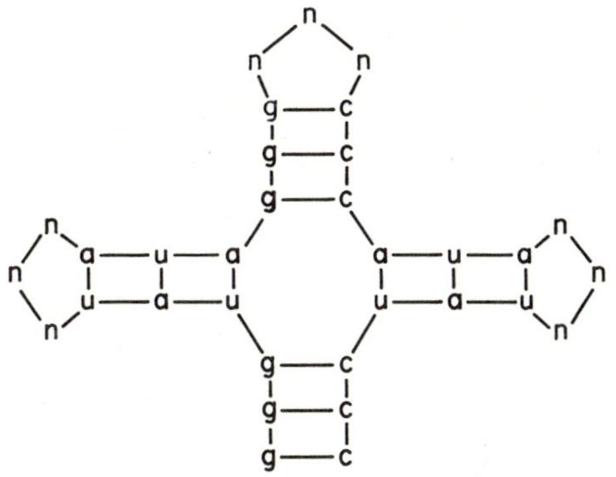

FIG. 6.1. Best second order secondary structure for **a**.

The loop has a hairpin from 5 to 13,

Base number	5 pairs with base number 13	:	*ua*
Base number	6 pairs with base number 12	:	*au*
Base number	7 pairs with base number 11	:	*ua*

another hairpin from 23 to 31,

Base number	23 pairs with base number 31	:	*au*
Base number	24 pairs with base number 30	:	*ua*
Base number	25 pairs with base number 29	:	*au*

and a final hairpin from 14 to 22.

Base number	14 pairs with base number 22	:	*gc.*
Base number	15 pairs with base number 21	:	*gc.*
Base number	16 pairs with base number 20	:	*gc.*

7. *N*th ORDER SECONDARY STRUCTURES

Finally, Nth order secondary structures are considered. This section is a direct generalization of Section 6. The structures of this section are not easy to visualize, but hopefully the work on second order structures provides a natural motivation. Also, at the conclusion of this section, a result is given to characterize the solution of the best secondary structure (of any order) for a given sequence. Due to the previous work of Section 6, the results of this section will be briefly stated and proofs omitted. It is assumed throughout that $N \geqslant 2$.

First, the components of an Nth order structure are classified.

DEFINITION 7.1. Let A be the adjacency matrix for an Nth order secondary structure and suppose $A = A^{(0)} \to A^{(1)} \to \cdots \to A^{(N-1)} \to A^{(N)}$ as in Definition 2.3. $A^{(N-1)}$ is first order and let LO be the set of loops, LA_0 the set of ladders, IL the set of interior loops, B' the set of bulges, J the set of joins, and T the set of tails, and let $B = B' \sim IL$.

(i) Assume $a_1 \cdots a_k \in LO$. If $a_1 \cdots a_k$ is a loop in $A^{(0)}$, $a_1 \cdots a_k$ is a *zero order loop in* $A^{(0)}$. Otherwise, $a_1 \cdots a_k$ is said to be an $(N-1)$*st order loop in* $A^{(0)}$. These relations are written as $a_1 \cdots a_k \in LO_0$ or $a_1 \cdots a_k \in LO_{N-1}$, respectively.

(ii) If $a_1 \cdots a_k \in B$, then, if $a_1 \cdots a_k$ is a bulge in $A^{(0)}$, $a_1 \cdots a_k$ is a *zero order bulge in* $A^{(0)}$ or $a_1 \cdots a_k \in B_0$. Otherwise, $a_1 \cdots a_k$ is called an $(N-1)$*st order bulge in* $A^{(0)}$ or $a_1 \cdots a_k \in B_{N-1}$.

(iii) Assume $(a_1 \cdots a_k, b_1 \cdots b_l) \in IL$. If the pair is an interior loop in $A^{(0)}$, it is said to be a *zero order interior loop in* $A^{(0)}$. Otherwise, the pair is said to be an $(N-1)$*st order interior loop in* $A^{(0)}$. The relations are written as $(a_1 \cdots a_k, b_1 \cdots b_l) \in IL_0$ or $(a_1 \cdots a_k, b_1 \cdots b_l) \in IL_{N-1}$, respectively.

(iv) Assume $a_1 \cdots a_k \in J$. If this sequence is a join in $A^{(0)}$, it is said to be a *zero order join in* $A^{(0)}$. Otherwise, it is said to be an $(N-1)$*st order join in* $A^{(0)}$. The relations are written as $a_1 \cdots a_k \in a_k \in J_0$ or $a_1 \cdots a_k \in J_{N-1}$, respectively.

(v) If $a_1 \cdots a_k \in T$ and $a_1 \cdots a_k$ is also a tail in $A^{(0)}$, then $a_1 \cdots a_k$ is said to be a *zero order tail in* $A^{(0)}$. Otherwise, $a_1 \cdots a_k$ is said to be an $(N-1)$*st order tail in* $A^{(0)}$. The relations are written as $a_1 \cdots a_k \in T_0$ or $a_1 \cdots a_k \in T_{N-1}$, respectively.

(iv) LA_0 will be called the set of *zero order ladders in* $A^{(0)}$.

THEOREM 7.1. *Any* N*th order secondary structure can be uniquely decomposed into the sets* LO_0, LO_{N-1}, B_0, B_{N-1}, IL_0, IL_{N-1}, J_0, J_{N-1}, T_0, T_{N-1}, *and* LA_0.

It is clear that for $N > 2$, J_{N-1}, for example, could be further decomposed. However, this will not be necessary for the algorithm given below and is therefore omitted.

The functions ζ_1, β_1, and γ_1 are extended in the following manner. Let $a_1 \cdots a_k$ be a sequence in B_{N-1}, say. Then define

$$\beta_{N-1}(a_1 \cdots a_k) = \beta_{N-2}(R(a_1 \cdots a_k)),$$

where R is defined on $a_1 \cdots a_k$ considered as a sequence in $A^{(N-2)}$. The remaining functions ζ_{N-1} and γ_{N-1} are similarly defined.

DEFINITION 7.2. Let $\mathbf{a} \in \mathscr{A}^N$ where ξ is a join function, p, α, β, γ, ζ are regular functions, and r is a replacement function. Let $S(\mathbf{a})$ be a secondary structure of no more than Nth order. Then the *Nth order free energy for* $S(\mathbf{a})$ is defined by

$$\begin{aligned}F_N(S(\mathbf{a})) = &\sum_{w_0 \in LO_0} \zeta_0(w_0) + \sum_{w_N \in LO_{N-1}} \zeta_N(w_N) + \sum_{t \in T_{N-1}} F_{N-1}(t) \\ &+ \sum_{v_0 \in J_0} \xi_0(v_0) + \sum_{v_N \in J_{N-1}} \xi_N(v_N) + \sum_{x_0 \in B_0} \beta_0(x_0) + \sum_{x_N \in B_{N-1}} \beta_N(x_N) \\ &+ \sum_{y_0 \in IL_0} \gamma_0(y_0) + \sum_{y_N \in IL_{N-1}} \gamma_N(y_N) + \sum_{z_0 \in LA_0} \alpha_0(z_0).\end{aligned}$$

If $S(\mathbf{a})$ has order less than N, let $S(\mathbf{a}) \in T_{N-1}$. The *Nth order free energy for* $\mathbf{a}$ is given by

$$F_N(\mathbf{a}) = \min_{S \in \mathscr{S}} F_N(S),$$

where $\mathscr{S} = \{S(\mathbf{a}): S(\mathbf{a})$ is no more than Nth order, w a loop in $S(\mathbf{a})$ implies w has no more than m elements, and z a ladder in $S(\mathbf{a})$ implies $\alpha(z) \leqslant \delta\}$. If $\mathscr{S}^* = \mathscr{S} \cap \{S(\mathbf{a}): A^{(N-1)}$ has no more than one loop$\}$, then define

$$F_N{}^*(\mathbf{a}) = \min_{S \in \mathscr{S}^*} F_N(S).$$

The proof of the next theorem follows that of Theorem 6.2.

THEOREM 7.2. *Let* $\mathbf{a} \in \mathscr{A}^n$ *and assume* $p, \alpha, \beta, \gamma, \zeta$ *are regular,* ξ *is a join function,* r *is a replacement function, and* δ, m *are as in the definition of* $F_N(\mathbf{a})$. *Define*

$$\begin{aligned} b_j &= a_{n-j+1}, & 1 \leqslant j \leqslant n, \\ \rho(0, j) &= \rho(i, 0) = 0, & 0 \leqslant i, j \leqslant n. \end{aligned}$$

Then inductively define $\rho(i,j)$ *for* $0 < ij$, $i + j \leqslant n - m$ *to be the minimum of zero and*:

(i)

$$\rho(i-k, j-k) + \alpha(a_{i-k+1} \cdots a_i, b_{j-k+1} \cdots b_j),$$

where $1 \leqslant k \leqslant \min\{i, j\}$ *and* $\alpha(a_{i-k+1} \cdots a_i, b_{j-k+1} \cdots b_j) \leqslant \delta$;

(ii)

$$\rho(i-k, j) + \min_{v=0, N} \beta_v(a_{i-k+1} \cdots a_i),$$

where $1 \leqslant k \leqslant i-1$ *and some alignment for* $\rho(i-k, j)$ *ends in a pair*;

(iii)

$$\rho(i, j-k) + \min_{v=0, N} \beta_v(b_{j-k+1} \cdots b_j),$$

where $1 \leqslant k \leqslant j-1$ *and some alignment for* $\rho(i, j-k)$ *ends in a pair*; and (iv)

$$\rho(i-k, j-l) + \min_{v=0, N} \gamma_v(a_{i-k+1} \cdots a_i, b_{j-l+1} \cdots b_j),$$

where $1 \leqslant k \leqslant i-1, 1 \leqslant l \leqslant j-1$, *and some alignment for* $\rho(i-k, j-l)$ *ends in a pair.*

Then

$$F_N{}^*(\mathbf{a}) = \min\{\rho(i, j) + \min_{v=0, N} \zeta_v(a_i b_j; a_{i+1} \cdots a_{n-j}):$$
$$i + j \leqslant n - m \text{ and some alignment for } \rho(i, j) \text{ ends in a pair}\},$$

where $F_N{}^*(\mathbf{a}) = 0$ *if the above set is empty.*

DEFINITION 7.3. Define the *Nth order hairpin matrix* $H^{(N)}$ to be the symmetric $n \times n$ matrix with

$$h_{ij}^{(N)} = \min_{S \in \mathscr{S}_{ij}} F_N(S), \qquad 1 \leqslant i \leqslant j' \leqslant n,$$

where $\mathscr{S}_{ij} = \{S(a_i \cdots a_j) = S : S$ is no more than Nth order, if w is a loop then w has at least m elements, if z is a ladder then $\alpha(z) \leqslant \delta < 0$. If S is first order, S has exactly one loop and $p(a_i, a_j) = 1$. If S is order $O > 1$, the structure corresponding to $\mathscr{A}^{(O-1)}$ has exactly one loop and $p(a_i, a_j) = 1$. If $\mathscr{S}_{ij} = \varnothing$, $h_{ij}^{(N)} = 0.\}$

THEOREM 7.3. *Theorem* 4.1 *holds for* $H^{(N)}$ *if* $\tau \in \{\beta, \gamma, \zeta\}$ *is replaced by* $\min\{\tau_0, \tau_N\}$ *and* h_{1n} *is replaced by* $h_{1n}^{(N)}$.

THEOREM 7.4. *Assume* $\xi = 0$ *in Theorem* 7.2. *Then*

$$F_N(a_1 \cdots a_i) = F_N(i) = \min\{F_N(i-j) + h_{i-j+1, i}^{(N)} : 1 \leqslant j \leqslant i\}.$$

Finally, the order of the best structure for **a** is characterized in terms of the algorithm.

THEOREM 7.5. *The order* N *of* $S(\mathbf{a})$ *such that* $F_N(S(\mathbf{a})) = \min_M F_M(\mathbf{a})$ *satisfies*

$$N = \min\{M : H^{(M)} = H^{(M+1)}\}.$$

8. CONCLUSION

A general algorithm for the evaluation of free energy can be obtained from the above work. (i) First, set $f = 0$. (ii) Then evaluate all hairpins in the structure and add the free energy to f. (iii) Replace the evaluated hairpins using the replacement functions. (iv) If the new structure is unpaired, f is the free energy of the original structure. Otherwise go to (ii).

If replacement functions are found to be the wrong approach, then (iii) can be altered. For example, it might be necessary to have a loop function of order 1, 2, ... The approach used in this paper, then, is easy to generalize to fit more accurate models. It does seem entirely adequate for current information.

A computer program is being developed, and a preliminary version was used for the naive example given above. As an example of its power, a portion of the first order loop in the Min Jou flower model [18] is examined using the same (combinatorial) energy assignments given in the earlier example. The sequence of length 68 is

$$\mathbf{a} = ucaaacgacgcuaacgacucccuuagcccaaagg$$

$$\cdot\ uagaaaaucccucuggaacguaacggaauuguuau,$$

which composes a first order structure of three loops in the flower model. There are 21 bases in the joins for this structure, and 16 base pairs (including two *gu* pairs and two ladders with only one base pair). The algorithm (with parameters as specified earlier) predicts a best single loop hairpin of $F_1{}^*(a) = -12$ with 12 bases in the tails, and 17 base pairs with *no gu* pairs. The best first order structure of $F_1(a) = -12$ has 12 bases in tails and joins and 17 base pairs. Of course, more precise energy functions would make the comparison more realistic, but the two structures given above seem to improve on that portion of the flower model. The three structures are displayed in Fig. 8.1.

A study is being planned in which good estimates of the energy functions will be used to study a set of tRNAs. Of course, that work will be useful in a final evaluation of this paper, and the results will be compared with other studies. We conjecture that, except for minor modifications, the algorithms proposed are the most efficient possible for determination of secondary structure on computers.

While the ladder function α given in this paper is strictly additive, it is important to note that it can be modified to handle general α. In particular, nearest neighbor effects [24] can be treated by adding a (negative) free energy for adjacent bonds. This allows the use of the best estimates of the energy functions.

Finally, the connection between the graph theory of secondary structure of single stranded nucleic acids and that for secondary structure in single stranded proteins is currently being considered.

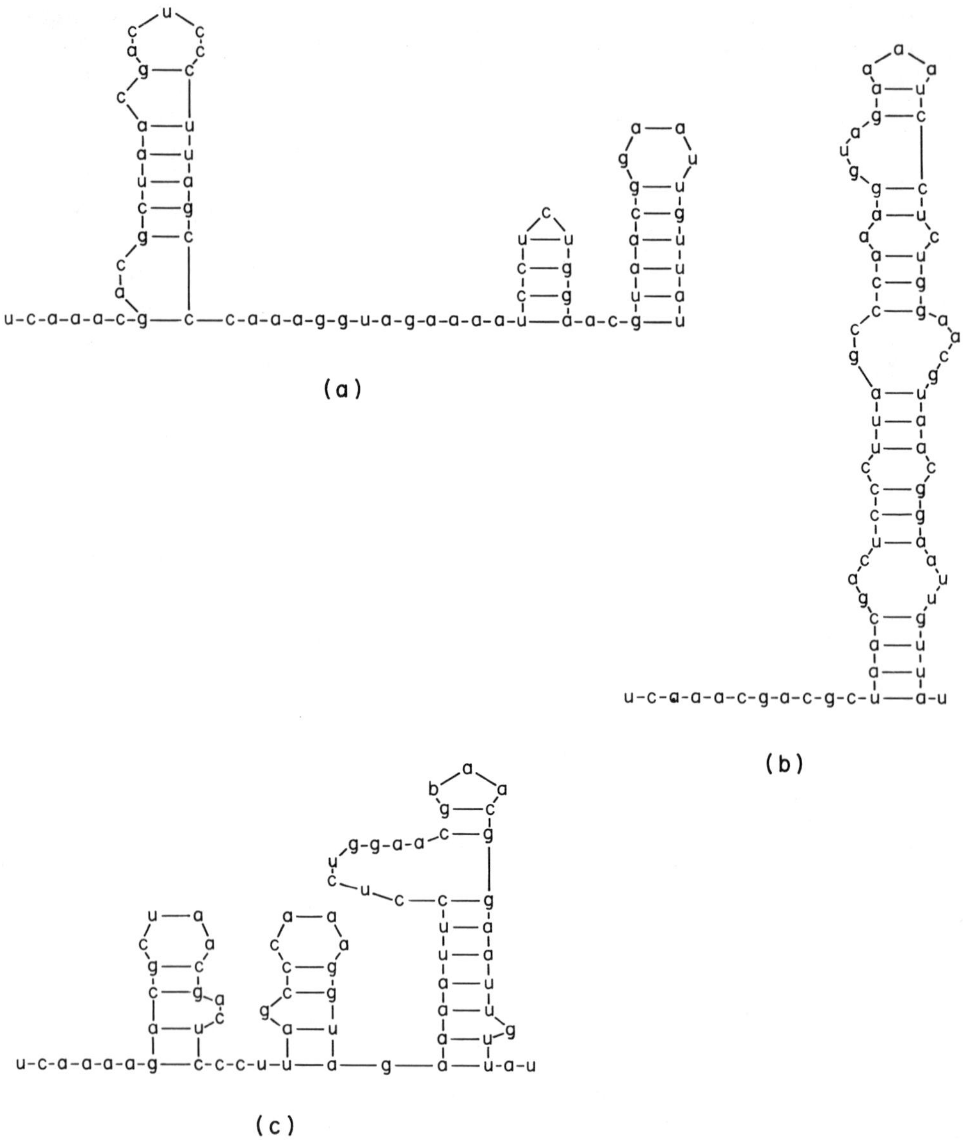

FIG. 8.1. Secondary structures for a portion of the coat protein gene. (a) Structure appearing in the flower model, (b) best hairpin, and (c) best first order structure.

ACKNOWLEDGMENTS

The author wishes to express his appreciation to Charles DeLisi of the National Cancer Institute who initially proposed the problem of secondary structure to me in such clear and straightforward terms and who provided much insight, encouragement, and advice during the course of this investigation. Thanks are also due William Beyer and Paul Stein of Los Alamos Scientific Laboratory and Temple F. Smith of Northern Michigan University for helpful discussions.

REFERENCES

1. V. A. BLOOMFIELD, D. M. CROTHERS, AND I. TINOCO, "Physical Chemistry of Nucleic Acids," Harper, New York, 1974.
2. N. G. DEBRUIJN AND P. ERDOS, Some linear and some quadratic recursion formulas II, *Indag. Math.* **55** (1952), 152–163.
3. C. DELISI AND D. M. CROTHERS, Prediction of RNA secondary structure, *Proc. Nat. Acad. Sci. U.S.A.* **68** (1971), 2682–2685.
4. C. DELISI, Conformational changes in transfer RNA. I. Equilibrium theory, *Biopolymers* **12** (1973), 1713–1728.
5. W. FELLER, "An Introduction to Probability Theory and Its Applications. Volume I" (3rd ed.), Wiley, New York 1967.
6. T. R. FINK AND D. M. CROTHERS, Free energy of imperfect nucleic acid helices. I. The bulge defect, *J. Mol. Biol.* **66** (1972), 1–12.
7. W. M. FITCH, Considerations regarding the regulation of gene transcription and messenger translation, *J. Mol. Evol.* **1** (1972), 185–207.
8. W. M. FITCH, The large extent of putative secondary nucleic acid structure in random nucleotide sequences or amino acid derived messenger-RNA, *J. Mol. Evol.* **3** (1974), 279–291.
9. J. R. FRESCO, B. M. ALBERTS, AND P. DOTY, Some molecular details of the secondary structure of ribonucleic acid, *Nature* **188** (1960), 98–101.
10. J. R. FRESCO, L. C. KLOTZ, AND E. G. RICHARDS, A new spectroscopic approach to the determination of helical secondary structure in ribonucleic acids, *Cold Spring Harbor Symp. Quant. Biol.* **28** (1963), 83.
11. J. GRALLA AND D. CROTHERS, Free energy of imperfect nucleic acid helices, II. Small hairpin loops, *J. Mol. Biol.* **73** (1973), 497–517.
12. J. GRALLA AND D. CROTHERS, Free energy of imperfect nucleic acid helices, III. Small internal loops resulting, *J. Mol. Biol.* **78** (1973), 301–319.
13. J. GRALLA AND C. DELISI, M-RNA is expected to form stable secondary structures, *Nature* **248** (1974), 330–332.
14. R. W. HOLLEY, J. APGAR, G. A. EVERETT, J. T. MADISON, M. MARQUISEE, S. H. MERRILL, J. R. PENSWICH, AND A. ZAMER, Structure of a ribonucleic acid, *Science* **147** (1965), 1462–1465.
15. W. N. HSIEH, Proportions of irreducible diagrams, *Studies in Applied Math.* **52** (1973), 277–283.
16. D. J. KLEITMAN, Proportions of irreducible diagrams, *Studies in Applied Math.* **49** (1970), 297–299.
17. A. J. MARK AND J. A. PETRUSKA, A method for predictory RNA structure from amino-acid sequence data, *J. Mol. Biol.* **72** (1972), 609–617.
18. W. MIN JOU, G. HAEGEMAN, M. YSEBAERT, AND W. FIERS, Nucleotide sequence of the gene coding for the bacteriophage MS2 coat protein, *Nature* **237** (1972), 82–88.

19. J. M. Pipas and J. E. McMahon, Method for predicting RNA secondary structure, *Proc. Nat. Acad. Sci. U.S.A.* **72** (1975), 2017–2021.
20. I. Tinoco, O. C. Uhlenbeck, and M. D. Levine, Estimation of secondary structure in ribonucleic acids, *Nature* **230** (1971), 362–367.
21. O. C. Uhlenbeck, P. N. Borer, B. Dengler, and I. Tinoco, Stability of RNA hairpin loops: A_6–C_m–U_6, *J. Mol. Biol.* **73** (1973), 483–496.
22. M. S. Waterman, T. F. Smith, and W. A. Beyer, Some biological sequence metrics, *Advances in Math.* **20** (1976), 367–387.
23. A. M. Lesk, A combinatorial study of the effects of admitting non-Watson–Crick base pairing and of base composition on the helix-forming potential of polynucleotides of random sequences, *J. Theor. Biol.* **44** (1974), 7–17.
24. D. Poland and H. A. Scheraga, "Theory of Helix-Coil Transitions in Biopolymers," Academic Press, New York, 1970.

AMS (MOS) subject classifications: 05A15, 05C99, 68A10, 92A05.

Limits of Zeros of Recursively Defined Families of Polynomials

S. BERAHA, J. KAHANE, AND N. J. WEISS†

Department of Mathematics
Queens College, City University of New York
Flushing, New York

1. INTRODUCTION

Let $\{P_n(z), n = 0, 1, 2, \ldots\}$ be a sequence of polynomials with complex coefficients satisfying the recursion of degree k

$$P_{n+k}(z) = -\sum_{j=1}^{k} f_j(z)P_{n+k-j}(z), \tag{1}$$

where the f_j are polynomials.

We define the complex number x to be a *limit of zeros* of the sequence $\{P_n\}$ if there is a sequence $\{z_n\}$ such that $P_n(z_n) = 0$ and $z_n \to x$ as $n \to \infty$. Note that a limit of zeros is not merely the limit of a sequence of zeros.

In this paper we give necessary and sufficient conditions, subject to two mild nondegeneracy conditions, that x be a limit of zeros. These conditions are given in terms of the coefficients f_j and the starting polynomials $P_0, P_1, \ldots, P_{k-1}$.

The next section contains a statement of the main result. Section 3 consists of preliminaries and is centered around some simple facts about algebraic functions. The main result is proved in Section 4, and the extent to which the result can be modified when the nondegeneracy conditions fail is discussed in Section 5. Section 6 describes an application to families of polynomials that arise in map-coloring problems.

We conclude this section with some remarks about the concept "limit of zeros." If x is a limit of zeros of $\{P_n\}$, then either $P_n(x) = 0$ for all sufficiently large n or x is an accumulation point of $Z(\{P_n\})$, the union of the zeros of the P_n, but the converse is not a priori true. As a byproduct of our main theorem we will see that, subject to our nondegeneracy conditions, a point which is a zero of P_n for large n or an accumulation point of $Z(\{P_n\})$ actually is a limit of zeros.

† Research partially supported by National Science Foundation Grant GP-31416.

ISBN 0-12-599101-0

It is routine to check that the statement that x is a limit of zeros is equivalent to the assertion that for every positive ϵ there is an N such that for all $n \geqslant N$, $P_n(z_n) = 0$ for some $z_n \in D_\epsilon(x) = \{z : |z - x| < \epsilon\}$. Finally, from this characterization of limits of zeros it follows immediately that x is a limit of zeros if there are limits of zeros arbitrarily close to x.

2. Statement of the Main Result

It is well known that for any fixed z, the solution of the recursion (1) given $P_0(z), \ldots, P_{k-1}(z)$ depends on the roots of the characteristic equation

$$Q_z(\lambda) = \lambda^k + \sum_{j=1}^{k} f_j(z)\lambda^{k-j} = 0. \tag{2}$$

Let these roots be $\lambda_1(z), \ldots, \lambda_k(z)$, with possible repetitions, where the numbering is temporarily arbitrary for each z. We call the $\lambda_j(z)$ the *root values* of z, and refer to the functions $z \to \lambda_j(z)$ as *roots*. If the $\lambda_j(z)$ are distinct for a particular z, then

$$P_n(z) = \sum_{j=1}^{k} \alpha_j(z)\lambda_j(z)^n. \tag{3}$$

If there are repeated root values at z, (3) is modified in the usual way, e.g., if $\lambda_1(z) = \lambda_2(z) \neq \lambda_j(z)$, $j > 2$, the term $\alpha_1\lambda_1{}^n + \alpha_2\lambda_2{}^n$ is replaced by a term $\alpha_1\lambda_1{}^n + n\alpha_2\lambda_1^{n-1}$. The $\alpha_j(z)$ are determined in any event by solving the k linear equations in the $\alpha_j(z)$ obtained by letting $n = 0, 1, \ldots, k-1$ in (3) or its variant.

We turn now to the nondegeneracy conditions required for our result.

A. $\{P_n\}$ does not satisfy a recursion of order less than k.

B. For no possible ordering of the λ_j is it true that $\lambda_1(z) \equiv \omega\lambda_2(z)$ for a constant ω with $|\omega| = 1$.

Condition B eliminates certain cases in which the $f_j(z)$ are constant together with other cases which can be called "constant coefficient type." If, for example, $k = 2$ and $\lambda_1(z) \equiv \omega\lambda_2(z)$, then setting $R_n(z) = P_n(z)/\lambda_2(z)^n$ we see that $\{R_n\}$ satisfies the constant coefficient recursion $R_{n+2} = (1 + \omega)R_{n+1} - \omega R_n$. The significance of the unimodularity of ω will appear below. Also, we shall see that A and B are (essentially) always satisfied when $Q_z(\lambda)$ is irreducible as a polynomial in z and λ.

Now we can state our main result.

Theorem. *Suppose that* $\{P_n(z)\}$ *satisfies* (1) *and the nondegeneracy requirements* A *and* B.

Then x is a limit of zeros of $\{P_n\}$ if and only if the $\lambda_j(x)$ can be ordered so that one of the following holds:

(i) $|\lambda_k(x)| > |\lambda_j(x)|$, $1 \leqslant j \leqslant k-1$, and $\alpha_k(x) = 0$.
(ii) $|\lambda_1(x)| = |\lambda_2(x)| = \cdots = |\lambda_l(x)| > |\lambda_j(x)|$, $l+1 \leqslant j \leqslant k$, for some $l \geqslant 2$.

We call a root *dominant* at z when its modulus is not exceeded by the modulus of any other root at z. The condition that x be a limit of zeros may be rephrased by saying that there is a single dominant root at x whose coefficient in (3) or its variant vanishes or there is more than one dominant root at x. It will appear that $\alpha_j(z)$ can have only finitely many zeros and that equimodularity of roots holds along entire curves; the two types of limits of zeros are thus referred to as *isolated* and *nonisolated.* The determination of nonisolated limits of zeros depends only on the λ_j, i.e., on the f_j, whereas determination of isolated limits of zeros depends on $P_0, \ldots, P_{k-1}$ as well.

We give now two simple examples to show that the nondegeneracy conditions A and B are necessary if the theorem is to hold. If $P_{n+2}(z) = (1+z)P_{n+1}(z) - zP_n(z)$, $P_0(z) = 1$, $P_1(z) = z$, then $P_n(z) = z^n$, so 0 is the only limit of zeros. But the roots are $\lambda_1(z) = z$, $\lambda_2(z) = 1$, and $\alpha_2(z) \equiv 0$, so (i) is satisfied for all z with $|z| < 1$. This shows that condition A is needed. (Note that $\{P_n\}$ satisfies the lower order recursion $P_{n+1} = zP_n$.)

If $P_{n+2}(z) = P_n(z)$, then $Z(\{P_n\})$ and thus the set of limits of zeros is finite. Since $\lambda_j(z) = \pm 1$, $j = 1, 2$, for every z, (ii) is satisfied for every z, and so we see that condition B is needed.

Further analysis of the conditions A and B will appear in Section 3 and Section 5.

3. Preliminaries

If $\lambda(z)$ is analytic in some neighborhood U in the complex plane and there is a polynomial $Q(z, \lambda) = Q_z(\lambda)$ such that $Q_z(\lambda(z)) \equiv 0$ in U, then λ is said to be an *algebraic function.* (For the simple facts about algebraic functions used here and below see, e.g., [1, Chapter VI].) Up to a constant multiple there is a unique $Q_z(\lambda)$ irreducible in z and λ such that $Q_z(\lambda(z)) \equiv 0$ for a given algebraic function $\lambda(z)$.

Conversely, given a polynomial $Q_z(\lambda)$ there are corresponding algebraic functions. (The discussion to follow could be made more elegant by using Riemann surfaces, but a cruder approach will suffice.) Let x be an arbitrary complex number and suppose that $Q_z(\lambda)$ is of degree k in λ. There is a disk U centered at x such that $Q_z(\lambda)$ has k distinct roots at every point of $U - \{x\}$. Let U' be U with a slit at x; for the sake of definiteness, we take

$$U' = \{z \in U : z \neq x, 0 < \arg(z - x) < 2\pi\}. \tag{4}$$

There are functions $\lambda_1(z), \ldots, \lambda_k(z)$ analytic in U' such that $Q_z(\lambda_j(z)) \equiv 0$ in U', and the λ_j are analytically continuable in all of $U - \{x\}$, with the relation $Q_z(\lambda_j(z)) \equiv 0$ persisting. If we define λ_j on the slit except at x by

$$\lambda_j(z) = \lim \lambda_j(\zeta) \qquad \text{as} \qquad \zeta \to z, \quad \arg(\zeta - x) \to 0, \tag{5}$$

then the roots of $Q_z(\lambda)$ for $z \in U - \{x\}$ are exactly the values $\lambda_1(z), \ldots, \lambda_k(z)$. If, moreover, the leading coefficient in $Q_z(\lambda)$ as a polynomial in λ is 1, which we assume from here on, then the λ_j can be defined on all of U so that they are continuous at x, and the roots of $Q_x(\lambda)$ are $\lambda_1(x), \ldots, \lambda_k(x)$.

If analytic continuation of some λ_j once around x (more precisely, once in, say, the counterclockwise direction around a circle centered at x) leads back to λ_j, then λ_j is analytic in all of U, and this is always the case if the k roots of $Q_x(\lambda)$ are distinct. Otherwise, repeated continuation around x of, say, λ_1, leads to a subset of the other λ_j, which we can take to be $\{\lambda_2, \ldots, \lambda_p\}$, and then back to λ_1. In this case $\lambda_1(x) = \cdots = \lambda_p(x)$, and there is a function F analytic in a neighborhood of 0, with $F(0) = 0$, such that $\lambda_j(z) = \lambda_j(x) + F((z - x)_j^{1/p})$, $z \in U'$, $j = 1, \ldots, p$, where $(z - x)_j^{1/p}$ is one of the p branches of $(z - x)^{1/p}$ analytic on U'.

Using this last representation of λ_j, it is easy to see that an analog of the maximum modulus principle for ordinary analytic functions still holds. In particular, if $\lambda = \lambda_1(x) = \cdots = \lambda_p(x)$, then, letting C be the boundary of U,

$$|\lambda| \leqslant \max_{1 \leqslant j \leqslant p} \sup\{|\lambda_j(z)| : z \in C\}.$$

An isolated fact we shall need is that the quotient of two algebraic functions is algebraic.

Consider now the coefficients $\alpha_j(z)$. The point of the nondegeneracy condition A is that condition A is equivalent to the assertion

$$\alpha_j(z) \not\equiv 0, \qquad j = 1, \ldots, k. \tag{6}$$

Roughly speaking, if $\{P_n\}$ satisfies a lower order recursion, then some of the α_j must vanish identically since (3) then holds with a smaller k. (This is not a proof, but see Section 5.) On the other hand, if, say, $\alpha_k(z) \equiv 0$, then (3) holds with k replaced by $k - 1$. And then by solving for the $g_j(z)$ in the equations

$$\lambda_i(z)^{k-1} + \sum_{j=1}^{k-1} g_j(z)\lambda_i(z)^{k-1-j} = 0, \qquad i = 1, \ldots, k-1,$$

we find that $P_{n+k-1}(z) = -\sum_{j=1}^{k-1} g_j(z)P_{n+k-1-j}(z)$, where

$$g_1(z) = -(\lambda_1(z) + \cdots + \lambda_{k-1}(z)), \quad \ldots, \quad g_{k-1}(z) = (-1)^{k-1}\lambda_1(z) \cdots \lambda_{k-1}(z)$$

are the appropriate symmetric functions. What is not clear is that the g_j are polynomials; to show this requires some analysis.

Suppose that $\alpha_1(z) \equiv 0$. Using Cramer's rule to solve for $\alpha_1(z)$ in (3) with $n = 0, \ldots, k-1$, we find that a determinant $\delta(z)$ must vanish identically. (The lengthy calculations needed to justify what is coming are omitted.) If the $\lambda_j(z)$ are distinct, then $\delta(z) = \pm\delta_1(z) \prod_{j>i=2}^{k} (\lambda_i(z) - \lambda_j(z))$, where $\delta_1 = P_0\lambda_2 \cdots \lambda_k \pm \cdots \pm P_{k-2}(\lambda_2 + \cdots + \lambda_k) \pm P_{k-1}$, the missing terms being products of $\pm P_j$ with symmetric functions in $\lambda_2, \ldots, \lambda_k$. We assume that $\lambda_i(z) = \lambda_j(z)$ for only finitely many $z, i \neq j$, which implies that $\delta_1(z) \equiv 0$. (Otherwise, an omitted variant of this argument is needed.) Since the $\lambda_j(z)$, $j = 1, \ldots, k$, are all the roots of $Q_z(\lambda)$, the elementary symmetric functions in the expression for $\delta_1(z)$ can be written as polynomials in $\lambda_1(z)$ and the coefficients $f_j(z)$ of $Q_z(\lambda)$, polynomials which are of degree at most $k-1$ in $\lambda_1(z)$.

Thus $\delta_1(z) = R_z(\lambda_1(z)) \equiv 0$, where $R_z(\lambda)$ is of degree at most $k-1$ in λ.

If $Q_z(\lambda)$ is irreducible in z and λ, this is impossible. Otherwise, $Q_z(\lambda) = S_z(\lambda)T_z(\lambda)$, where $S_z(\lambda)$ is irreducible and $S_z(\lambda_1(z)) \equiv 0$. The other roots of $S_z(\lambda)$ must also be of the form $\lambda_j(z)$; we can take them to be $\lambda_2(z), \ldots, \lambda_r(z)$. But then $R_z(\lambda_j(z)) \equiv 0$, $1 \leqslant j \leqslant r$, implying in turn that $\alpha_j(z) \equiv 0$, $1 \leqslant j \leqslant r$. And finally, we have $P_n = \sum_{j=r+1}^{k} \alpha_j\lambda_j^{\,n}$, implying as we have seen that $\{P_n\}$ satisfies a recursion of order $k - r$ whose coefficients are elementary symmetric functions of the λ_j, $r + 1 \leqslant j \leqslant k$, which *are* polynomials because these λ_j are exactly the roots of $T_z(\lambda)$. This completes the proof of our assertion that condition A holds only if $\alpha_j(z) \not\equiv 0, j = 1, \ldots, k$.

From the preceding discussion it is easy to see more precisely why condition A is necessary for the sufficiency of (i) in the theorem. In fact, we can add a "spurious" root $\lambda_{k+1}(z)$ to any recursion for $\{P_n\}$ of order k by letting $R_z(\lambda) = Q_z(\lambda)[\lambda - \lambda_{k+1}(z)]$ and using the coefficients of $R_z(\lambda)$ to generate a recursion of order $k+1$ satisfied by $\{P_n\}$. But then $\alpha_{k+1}(z) \equiv 0$, while on the other hand $\lambda_{k+1}(z)$ can be chosen so that it is the single dominant root at one or more points which are not limits of zeros.

Notice now that everything that has been said about algebraic functions applies to the roots $\lambda_j(z)$. Since the $\alpha_j(z)$ can be determined by using Cramer's rule to solve the equations in (3) for $n = 0, \ldots, k-1$, they have the same analyticity properties as the $\lambda_j(z)$, except that $\alpha_j(z)$ may not converge as z approaches a point where there are repeated roots. (The denominator in the expressions for the $\alpha_j(z)$ is the van der Monde determinant $\Delta(z) = \pm\prod_{i<j} [\lambda_i(z) - \lambda_j(z)]$.) Nevertheless, a routine calculation with determinants shows that $\alpha_j(z) \to \alpha_j(x)$ as $z \to x$ so long as λ_j is not a repeated root at x, even if other roots are repeated.

We must also consider more carefully the behavior of $\alpha_j(z)$ near x when $\lambda_j(x)$ is a repeated root value. To be more precise, we consider $\sum_{j=1}^{p} \alpha_j(z)\lambda_j(z)^n$

in U, where U is the usual small disk about x and

$$\lambda_1(x) = \cdots = \lambda_p(x) \neq \lambda_j(x), \qquad p+1 \leqslant j \leqslant k.$$

Let U' be the slit disc defined by (4) in which the λ_j can be taken to be analytic, and let the λ_j be defined on all of U by (5) and continuity at x. Then, assuming the equality of $\lambda_1(x), \ldots, \lambda_p(x)$, we have:

LEMMA 3.1. *For $z \in U - \{x\}$, $n \geqslant p$,*

$$\sum_{j=1}^{p} \alpha_j(z)\lambda_j(z)^n = \sum_{j=1}^{p} \beta_j(z)[n!/(n-j+1)!]\lambda(z,n,j)^{n-j+1}, \tag{7}$$

where $|\beta_j(z)| \leqslant C$, $\lambda(z, n, 1) = \lambda_1(z)$, and $\lambda(z, n, j) \to \lambda_1(x)$ uniformly in n as $z \to x$, $j = 2, \ldots, p$.

Note. It follows from the continuity of $P_n(z)$ that the LHS of (7) converges to $\sum_{j=1}^{p} \beta_j[n!/(n-j+1)!]\lambda_1(x)^{n-j+1}$ as $z \to x$ for each n; the point of the lemma is the uniformity in n of the convergence of $\lambda(z, n, j)$ to $\lambda_1(x)$.

Proof. We prove the lemma in the case $k = 3$, $p = 2$, which is sufficiently typical to exemplify the proof for any k, p. Let $t(z) = \lambda_2(z) - \lambda_1(z)$. Then, suppressing the argument z, we have $\alpha_1 = d_1/td$, $\alpha_2 = d_2/td$, where $d = (\lambda_1 - \lambda_3)(\lambda_2 - \lambda_3)$,

$$d_1 = \begin{vmatrix} P_0 & 1 & 1 \\ P_1 & \lambda_2 & \lambda_3 \\ P_2 & \lambda_2^2 & \lambda_3^2 \end{vmatrix}, \qquad d_2 = \begin{vmatrix} 1 & P_0 & 1 \\ \lambda_1 & P_1 & \lambda_3 \\ \lambda_1^2 & P_2 & \lambda_3^2 \end{vmatrix}.$$

Note that $|d| = |d(z)|$ is bounded away from 0 as $z \to x$. Now $d_1 = -d_2 + te_1$, where

$$e_1 = \begin{vmatrix} P_0 & 0 & 1 \\ P_1 & 1 & \lambda_3 \\ P_2 & 2\lambda_1 + t & \lambda_3^2 \end{vmatrix},$$

and so $\alpha_1\lambda_1{}^n + \alpha_2\lambda_2{}^n = [e_1\lambda_1{}^n + d_2((\lambda_1 + t)^n - \lambda_1{}^n)/t)]/d$. By an easy mean value type result, $\{[\lambda_1(z) + t(z)]^n - \lambda_1(z)^n\}/t(z) = n\lambda(z, n)^{n-1}$, where $\lambda(z, n) \to \lambda_1(x)$ as $z \to x$ uniformly in n because $t(z) \to 0$ as $z \to x$.

Letting $\beta_1 = e_1/d$, $\beta_2 = d_2/d$, $\lambda(z, n, 2) = \lambda(z, n)$, we have established (7).

One more preliminary result is needed in the case of repeated roots. As soon as one of the λ_j fails to be single valued near x, it becomes possible that all of the α_j fail to be single valued near x. But we have the following, where U is as above.

LEMMA 3.2. *If $\lambda_j(z)$ is single valued in U, so is $\alpha_j(z)$.*

Proof. Consider the effect of analytic continuation of $\alpha_j(z)$ once around x in U. Continuation around x has the effect of acting on the $\lambda_i(x)$, $i \neq j$, by a permutation π. The effect of the continuation on the determinants in both the numerator and denominator of the expression obtained by using Cramer's rule to solve for $\alpha_j(z)$ in (3) is thus multiplication by sgn π. In particular, $\alpha_j(z)$ is taken back to itself and must be single valued.

4. PROOF OF THE THEOREM

We begin by showing that one of the conditions (i), (ii) must hold if x is a limit of zeros. The proof will be given for $k = 3$; the extension to other k will be evident. We must prove that if x is a limit of zeros and one of the roots, say λ_3, is dominant at x, then $\alpha_3(x) = 0$.

Two estimates for numerical sequences, whose routine proofs are omitted, will be needed.

LEMMA 4.1. *Suppose that $\gamma_n, \delta_n, \mu_n, v_n, \sigma_n, \lambda_j > 0$, $\gamma_n, \delta_n \leqslant M$, $n = 1, 2, 3, \ldots, j = 1, 2$, and that $\mu_n \to \lambda_1$, $v_n \to \lambda_1$, $\sigma_n \to \lambda_2$.*

(a) $\limsup_{n\to\infty}(\gamma_n\mu_n{}^n + \delta_n\sigma_n{}^n)^{1/n} \leqslant \max(\lambda_1, \lambda_2)$.
(b) $\limsup_{n\to\infty}(\gamma_n\mu_n{}^n + n\delta_n v_n^{n-1})^{1/n} \leqslant \lambda_1$.

Now to the proof. We have $x = \lim z_n$ with $P_n(z_n) = 0$ and $|\lambda_3(x)| > |\lambda_1(x)|$, $|\lambda_2(x)|$ and want to show that $\alpha_3(x) = 0$.

Case I: $\lambda_1(x) \neq \lambda_2(x)$. Take U as usual so that the λ_j are analytic in U; we can assume that the z_n are in U. Referring to (3), we see that the condition $P_n(z_n) = 0$ amounts to

$$\alpha_3(z_n)\lambda_3(z_n)^n = -[\alpha_1(z_n)\lambda_1(z_n)^n + \alpha_2(z_n)\lambda_2(z_n)^n] \tag{8}$$

and thus

$$|\alpha_3(z_n)|^{1/n}|\lambda_3(z_n)| = |\alpha_1(z_n)\lambda_1(z_n)^n + \alpha_2(z_n)\lambda_2(z_n)^n|^{1/n}. \tag{9}$$

If $\alpha_3(x) \neq 0$, the LHS of (9) converges to $|\lambda_3(x)|$ as $n \to \infty$. On the other hand, it follows from Lemma 4.1a that the lim sup of the RHS of (9) is bounded by $\max(|\lambda_1(x)|, |\lambda_2(x)|)$, a contradiction since $|\lambda_3(x)| > |\lambda_1(x)|, |\lambda_2(x)|$. Thus $\alpha_3(x)$ must vanish.

Case II: $\lambda_1(x) = \lambda_2(x)$. In this case (3) does not hold at x. If $z_n = x$ for infinitely many n, then a look at the variant of (3) that does hold at x together with the recollection that λ_3 is the single dominant root at x reveal that $\alpha_3(x) = 0$. So we may as well assume that $z_n \neq x$ for all n. Defining λ_1 and

λ_2 in $U - \{x\}$ by (5), we apply Lemma 3.1 to conclude that for $z \in U - \{x\}$,

$$\alpha_1(z)\lambda_1(z)^n + \alpha_2(z)\lambda_2(z)^n = \beta_1(z)\lambda_1(z)^n + n\beta_2(z)\lambda(z, n, 2)^{n-1}, \tag{10}$$

where the β_j are bounded and $\lambda(z, n, 2) \to \lambda_1(x)$ as $z \to x$, uniformly in n. Since $z_n \to x$, the uniformity of convergence implies that $\lambda(z_n, n, 2) \to \lambda_1(x)$ as $n \to \infty$. Applying Lemma 4.1b to (10) after taking absolute values and setting $z = z_n$, we see that the lim sup of the RHS of (9) is at most $|\lambda_1(x)|$.

Finally, $\alpha_3(x)$ is single valued in U by Lemma 3.2, and in particular is continuous at x, so the LHS of (9) again tends to $|\lambda_3(x)|$ as long as $\alpha_3(x) \neq 0$. And again we have a contradiction unless $\alpha_3(x) = 0$.

Remark. The proof just given works equally well under the milder hypothesis that either $P_n(x) = 0$ for infinitely many n or x is an accumulation point of $Z(\{P_n\})$, the set of zeros of the P_n. Therefore, once we have proved the sufficiency of conditions (i) and (ii) for x to be a limit of zeros, we shall have proved also that x satisfying the milder hypotheses are limits of zeros.

Returning to the proof of the theorem, we show that the condition (i) is sufficient to guarantee that x is a limit of zeros. Again supposing for convenience that $k = 3$, we show that if $|\lambda_3(x)| > |\lambda_1(x)|, |\lambda_2(x)|$ and $\alpha_3(x) = 0$, then x is a limit of zeros. As in the proof of necessity above, the situation is more complicated when $\lambda_1(x) = \lambda_2(x)$, but the basic idea of the proof does not change.

Case I: $\lambda_1(x) \neq \lambda_2(x)$. As was remarked in Section 1, to show that x is a limit of zeros we must show that for every $\epsilon > 0$ there is a sufficiently large N so that for every $n \geqslant N$, $P_n(z_n) = 0$, i.e., (8) holds, for $z_n \in D_\epsilon(x)$.

Taking ϵ sufficiently small, we can suppose that the λ_j are analytic in $U = D_\epsilon(x)$, and that $\lambda_3(z) \neq 0$, $z \in U$. Thus we can define

$$w(z) = w_n(z) = -[\alpha_1(z)\lambda_1{}^n(z) + \alpha_2(z)\lambda_2{}^n(z)]/\lambda_3(z)^n, \tag{11}$$

analytic in U, and are reduced to showing that for n sufficiently large, $\alpha_3(z) = w(z)$ for some $z = z_n \in U$.

By making ϵ smaller yet, if necessary, and recalling that $\alpha_3(z) \not\equiv 0$ (see (6)), we may suppose that $|\alpha_3(z)| \geqslant m > 0$ and also that $|\lambda_j(z)| / |\lambda_3(z)| \leqslant r < 1$, $j = 1, 2$, if $|z - x| = \epsilon$, i.e., if $z \in C$, the boundary of U. Now let

$$M = \sup\{\max(|\alpha_1(z)|, |\alpha_2(z)|) : z \in C\},$$

and choose N sufficiently large so that $2Mr^N < m$. If $n \geqslant N$, we see by referring to (11) that $|w(z)| < |\alpha_3(z)|$ on C, and so by Rouché's theorem [1, p. 124], $\alpha_3(z) - w(z)$ and $\alpha_3(z)$ have the same number of zeros in U. But $\alpha_3(x) = 0$, so this number is positive, proving that $\alpha_3(z) = w(z)$ for at least one $z \in U$, which was to be shown.

Case II: $\lambda_1(x) = \lambda_2(x)$. Let U be an arbitrarily small disk centered at x, which can be supposed sufficiently small so that the results of Section 3 apply. By Lemma 3.2, $\alpha_3(z)$ is single valued in U. Since $\lambda_3(z)^n$ and $P_n(z)$ are also single valued, the same must be true of $\alpha_1(z)\lambda_1(z)^n + \alpha_2(z)\lambda_2(z)^n$, and thus the function w defined by (11) is analytic in U.

Now applying Lemma 3.1, we again obtain (10). Since $\lambda(z, n, 2) \to \lambda_1(x)$ uniformly in n as $z \to x$, we can assume not only that $|\lambda_1(z)|/|\lambda_3(z)| \leqslant r < 1$ but that $|\lambda(z, n, 2)|/|\lambda_3(z)| \leqslant r$, $|\lambda(z, n, 2)| \geqslant \epsilon > 0$, if $z \in C$, the boundary of U. Letting

$$M = \sup\{\max(|\beta_1(z)|, |\beta_2(z)|): z \in C\},$$

and supposing that $|\alpha_3(z)| \geqslant m > 0$ on C, we choose N so that $M[1 + (n/\epsilon)]r^n < m$ for $n \geqslant N$; for such n, we again have $|w(z)| < |\alpha_3(z)|$ on C. The rest of the proof is exactly as in case I.

To complete the proof of the theorem, we establish the sufficiency of condition (ii). Consider the case of more than one dominant root at x; we number the roots so that

$$|\lambda_1(x)| = \cdots = |\lambda_l(x)| > |\lambda_{l+1}(x)|, \ldots, |\lambda_k(x)|, \tag{12}$$

where $l \geqslant 2$.

The first step consists of showing that such points x are not isolated, i.e., that after possible renumbering (12) holds arbitrarily close to x (with perhaps a different value, still exceeding 1, for l). Suppose the contrary. Then there is a disk U about x such that there is a single dominant root value at every point of $U - \{x\}$. By making U smaller if necessary, we may remove from it a slit as in (4), obtaining the slit disk U' in which the roots λ_j may be taken to be analytic. Defining the λ_j in all of U as in (5), we can suppose that $|\lambda_1(z)| > |\lambda_j(z)|$, $2 \leqslant j \leqslant k$, for $z \in U - \{x\}$.

Let $z_0 \neq x$ be on the slit. Then the k distinct root values at z_0 are given by the $\lambda_j(z_0)$. But the root values are also given by the numbers

$$\tilde{\lambda}_j(z_0) = \lim \lambda_j(\zeta) \qquad \text{as} \quad \zeta \to z_0, \quad \arg(\zeta - x) \to 2\pi.$$

By continuity, $\lambda_1(z_0)$ and $\tilde{\lambda}_1(z_0)$ are both dominant root values; since there is but one such at each point of $U - \{x\}$, $\tilde{\lambda}_1(z_0) = \lambda_1(z_0)$. Referring to (5), we see that continuation of λ_1 once around x leads again to λ_1, and thus λ_1 is single valued in U.

Now consider the analytic continuations of λ_2 around x; we can suppose them to be $\lambda_2, \ldots, \lambda_p$, $2 \leqslant p \leqslant l$. If $\mu_j = \lambda_j/\lambda_1$, $2 \leqslant j \leqslant p$, the μ_j are algebraic and have a common limit ω of modulus 1 at x. Moreover, since λ_1 is single valued, the μ_j are exactly the analytic continuations of μ_2 around x. Extending the μ_j to all of U by (5) and applying the maximum principle referred to in Section 3 to a disk V with boundary C centered at x and contained in

U, we find that

$$1 = |\omega| \leqslant \max_{2 \leqslant j \leqslant p} \{\sup\{|\mu_j(z)| : z \in C\}\},$$

and in particular that $|\mu_j(z)| \geqslant 1$ for some $z \in U - \{x\}$ and $j \geqslant 2$. But we have been supposing that $|\lambda_1(z)| > |\lambda_j(z)|$ in $U - \{x\}$, i.e., that $|\mu_j(z)| < 1$, $j > 1$, and so we have obtained a contradiction. Our assertion that the set of x satisfying (12) is nonisolated is thus proved.

The nonisolated nature of such x simplifies our analysis considerably. For we saw at the end of Section 1 that if there are limits of zeros arbitrarily near x, then x is a limit of zeros; and thus in showing that a point x satisfying (12) is a limit of zeros, we may assume that x has any property that is known to hold at every point other than x sufficiently close to x, even if x does not have the property originally. Using the nondegeneracy condition B and (6) we thus can, and do, assume in addition to (12) that

(a) the $\lambda_j(x)$ are distinct, $1 \leqslant j \leqslant k$,
(b) $(\lambda_i/\lambda_j)'(x) \neq 0$, $1 \leqslant i \neq j \leqslant l$,
(c) $\alpha_j(x) \neq 0$, $1 \leqslant j \leqslant k$.

Because of (a), which guarantees that the λ_j are analytic near x and (b), we may [1, p. 107], making U smaller if necessary, suppose that the $\mu_{ij} = \lambda_i/\lambda_j$ are analytic 1–1 maps from U onto a neighborhood of $\mu_{ij}(x)$, $1 \leqslant i \neq j \leqslant l$. In particular, since $|\mu_{ij}(x)| = 1$, the set of z in U for which $|\lambda_i(z)| = |\lambda_j(z)|$ consists of an arc Γ_{ij} through x and, in particular, every point in U at which there is more than one dominant root must lie on some Γ_{ij}. By going out an arbitrarily small distance from x along one of these Γ_{ij} (which must either coincide or meet only at x) we may assume, again making U smaller, renumbering, and changing l, if necessary,

(d) there is an arc Γ through x such that (12) holds (with x replaced by z) for $z \in \Gamma$ and $|\lambda_i(z)| \neq |\lambda_j(z)|$, $z \in U - \Gamma$, $1 \leqslant i \neq j \leqslant l$.

If U is small enough, the arc Γ in (d) will divide U into two parts, U_1 and U_2, and the roots can be numbered so that

$$|\lambda_1(z)| > |\lambda_2(z)| > \cdots > |\lambda_l(z)|, \qquad z \in U_1, \tag{13}$$

and the opposite ordering holds in U_2. Now $\mu = \mu_{1l} = \lambda_1/\lambda_l$ maps U 1–1 onto a neighborhood V of $\omega_0 = \mu(x)$, where $|\omega_0| = 1$, and μ takes Γ onto an arc $\tilde{\Gamma}$ of the unit circle through ω_0. If ν is the analytic map on V inverse to μ, then for $z \in V$,

$$P_n(\nu(z)) = \alpha_1(\nu(z))z^n\lambda_l(\nu(z))^n + \sum_{j=2}^{k} \alpha_j(\nu(z))\lambda_j(\nu(z))^n.$$

Dividing through by $\alpha_1 \lambda_l(v(z))^n$ and setting

$$\varphi_j(z) = \lambda_j(v(z))/\lambda_l(v(z)), \qquad \beta_j(z) = \alpha_j(v(z))/\alpha_1(v(z)),$$

(cf. (c)) we see that the existence of a zero of P_n in U amounts to the existence in V of a zero of

$$\Pi_n(z) = [P_n/\alpha_1 \lambda_l^n](v(z)) = z^n + \sum_{j=2}^{k} \beta_j(z)\varphi_j(z)^n. \tag{14}$$

Reviewing our construction so far, we see that $\varphi_l(z) \equiv 1$, $|\varphi_j(z)| = 1$, $2 \leqslant j \leqslant l-1$, if $|z| = 1$, and that V is divided into two parts, V_1 and V_2, by the arc $\tilde{\Gamma}$ of the unit circle such that

$$|z| > |\varphi_2(z)| > \cdots > |\varphi_{l-1}(z)| > 1, \qquad z \in V_1, \tag{15}$$

because of (13), and the opposite inequality holds on V_2. Moreover, by making U smaller if necessary, we have from (12) that

$$|\varphi_j(z)| \leqslant \delta < 1, \qquad z \in V, \quad l+1 \leqslant j \leqslant k. \tag{16}$$

Now define f_n on V by

$$\begin{aligned} f_n(z) &= \Pi_n(z) - z^n - \sum_{j=2}^{l-1} \beta_j(z)\varphi_j(z)^n \\ &= \beta_l(z) + \sum_{j=l+1}^{k} \beta_j(z)\varphi_j(z)^n, \end{aligned} \tag{17}$$

where the second equality follows from (14). Because of (16), we can, by making V small and n large, assume that $f_n(z)$ is arbitrarily close to $\beta_l(\omega_0)$ for $z \in V$. This provides the key to our proof, which we can now sketch roughly.

By taking a smaller (noncircular) neighborhood of U, we can assume that for some $r_0 > 1$ and $\theta_0 > 0$,

$$V = N_{r_0,\theta_0}(\omega_0) = \{re^{i\theta}\omega_0 : r_0^{-1} < r < r_0, \ -\theta_0 < \theta < \theta_0\}. \tag{18}$$

The boundary of V is then $C = C_1 + C_2 + C_3 + C_4$, where $|z| = r_0$ on C_1, $|z| = r_0^{-1}$ on C_3, $z = re^{i\theta_0}\omega_0$ on C_2, and $z = re^{-i\theta_0}\omega_0$ on C_4.

From (17), we see that

$$\Pi_n(z) = z^n + \sum_{j=2}^{l-1} \beta_j(z)\varphi_j(z)^n + f_n(z). \tag{19}$$

Consider the image under Π_n of C. On C_1, z^n dominates for n large and so the image of C_1 winds around 0 a number of times on the order of $n\theta_0$. Also, $\beta_l(\omega_0) \neq 0$, $f_n(z)$ stays near $\beta_l(\omega_0)$, and the inequality opposite to (15) holds on C_3, so f_n dominates on C_3 and the image of C_3 does not wind around 0 at all, if n is large. To conclude that the net effect of traversing C is to wind around 0 a positive number of times and thus that Π_n has a positive number of zeros in $V = N_{r_0,\theta_0}(\omega_0)$, we are thus reduced to controlling the winding around 0 of the images under Π_n of C_2 and C_4, which is the most complicated part of the proof.

It is worth noting that in the "most typical" case, when there are exactly two dominant roots at x, i.e., when $l = 2$, the complication disappears. For then the middle term in the RHS of (19) disappears, we can assume that $f_n(z)$ lies in a half-plane, say the right half-plane, for $z \in V$, and by adjusting θ_0 according to the value of n we can suppose that $z^n = r^n e^{\pm in\theta_0} \omega_0{}^n$ is in the right half-plane as z traverses C_2 or C_4. In particular, in these circumstances the images under Π_n of C_2 and C_4 will be contained in the right half-plane and will thus not wind around 0. (Unfortunately, the complication cannot be avoided, as may be seen by considering the case $\lambda_1(z) = z$, $\lambda_2(z) = z^2$, $\lambda_3(z) = z^3$.)

To begin making our argument precise we recall that if a function f is analytic in a region containing the neighborhood V with boundary C, the number of zeros of f in V is the number of times the image under f of C winds around 0, i.e.,

$$(1/2\pi i) \int_C [f'(z)/f(z)]\, dz.$$

Also, we prove the following "partial Rouché theorem," where for any arc Γ in the plane and analytic function φ defined on Γ,

$$I_\Gamma(\varphi) = \mathrm{Re}\left\{(1/2\pi i) \int_\Gamma [\varphi'(z)/\varphi(z)]\, dz\right\}. \tag{19a}$$

LEMMA 4.2. *Suppose that f and g are analytic in a region R, that Γ is an arc contained in R, and that $|f(z)| > |g(z)|$, $z \in \Gamma$. Then*

$$I_\Gamma(f) - \tfrac{1}{2} < I_\Gamma(f + g) < I_\Gamma(f) + \tfrac{1}{2}.$$

Proof. If $|\zeta_1| > |\zeta_2|$ then $|\arg(\zeta_1 + \zeta_2) - \arg \zeta_1| = 2\pi k + \delta$, where k is integral and $0 < \delta < \pi/2$. Thus we can choose branches of $\log f$ and $\log(f + g)$ analytic on a neighborhood of Γ so that

$$|\mathrm{Re}\{(1/2\pi i)[\log(f(z) + g(z)) - \log f(z)]\} < \tfrac{1}{4}, \qquad z \in \Gamma.$$

If z_0 and z_1 are the initial and final points of Γ, then

$$I_\Gamma(f) = \mathrm{Re}\{(1/2\pi i)[\log f(z_1) - \log f(z_0)]\},$$

and similarly for $I_\Gamma(f + g)$, so $|I_\Gamma(f + g) - I_\Gamma(f)| < \tfrac{1}{2}$, and the lemma is proved.

Now we turn back to the function Π_n given in $V = N_{r_0,\theta_0}(\omega_0)$ by (19). At this stage we simplify matters by assuming that $l = 3$, a case that is sufficiently typical. Letting $\varphi(z) = \varphi_2(z)$, $\beta(z) = \beta_2(z)$, and, comparing with (19), we see that

$$\Pi_n(z) = z^n + \beta(z)\varphi(z)^n + f_n(z), \tag{20}$$

where f_n is given by (17).

We start by setting out the properties of f_n that we need. Because $\beta_l(\omega_0) \neq 0$ (by (c)), it follows from (16) and (17) that if V is sufficiently small and n sufficiently large,

$$0 < m_1 \leqslant |f_n(z)| \leqslant m_2, \qquad z \in V, \tag{21}$$

for some m_1, m_2, and, moreover, that $f_n(z)$ is contained in some fixed half-plane for $z \in V$. In particular, if Γ is any arc contained in V,

$$|I_\Gamma(f_n)| < \tfrac{1}{2}. \tag{22}$$

The heart of the argument consists of an analysis of the local behavior of $\varphi = \varphi_2 = \lambda_2/\lambda_3$. It follows from (d) and (15) that $|\varphi(z)| = 1$ if $|z| = 1$, and either $1 < |\varphi(z)| < |z|$ or $1 > |\varphi(z)| > |z|$ if $|z| \neq 1$, and it follows from (b) that $\varphi'(z) \neq 0$ in V. Using the polar coordinate version of the Cauchy–Riemann equations and the vanishing of $\partial|\varphi(z)|/\partial\theta$ for $|z| = 1$, we find that $\varphi(z) = R(z)e^{i\psi(z)}$, with

$$\frac{\partial R}{\partial r}(e^{i\theta}) = \frac{\partial \psi}{\partial \theta}(e^{i\theta}) = [e^{i\theta}/\varphi(e^{i\theta})]\varphi'(e^{i\theta}).$$

Put equivalently, $\varphi(z)$ for z near $e^{i\theta} \in V$ is approximated to the first order by cz^σ, where

$$\sigma = \frac{\partial R}{\partial r}(e^{i\theta}), \qquad c = \varphi(e^{i\theta})/e^{i\sigma\theta}.$$

Moreover, $\sigma = \sigma(\theta) \neq 0$ because $\varphi'(e^{i\theta}) \neq 0$, $e^{i\theta} \in V$, and $\sigma(\theta) \neq 1$ since (using (b) again) $(\varphi(z)/z)'(e^{i\theta}) \neq 0$. Also, since $1 < |\varphi(z)| < |z|$ if $|z| > 1$, it follows that $0 < \sigma(\theta) < 1$. Changing variables, we have that $\varphi(re^{i\theta}\omega_0)$ is approximated by $cr^{\sigma(\theta)}e^{i\sigma(\theta)\theta}$ near $e^{i\theta}\omega_0 \in N_{r_0,\theta_0}(\omega_0)$ (see (18)), where $|c| = |c(\theta)| = 1$. In particular,

$$\begin{aligned} |\varphi((1+\epsilon)e^{i\theta}\omega_0)| &= (1+\epsilon)^{\sigma(\theta)} + O(\epsilon^2) \\ &= 1 + \epsilon\sigma(\theta) + O(\epsilon^2). \end{aligned} \tag{23}$$

Furthermore, since $0 < \sigma(\theta) < 1$, a compactness argument shows that

$$0 < \sigma_0 \leqslant \sigma(\theta) \leqslant \sigma_1 < 1, \qquad -\theta_0 \leqslant \theta \leqslant \theta_0. \tag{24}$$

We now exploit the easily verifiable estimate

$$(1 + a/n + O(n^{-2}))^n = e^a + O(n^{-1}). \tag{25}$$

Let m_1 and m_2 be as in (21), and let $m_3 = \sup\{|\beta(z)|: z \in V\}$, recall (24), and choose A sufficiently large so that

$$e^A > 2(m_3 e^{A\sigma_1} + m_2), \qquad m_1 > 2(e^{-A} + m_3 e^{-A\sigma_0}).$$

The point of these inequalities is that if n is sufficiently large, and in particular large enough so that $1 + A/n < r_0$, we can apply (21), (23), (24), and (25) to conclude that for $z \in V$,

$$|z|^n > |\beta(z)\varphi(z)^n + f_n(z)|, \qquad |z| = 1 + A/n, \tag{26}$$

$$|f_n(z)| > |z^n + \beta(z)\varphi(z)^n|, \qquad |z| = (1 + A/n)^{-1}. \tag{27}$$

Of course, (26) and (27) hold for $|z| = r_0, r_0^{-1}$, respectively, for large n. But we also have the following, which is immediate from (25) but not true if $|z| = r_0$:

$$|z|^n \leqslant m_4 \qquad \text{if} \quad |z| \leqslant (1 + A/n). \tag{28}$$

We shall prove that for sufficiently large n, Π_n has a zero in a subneighborhood V_n of $V = N_{r_0,\theta_0}(\omega_0)$ of the form

$$V_n = \{re^{i\theta}\omega_0 : (1 + A/n)^{-1} < r < (1 + A/n),\ \theta_{n,1} < \theta < \theta_{n,2}\} \tag{29}$$

where the $\theta_{n,j}$ will be chosen below and satisfy $-\theta_0 \leqslant \theta_{n,1} < \theta_{n,2} \leqslant \theta_0$. Notice that V_n is indeed contained in V for large n. The point of restricting ourselves to V_n is that we shall be able to control the behavior of Π_n on C_2 and C_4, the parts of the boundary of V_n where $\theta = \theta_{n,j}$.

Our goal is an estimate of $I_{C_2}(\Pi_n)$ and $I_{C_4}(\Pi_n)$, and we begin with an estimate for Π_n'. Recall that $|\varphi(z)| < |z|$ if $|\varphi(z)| > 1$, $z \in V$; together with (28) this implies that $|\varphi(z)|^n \leqslant m_4$ for $z \in V_n$. Using (16) also it follows that

$$|\Pi_n'(z)| = |(z^n + \beta(z)\varphi(z)^n + f_n(z))'| \leqslant m_5 n, \qquad z \in V_n. \tag{30}$$

We claim that there is a constant m_6 such that for every sufficiently large n it is possible to choose $\theta_{n,1}, \theta_{n,2}$ so that

$$-\theta_0 \leqslant \theta_{n,1} \leqslant -\theta_0 + (2\pi/n) < \theta_0 - (2\pi/n) \leqslant \theta_{n,2} \leqslant \theta_0, \tag{31}$$

$$|\Pi_n[r\exp(i\theta_{n,j})\omega_0]| \geqslant m_6 > 0, \qquad (1 + A/n)^{-1} \leqslant r \leqslant (1 + A/n), \qquad j = 1, 2. \tag{32}$$

Before proving our claim we show that together with the above it implies the existence of a zero of Π_n in V_n, and thus the theorem. Let C_n be the boundary of V_n, and let C_1, C_2, C_3, C_4 be the parts of C_n containing $z = re^{i\theta}\omega_0$ such that, respectively, $r = (1 + A/n)$, $\theta = \theta_{n,2}$, $r = (1 + A/n)^{-1}$, $\theta = \theta_{n,1}$.

What must be shown in (see (19a)) that

$$I_{C_n}(\Pi_n) = \sum_{j=1}^{4} I_{C_j}(\Pi_n) > 0. \tag{33}$$

From (26) and Lemma 4.2 it follows that

$$I_{C_1}(\Pi_n(z)) > I_{C_1}(z^n) - \tfrac{1}{2} = (n/2\pi)(\theta_{n,2} - \theta_{n,1}) - \tfrac{1}{2}.$$

Taking into account (31), we see that

$$I_{C_1}(\Pi_n) > (n\theta_0/\pi) - \tfrac{5}{2}. \tag{34}$$

Similarly, (27), Lemma 4.2 and (22) imply that

$$I_{C_3}(\Pi_n) > -1. \tag{35}$$

A direct estimate of $I_{C_2}(\Pi_n)$ and $I_{C_4}(\Pi_n)$ using (30) and (32) shows that

$$I_{C_j}(\Pi_n) \geqslant -\{(2\pi)^{-1}[(1 + A/n) - (1 + A/n)^{-1}]m_5 n m_6^{-1}\} > -m_7, \qquad j = 2, 4. \tag{36}$$

Finally, (34), (35), and (36) together certainly show that (33) holds if n is sufficiently large.

The conclusion of the proof of the theorem consists in verifying the claim involving (31) and (32). We do this first for $j = 2$. Suppose that $z \in V$, $|z - e^{i\theta_0}\omega_0| = O(n^{-1})$. Then from the discussion preceding (23) we see that

$$\varphi(z) = e^{i\gamma}z^{\sigma} + O(n^{-2}), \qquad \text{where} \quad \sigma = \sigma(\theta_0),$$
$$e^{i\gamma} = c(\theta_0) = \varphi(e^{i\theta_0}\omega_0)/(e^{i\theta_0}\omega_0)^{\sigma}.$$

Also $\beta(z) = \beta + O(n^{-1})$, where $\beta = \beta(e^{i\theta_0}\omega_0)$, and we have from (16) and (17) with $l = 3$ that $f_n(z) = \delta + O(n^{-1})$, where $\delta = \beta_3(e^{i\theta_0}\omega_0)$. In light of these estimates, (20) and (25), it is enough to establish the existence of m_6 such that for every large enough n one can choose $\bar\theta = \theta_{n,2}$ satisfying (31) so that (32) holds for $j = 2$ with $\Pi_n(z)$ replaced by $z^n + \beta e^{in\gamma}z^{n\sigma} + \delta$, i.e., so that

$$|r^n(e^{i\bar\theta}\omega_0)^n + \beta e^{in\gamma}r^{n\sigma}(e^{i\bar\theta}\omega_0)^{n\sigma} + \delta| \geqslant m_6 > 0 \tag{37}$$

if $(1 + A/n)^{-1} \leqslant r \leqslant (1 + A/n)$.

Now the last inequality in r implies that $\frac{1}{2}e^{-A} < r^n < 2e^A$ for large n; also, as $\bar\theta$ varies from $\theta_0 - 2\pi/n$ to θ_0, the values of $(e^{i\bar\theta}\omega_0)^n$ vary over the entire unit circle. Therefore, the assertion that (37) holds for large n for the appropriate $\bar\theta$ will be proved if it can be shown that

$$\inf_{0\leqslant\tau\leqslant 2\pi}\ \sup_{0\leqslant\psi\leqslant 2\pi}\ \inf_{\frac{1}{2}e^{-A}\leqslant u\leqslant 2e^A} |\Phi(\tau, \psi, u)| = m_6 > 0 \tag{38}$$

where

$$\Phi(\tau, \psi, u) = ue^{i\psi} + \beta e^{i\tau}u^{\sigma}e^{i\sigma\psi} + \delta.$$

For fixed τ, $\Phi(\tau, \psi, u)$ is analytic in $\zeta = ue^{i\psi}$; in particular, $\Phi(\tau, \psi, u)$ vanishes for some u for only finitely many values of ψ. Thus $\Psi(\tau) = \sup_\psi \inf_u |\Phi(\tau, \psi, u)| > 0$ for every τ. Moreover, since $|\Phi|$ is continuous on the product of three closed bounded intervals, it is an easy exercise to show that Ψ is continuous in τ, from which it follows that $\inf_{0 \leqslant \tau \leqslant 2\pi} \Psi(\tau) = m_6 > 0$. This establishes (38) and thus proves the assertion involving (31) and (32) for $j = 2$. An identical argument handles the case $j = 1$, and the proof of the theorem is complete.

5. The Nondegeneracy Conditions

In this section we consider the nondegeneracy conditions A and B. In particular, we give algorithms for determining whether A and B hold for a given $\{P_n\}$ and discuss the extent to which the theorem can be modified in case A or B fail.

We examine first the question whether A fails, i.e., whether $\{P_n\}$ satisfies a lower order recursion. Let $H(z) = \det((a_{ij}(z)))$, where $a_{ij}(z) = P_{i+j-2}(z)$, $1 \leqslant i, j \leqslant k$. If $\{P_n\}$ satisfies a recursion of order less than k, then the last column of $a_{ij}(z)$ can, for each z, be expressed as a linear combination of the first $k-1$ columns, and so $H(z) \equiv 0$. In particular, if $H(z) \not\equiv 0$, condition A must hold.

It is easy to check that

$$(a_{ij}(z)) = (c_{ij}(z))(d_{ij}(z))(e_{ij}(z)),$$

where $c_{ij}(z) = \lambda_j(z)^{i-1}$, $d_{ij}(z) = \delta_{ij}\alpha_j(z)$, $e_{ij}(z) = \lambda_i(z)^{j-1}$, whenever the $\lambda_i(z)$ are distinct. When the $\lambda_j(z)$ are distinct, which they are for all but finitely many z if condition B holds, we have $\det((c_{ij}(z))) \neq 0$, $\det((e_{ij}(z))) \neq 0$, and so $H(z) \equiv 0$ only if $\alpha_j(z) \equiv 0$ for some j. This proves the assertion stated without proof in Section 3 that the failure of condition A implies the identical vanishing of some $\alpha_j(z)$.

We now turn to the nondegeneracy condition B. Suppose that $\lambda_1(z) \equiv \omega\lambda_2(z)$. Let $R_z(\lambda)$ be the unique irreducible polynomial such that $R_z(\lambda_1(z)) \equiv 0$; $R_z(\lambda)$ must be a factor of $Q_z(\lambda)$. It is known [1, p. 223] that $R_z(\lambda)$ has repeated roots for only finitely many z, so if $\omega = 1$, $R_z(\lambda)$ must appear at least twice in the factorization of $Q_z(\lambda)$. If $\omega \neq 1$, let $S_z(\lambda) = R_z(\omega\lambda)$. Then $S_z(\lambda)$ must be the unique factor of $Q_z(\lambda)$ such that $S_z(\lambda_2(z)) \equiv 0$. Moreover, $S_z(\lambda) \not\equiv R_z(\lambda)$ unless ω is a primitive eth root of unity, the degree of R_z is of the form qe, and

$$R_z(\lambda) = T_z(\lambda^e) = (\lambda^e)^q + \sum_{j=1}^{q} g_j(z)(\lambda^e)^{q-j}.$$

In summary, condition B asserts exactly that $Q_z(\lambda)$ has no irreducible factors of the form $T_z(\lambda^e)$ and no pairs of irreducible factors of the form $R_z(\lambda)$, $R_z(\omega\lambda)$. And, in particular, if $Q_z(\lambda)$ is irreducible and not of the form $T_z(\lambda^e)$, condition B must hold.

We consider now the question whether condition B holds for a given recursion. For fixed ω, let

$$Q_{z,\omega}(\lambda) = [Q_z(\lambda) - Q_z(\omega\lambda)]/\lambda, \qquad \omega \neq 1,$$
$$Q_{z,1}(\lambda) = \partial Q_z(\lambda)/\partial\lambda,$$

and note that $\tilde{\lambda}$ and $\omega\tilde{\lambda}$ are both roots of $Q_z(\lambda)$ exactly if $Q_z(\lambda)$ and $Q_{z,\omega}(\lambda)$ have the common root $\tilde{\lambda}$. In particular, $\lambda_1(z) \equiv \omega\lambda_2(z)$ with respect to some numbering if and only if $Q_z(\lambda)$ and $Q_{z,\omega}(\lambda)$ have a common root for every z. An algorithm, admittedly not very practical, exists for determining whether this occurs. (Unfortunately, the algorithm does not distinguish the case when $|\omega| = 1$, and so can be used to show that B holds, but not necessarily that it fails to hold.)

The algorithm is based on the resultant $\mathrm{Res}(p, q)$ of two polynomials $p(\lambda)$ and $q(\lambda)$ of respective degrees m and n [6, p. 135]. $\mathrm{Res}(p, q)$ vanishes exactly when p and q have common roots and is given by the $(m + n) \times (m + n)$ determinant whose jth row consists of the coefficients in λ^{m+n-1}, $\lambda^{m+n-2}, \ldots, 1$ of $\lambda^{n-j}p(\lambda)$ if $1 \leqslant j \leqslant n$ and the coefficients of $\lambda^{m+n-j}q(\lambda)$ if $n + 1 \leqslant j \leqslant m + n$.

Now, $\lambda_1(z) \equiv \omega\lambda_2(z)$, i.e., $Q_z(\lambda)$ and $Q_{z,\omega}(\lambda)$ have a common root for all z if and only if $\mathrm{Res}(Q_z, Q_{z,\omega}) \equiv 0$. The first step in the algorithm is to check whether or not $\mathrm{Res}(Q_z, Q_{z,1}) \equiv 0$. For $\omega \neq 1$, $\mathrm{Res}(Q_z, Q_{z,\omega}) = K_z(\omega)$, a polynomial in z and ω, and the identical vanishing in z of $K_z(\omega)$ for some constant ω amounts to the polynomials $K_{z_1}(\varphi)$, $K_{z_2}(\varphi)$ having the common root ω for all z_1 and z_2, i.e., to the identity $\mathrm{Res}(K_{z_1}, K_{z_2}) \equiv 0$. (The authors thank Hyman Bass for this algorithm.)

We now show how the theorem can be modified in cases when the nondegeneracy conditions fail. As for condition A, if $\{P_n\}$ satisfies a lower order recursion, all that is necessary is to solve the appropriate linear equations and determine the (unique) lowest order recursion which $\{P_n\}$ does satisfy and then apply the theorem.

The situation is more complicated when condition B fails. We define a *root class* to be a maximal set of roots of the form $\omega_j\lambda(z)$, $j = 1, \ldots, p$, where $\omega_1 = 1$, $|\omega_j| = 1$, and λ is a root. If $p = 1$, we say that the root class is *ordinary*; the condition B fails exactly when there is at least one nonordinary root class. We say that a root class is dominant if the roots it contains are dominant.

For simplicity, we consider only root classes such that either $\omega_i \neq \omega_j$, $1 \leqslant i, j \leqslant p$, or $\omega_j = 1$, $1 \leqslant j \leqslant p$; we call such root classes type 1 and type 2, respectively, considering ordinary classes as being of either type. (Here and below, untreated cases can be analyzed easily by the interested reader.) If $\{\omega_j\lambda(z)\}$ is a root class of type 1, we let $\alpha_j(z)$ be the coefficient corresponding to $\omega_j\lambda(z)$ in the usual way. If the class is of type 2, i.e., if the root λ is repeated p times, the contribution of the class to the variant of (3) required to determine $P_n(z)$ is, for $n \geqslant p$,

$$\beta_1(z)\lambda(z)^n + n\beta_2(z)\lambda(z)^{n-1} + \cdots + [n!/(n-p+1)!]\beta_p(z)\lambda(z)^{n-p+1}.$$

It is easy to see that in the case when all root classes are of type 1 or type 2, one or the other of the following two conditions is necessary for x to be a limit of zeros of $\{P_n\}$.

(i)′ If $\{\omega_j\lambda\}$ is the unique dominant root class at x, then $\alpha_j(x) = 0$, $1 \leqslant j \leqslant p$, if $\{\omega_j\lambda\}$ is type 1, $\beta_p(x) = 0$ if $\{\omega_j\lambda\}$ is type 2.

(ii)′ There is more than one dominant root class.

In fact, turning back to the proof of the necessity of conditions (i) or (ii), we see that the necessity of (i)′ or (ii)′ follows from Lemma 4.1 once we have the equivalent of the elementary fact (see (9)) that $|\alpha_3(z_n)|^{1/n} \to 1$ as $z_n \to x$ if $\alpha_3(x) \neq 0$. The equivalent assertion is exactly the following:

LEMMA 5.1. (a) *If the α_j do not all vanish at x, $|\omega_j| = 1$, $\omega_i \neq \omega_j \neq 1$, and $z_n \to x$, then*

$$\limsup_{n\to\infty} |\alpha_1(z_n) + \omega_2{}^n\alpha_2(z_n) + \cdots + \omega_p{}^n\alpha_p(z_n)|^{1/n} = 1.$$

(b) *If $\beta_p(x) \neq 0$, $\lambda(x) \neq 0$, and $z_n \to x$, then*

$$\lim_{n\to\infty} |\beta_1(z_n) + n\beta_2(z_n)\lambda(z_n)^{-1} + \cdots + [n!/(n-p+1)!)\beta_p(z_n)\lambda(z_n)^{-p+1}|^{1/n} = 1.$$

Part (a) of the lemma follows easily from the condition $\omega_i \neq \omega_j \neq 1$. Part (b) is a consequence of the domination of the other terms by the term involving β_p, which is on the order of n^{p-1}. Exploiting this domination, one can mimic the proof of the sufficiency of (i) and (ii) to show that (i)′ is sufficient if $\{\omega_j\lambda\}$ is of type 2 and (ii)′ is sufficient if the two or more dominant root classes are of type 2.

The case of type 1 root classes is more interesting. If the ω_j are roots of unity, nothing is really new. Suppose, for instance, that $\omega_j{}^e = 1, j = 1, \ldots, p$. Then, referring to (3) once more, we see that $\{P_n\}$ can be regarded as the union of e dinstinct subsequences $\{P_{me+r}: m = 0, 1, 2, \ldots\}$, $0 \leqslant r \leqslant e-1$, where the roots in the recursion satisfied by the subsequences are the eth powers of the roots of the original recursion, except that the root class $\{\omega_j\lambda\}$ coalesces into a single root λ^e.

Clearly, x is a limit of zeros of $\{P_n\}$ if and only if x is a limit of zeros of each $\{P_{me+r}\}$. Thus we can apply the theorem to each of subsequences to conclude that (i)′ and (ii)′ are sufficient to guarantee that x is a limit of roots of $\{P_n\}$, as long as conditions A and B are satisfied by each subsequence. If $\{\omega_j\lambda\}$ is the only nonordinary root class, then B is automatically satisfied by each subsequence. On the other hand, condition A and the sufficiency of (i)′ and (ii)′ may fail for one or more subsequences even in cases when condition A holds for $\{P_n\}$. For example, if

$$P_n(z) = z(2^n) + z((-2)^n) + (z-1)^n,$$

then A holds for $\{P_n\}$ and (i)′ is satisfied at $x = 0$, but 0 is not a limit of zeros, since the only zero of $P_{2n+1}(z)$ is $z = 1$.

Finally, we turn to type 1 root classes in which the ω_j are not roots of unity. Assume that $p = 2$, so the root class consists of $\lambda(z)$ and $\omega\lambda(z)$, where $\omega = e^{2\pi i\theta}$ and θ is irrational. We will give examples to show that the sufficiency of (i)′ and (ii)′ may depend on the degree of transcendence of θ.

In our examples, we use $\|y\|$ to denote the distance from y to the nearest integer. If θ is algebraic, it follows from the theorem of Roth [7, p. 123] that $|\theta - m/n| \geqslant C_\theta n^{-3}$ for all integers m if the integer n is sufficiently large, and thus $\|n\theta\| = n|\theta - m/n| \geqslant C_\theta n^{-2}$. On the other hand, if $\theta = \theta_0 = \sum_{j=1}^{\infty} 10^{-a_j}$, where $a_1 = 1$, $a_{j+1} = 10^{a_j}$, then $\|10^{a_j}\theta\| \leqslant 2 \cdot 10^{-a_{j+1}}$. In any event, notice that $|\omega^n - 1| = |e^{2\pi i n\theta} - 1|$ is bounded above and below by a multiple of $\|n\theta\|$.

Now for the examples. If $P_n(z) = z \cdot 2^n - z(2\omega)^n + 1^n$, condition (i)′ is satisfied at $x = 0$. Since the only zero of P_n is $z_n = 1/[2^n(\omega^n - 1)]$, 0 is a limit of zeros if and only if $z_n \to 0$. If θ is algebraic, $|2^n(\omega^n - 1)| > C_1 2^n n^{-2}$ so $z_n \to 0$, but if $\theta = \theta_0$ and $n = 10^{a_j} = a_{j+1}$, $|2^n(\omega^n - 1)| < C_2 2^n 10^{-n} = C_2 5^{-n}$, and $z_n \not\to 0$.

If $P_n(z) = z^n - \omega^n + 1^n$, (ii)′ is satisfied at $x = 1$. The zeros of $P_n(z)$ are $(\omega^n - 1)^{1/n}\omega_{n,j}$, where the $\omega_{n,j}$ are the nth roots of unity, so 1 is a limit of zeros if and only if $|\omega^n - 1|^{1/n} \to 1$. If θ is algebraic, then $|\omega^n - 1|^{1/n} \geqslant |C_1 n^{-2}|^{1/n} \to 1$, so $|\omega^n - 1|^{1/n} \to 1$; but if $\theta = \theta_0$, $n = 10^{a_j}$, then $|\omega^n - 1|^{1/n} \leqslant (C_2 10^{-n})^{1/n} = C_2^{1/n}/10$ and so $|\omega^n - 1|^{1/n} \not\to 1$.

The conjecture that (i)′ and (ii)′ are always sufficient in case θ is algebraic is inviting, but we have not attempted it.

6. An Application

Corresponding to any map is a polynomial whose degree is the number of countries of the map and whose value at a positive integer l is the number of ways the map can be colored using l or fewer colors. A polynomial arising in this way is called a chromatic polynomial; the study of chromatic polynomials was initiated by G. D. Birkhoff in [5]. The four-color conjecture, of course, amounts to the nonexistence of a chromatic polynomial with a zero at 4.

One of the authors has conjectured [2] that among the limits of zeros of all (not necessarily recursive) families of chromatic polynomials are found all numbers of the form $B_n = 2 + 2\cos(2\pi/n)$ for integral n. Notice that $B_1 = \lim_{n\to\infty} B_n = 4$.

Attempts at verifying the conjecture have so far centered on producing recursively defined families of chromatic polynomials corresponding to recursively defined families of maps. Applying the theorem to the families found in [2] and [4], one finds that several of the B_n are indeed limits of zeros of families of chromatic polynomials; applying the theorem to a family described in [3], one finds that 4 is such a limit of zeros.

References

1. Lars Ahlfors, "Complex Analysis," McGraw–Hill, New York, 1953.
2. S. Beraha, Infinite non-trivial families of maps and chromials, Ph.D. Thesis submitted to Johns Hopkins University.
3. S. Beraha and J. Kahane, Is the four-color conjecture almost false, *J. Combinatorial Theory Ser. B*, in press.
4. S. Beraha, J. Kahane, and R. Reid, B_7 and B_{10} are limit points of chromatic zeros, *Notices Amer. Math. Soc.* **20** (1973), 45.
5. G. D. Birkhoff, A determinant formula for the number of ways of coloring a map, *Ann. of Math.* **14** (2) (1912), 42–46.
6. Serge Lang, "Algebra," Addison–Wesley, Reading, Massachusetts, 1965.
7. William Leveque, "Topics in Number Theory," Vol. 2, Addison–Wesley, Reading, Massachusetts, 1956.

AMS (MOS) subject classifications: primary 39A10; secondary, 05C15.

STUDIES IN FOUNDATIONS AND COMBINATORICS
ADVANCES IN MATHEMATICS SUPPLEMENTARY STUDIES, VOL. 1

Time-Varying Linear Discrete-Time Systems: Realization Theory

BOSTWICK F. WYMAN[†]

Mathematics Department
Ohio State University
Columbus, Ohio

Generalized difference systems are defined, and the fundamental realization theorem for these systems is proved. Time-varying linear systems on partially ordered sets are studied in this context. Time-varying linear systems on the discrete line are studied, and a realization theory is developed for the pure autoregressive equation $y(t) + a_1(t)y(t-1) + \cdots + a_n(t)y(t-n) = u(t)$.

1. INTRODUCTION

This paper keeps some of the promises made in my previous papers [W75, W76, W76b]. Section 2 gives definitions and the basic realization theorem for *generalized difference systems*. The next part, Sections 3, 4, and 5, applies the general theory to the study of time-varying discrete-time linear systems over a locally finite partially ordered set. Although the treatment here is logically self-contained, the reader should consult [W76] for an extended motivation. The rest of the paper (Sections 6–9) is restricted still further to the case of time-varying linear systems on the discrete line $(\mathbb{Z}, \leqslant)$. This part of the paper, and to some extent the whole treatment of systems on posets, was inspired by the work of E. Kamen and his student K. Hafiz [K74, H75]. However, the methods used here are quite different and realization theory is emphasized.

2. GENERALIZED DIFFERENCE SYSTEMS

In this section we give a formal definition of a "generalized difference system," following [W75, W76], and prove the expected existence and uniqueness theorem for canonical realizations.

[†] Preparation of this paper was supported in part by a grant from the Ohio State University.

ISBN 0-12-599101-0

Throughout, we consider a base ring $\mathscr{K}$ which is assumed commutative with identity. We denote various right $\mathscr{K}$-modules as $\mathscr{X}$, $\mathscr{U}$, and $\mathscr{Y}$. Our first definition gives a suggestive name to a standard object.

2.1. Definition. An *algebra* of *generalized difference operators* over a commutative ring $\mathscr{K}$ with identity 1 is a ring R and a ring homomorphism $\varphi:\mathscr{K} \to R$ such that $\varphi(1)$ is the identity in R.

Note that R need not be commutative, and the image of $\mathscr{K}$ need not be central. We consider R as a $\mathscr{K}-\mathscr{K}$ bimodule by $ar = \varphi(a)r$ and $rb = r\varphi(b)$ for all $r \in R$, $a,b \in \mathscr{K}$. The map φ will usually be omitted from the notation.

We denote by ((mod-R)) and ((mod-$\mathscr{K}$)) the categories of right R- and right $\mathscr{K}$-modules. The map $\varphi:\mathscr{K} \to R$ induces an unnamed "forgetful functor":((mod-R)) $\to$ ((mod-$\mathscr{K}$)) which has a left adjoint Ω and a right adjoint Γ. (See, for example, [M71, AM75] for discussions of adjoints.) In particular, if $\mathscr{X}$ denotes an R-module (and the underlying $\mathscr{K}$-module) then every $\mathscr{K}$-linear map $G:\mathscr{U} \to \mathscr{X}$ yields a unique R-linear map $\tilde{G}:\Omega\mathscr{U} \to \mathscr{X}$, while a $\mathscr{K}$-linear map $H:\mathscr{X} \to \mathscr{Y}$ yields an R-linear $\tilde{H}:\mathscr{X} \to \Gamma\mathscr{Y}$ such that the following diagrams commute.

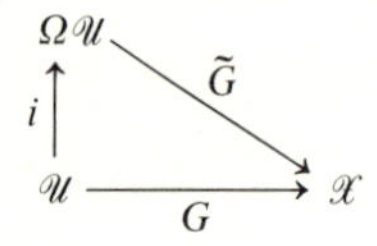

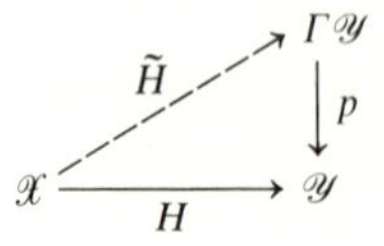

Here i and p are $\mathscr{K}$-linear maps which "come with" the functors Ω and Γ. In the present case, we can take $\Omega(\text{-}) = \text{-} \otimes_{\mathscr{K}} R$ and $\Gamma(\text{-}) = \operatorname{Hom}_{\mathscr{K}}(R, \text{-})$, where $\operatorname{Hom}_{\mathscr{K}}$ denotes the set of all right $\mathscr{K}$-linear maps. The right R-module structure on $\Omega(\mathscr{U})$ is familiar, and with our conventions the right R-structure on $\operatorname{Hom}_{\mathscr{K}}(R, \mathscr{Y})$ is given by $(h\alpha)(\beta) = h(\alpha\beta)$ for $h \in \operatorname{Hom}_{\mathscr{K}}(R, \mathscr{Y})$ and $\alpha, \beta \in R$.

If $G:\mathscr{U} \to \mathscr{X}$ is a right $\mathscr{K}$-module map, then the R-linear extension $\tilde{G}:\Omega\mathscr{U} \to \mathscr{X}$ given by the adjunction has the explicit formula $(u \otimes \alpha)\tilde{G} = uG \cdot \alpha$ for simple tensors. If $H:\mathscr{X} \to \mathscr{Y}$ is a right $\mathscr{K}$-module map, the corresponding R-linear map $\tilde{H}:\mathscr{X} \to \Gamma\mathscr{Y}$ is given by $(x\tilde{H})(\alpha) = (x\alpha)H$ for $x \in \mathscr{X}$ and $\alpha \in R$. The map $i:\mathscr{U} \to \Omega\mathscr{U}$ above is given by $i(u) = u \otimes 1$, while $p:\Gamma\mathscr{Y} \to \mathscr{Y}$ is given by $p(h) = h(1)$, where 1 is the identity of R. (See also [CE56, p. 28].)

Now we are ready to define systems and their corresponding input/output maps in this general context.

2.2. Definition. Let $\mathscr{K}$ be a commutative ring and $\varphi:\mathscr{K} \to R$ an algebra of generalized difference operators. Then a *generalized difference system over* $\mathscr{K}$ *with algebra* R, or an $R/\mathscr{K}$*-system* for short, is a quintuple $\Sigma = (\mathscr{X}, \mathscr{U}, \mathscr{Y}; G, H)$ where $\mathscr{X}$ is a right R-module, $\mathscr{U}$ and $\mathscr{Y}$ are right $\mathscr{K}$-modules, and $G: \mathscr{U} \to \mathscr{X}$, $H: \mathscr{X} \to \mathscr{U}$ are $\mathscr{K}$-module maps.

The *input/output* (or i/o) *diagram* for $\Sigma = (\mathscr{X}, \mathscr{U}, \mathscr{Y}; G, H)$ is the commutative R-module diagram

$$\begin{array}{ccc} \Omega\mathscr{U} & \xrightarrow{f_\Sigma} & \Gamma\mathscr{Y} \\ & {}_{\tilde{G}}\searrow \quad \nearrow_{\tilde{H}} & \\ & \mathscr{X} & \end{array}$$

where $\tilde{G}$ and $\tilde{H}$ are defined above, and $f_\Sigma = \tilde{H} \circ \tilde{G}$. The map $\tilde{G}$ is called the *reachability map* for Σ; $\tilde{H}$ is called the *observability map*; and f_Σ is called the i/o-*map*.

The system Σ is called *reachable* if $\tilde{G}$ is surjective, *observable* if $\tilde{H}$ is injective, and *canonical* if it is both reachable and observable.

We can also present the notion of an *abstract* i/o-*map* and define *realizations*. The usual existence and uniqueness results will follow.

2.3. Definition. Suppose given an algebra of generalized difference operators $\varphi: \mathscr{K} \to R$ with input functor Ω and output functor Γ. An *i/o-map* is an R-module homomorphism $f: \Omega\mathscr{U} \to \Gamma\mathscr{Y}$ for any $\mathscr{K}$-modules $\mathscr{U}$ and $\mathscr{Y}$.

An $R/\mathscr{K}$ system $\Sigma = (\mathscr{X}, \mathscr{U}, \mathscr{Y}; G, H)$ is a *realization* of f if $f = f_\Sigma$ or, equivalently, if the R-module diagram

$$\begin{array}{ccc} \Omega\mathscr{U} & \xrightarrow{f} & \Gamma\mathscr{Y} \\ & {}_{\tilde{G}}\searrow \quad \nearrow_{\tilde{H}} & \\ & \mathscr{X} & \end{array}$$

commutes. The system Σ is called a *canonical realization* if it is canonical as a system.

Note that no finiteness condition on $\mathscr{X}$ is required in the definition of realization. In fact, the appropriate "finiteness" conditions are quite elusive in this general context. Every i/o-map $f: \Omega\mathscr{U} \to \Gamma\mathscr{Y}$ has at least one realization. We give two basic (but rather uninteresting) realizations as a starting point. The "Ω-realization" is given by

$$\begin{array}{ccc} \Omega\mathscr{U} & \xrightarrow{f} & \Gamma\mathscr{Y} \\ & {}_{\mathrm{id}}\searrow \quad \nearrow_{f} & \\ & \Omega\mathscr{U} & \end{array}$$

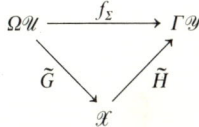

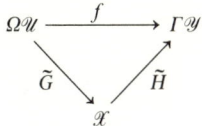

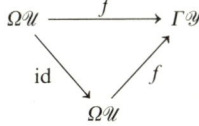

where id denotes the identity map. The "Γ-realization" is

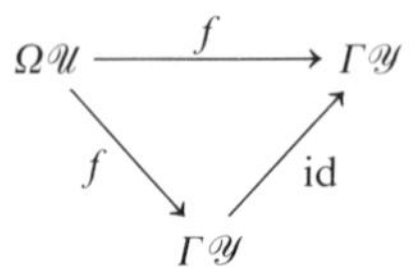

We need to show that every i/o-map has a canonical realization unique up to system isomorphism, and we begin by defining morphisms of systems.

2.4. DEFINITION. Let $\Sigma_i = (\mathscr{X}_i, \mathscr{U}, \mathscr{Y}; G_i, H_i)$, $i = 1, 2$, be two $R/\mathscr{K}$-systems with the same input space $\mathscr{U}$ and the same output space $\mathscr{Y}$. A morphism $T: \Sigma_1 \to \Sigma_2$ is an R-module homomorphism $T: \mathscr{X}_1 \to \mathscr{X}_2$ such that $T \circ G_1 = G_2$ and $H_2 \circ T = H_1$. That is,

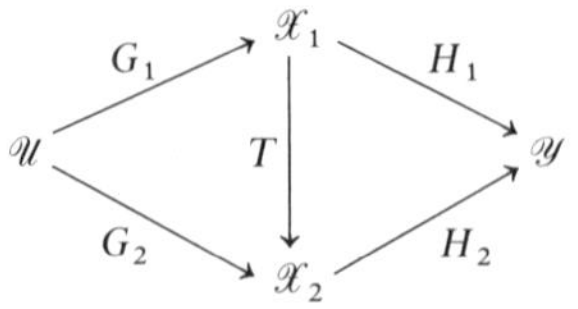

is a commutative diagram (of $\mathscr{K}$-modules, since $\mathscr{U}$ and $\mathscr{Y}$ are only $\mathscr{K}$-modules). The morphism T is called an *isomorphism* if $T: \mathscr{X}_1 \to \mathscr{X}_2$ is an R-module isomorphism.

The next proposition is perhaps a little surprising.

2.5. THEOREM. *Let $T: \Sigma_1 \to \Sigma_2$ be any morphism of systems as in Definition 2.4. Then $f_{\Sigma_1} = f_{\Sigma_2}$.*

Proof. The commutative $\mathscr{K}$-module diagram of the definition induces a commutative R-module diagram

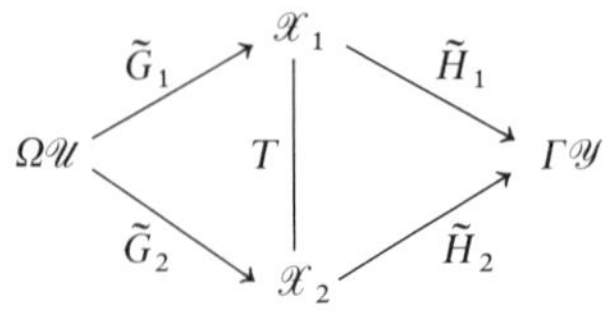

so that $f_{\Sigma_1} = \tilde{H}_1 \circ \tilde{G}_1 = \tilde{H}_2 \circ \tilde{G}_2 = f_{\Sigma_2}$. Q.E.D.

This proposition says that if two systems are related by *any* morphism (not just an isomorphism), then they have the same i/o-map. This motivates the next definition.

2.6. Definition. Let $f:\Omega\mathscr{U} \to \Gamma\mathscr{Y}$ be a fixed i/o-map. The *realization category of* f ((REAL-f)) has systems $\Sigma = (\mathscr{X}, \mathscr{U}, \mathscr{Y}; G, H)$ as objects, with morphisms defined as in 2.4. We denote by ((RCH-f)), ((OBS-f)), and ((CAN-f)) the full subcategories of ((REAL-f)) consisting of reachable, observable, and canonical systems, respectively.

The next "existence and uniqueness" theorem collects the main results of this section.

2.7. Theorem. *Suppose given an i/o-map* $f:\Omega\mathscr{U} \to \Gamma\mathscr{Y}$. *Then*

(a) *the categories* ((RCH-f)), ((OBS-f)), *and* ((CAN-f)) *are all nonempty,*

(b) *the systems in* ((CAN-f)) *are exactly the terminal objects in the category* ((RCH-f)),

(c) *the systems in* ((CAN-f)) *are exactly the initial objects in the category* ((OBS-f)),

(d) *if* Σ_1 *and* Σ_2 *are any two systems in* ((CAN-f)), *there exists a unique system isomorphism* $T:\Sigma_1 \to \Sigma_2$.

Proof. The Ω-realization discussed above is reachable, while the Γ-realization is observable. If $\Sigma = (\mathscr{X}, \mathscr{U}, \mathscr{Y}; G, H)$ is reachable, let $\mathscr{X}_0 = \mathscr{X}/\ker H$ and consider the S-module diagram

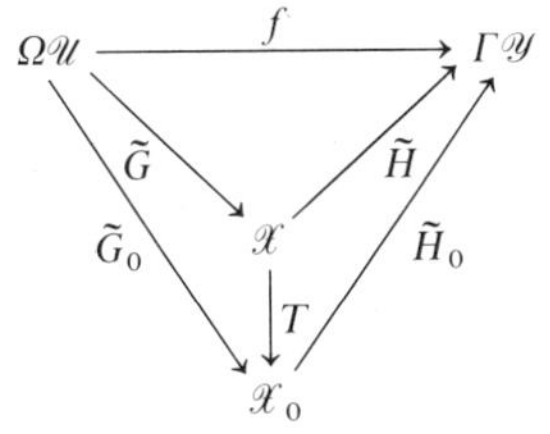

(*)

Here $T:\mathscr{X} \to \mathscr{X}_0$ is the natural map, and $\tilde{G}_0 = T_0\tilde{G}$ is surjective since $\tilde{G}$ and T are. The map $H_0:\mathscr{X}_0 \to \Gamma\mathscr{Y}$ exists and is injective by the standard isomorphism theorems. Therefore, $\Sigma_0 = (\mathscr{X}_0, \mathscr{U}, \mathscr{Y}; G_0, H_0)$ is canonical, and clearly $f_{\Sigma_0} = f$. We could just as well have started with an observable system $\Sigma = (\mathscr{X}, \mathscr{U}, \mathscr{Y}; G, H)$ and defined a canonical system Σ_r using the state module $\mathscr{X}_r = G(\Omega\mathscr{U}) \subseteq \mathscr{X}$. Details are left to the reader.

Next we show that canonical systems are terminal in ((RCH-f)). In diagram (*) above, assume $\Sigma = (\mathscr{X}, \mathscr{U}, \mathscr{Y}; G, H)$ is a given reachable system and $\Sigma_0 = (\mathscr{X}_0, \mathscr{U}, \mathscr{Y}; G_0, H_0)$ is any canonical system. We must show that there exists a *unique* S-module map $T: X \to X_0$ making the whole diagram commute. Now $\ker f = \ker \tilde{G}_0$ since $\tilde{H}_0$ is injective, and $\ker f \supseteq \ker \tilde{G}$.

Therefore, $\ker \tilde{G} \supseteq \ker \tilde{G}_0$, and T exists by an isomorphism theorem. The uniqueness of T is assured by the fact that $\tilde{G}$ is onto (because Σ is reachable).

The reader should check that canonical systems are initial objects in Obs(f).

Since any two initial objects (or any two terminal objects) in a category are related by a *unique* isomorphism in the category, part (d) follows immediately from consideration of either ((RCH-f)) or ((OBS-f)).

A proof along these lines appears in [W72]. For details on initial and terminal objects, see [M71, p. 20].

3. Posets, Incidence Algebras, and Dynamical Modules

In the next section we begin the study of time-varying linear difference systems as a special case of the abstract theory of Section 2. Here we define the rings we shall use and develop some of the algebraic theory. The treatment here is mostly self-contained, modulo references to the poset literature, but for motivation the reader should consult [W76].

Suppose given a partially ordered set ("poset") $(T, \leqslant)$ which is *locally finite* in the sense that segments $[s, t] = \{r : s \leqslant r \leqslant t\}$ are finite sets. Let k be a fixed field of scalars and denote by $\mathscr{A}$ the *incidence algebra* of T with coefficients in k [R64, DRS72]. An incidence function α in $\mathscr{A}$ is a partial function $\alpha : T \times T \to k$ with $\alpha(s, t)$ defined only for $s \leqslant t$ in T. Addition in $\mathscr{A}$ is functional addition, and multiplication in $\mathscr{A}$ is *convolution*:

$$(\alpha\beta)(s, t) = \sum_{r : s \leqslant r \leqslant t} \alpha(s, r)\beta(r, t).$$

Next, we define a subring of $\mathscr{A}$ which is suitable for our applications to system theory.

3.1. Definition. Let T be a locally finite poset with incidence algebra $\mathscr{A}$. An incidence function α in $\mathscr{A}$ is said to have *finite memory* if for every t in T,

$$\{s : s \leqslant t \text{ and } \alpha(s, t) \neq 0\}$$

is a finite set. The subring of all functions with finite memory is denoted by $\mathscr{A}_{\mathrm{f}}$ or $\mathscr{A}_{\mathrm{f}}(T)$ and is called the *finite memory algebra* of T.

We need to give some particular elements of $\mathscr{A}_{\mathrm{f}}$ which will be important. For fixed $s \leqslant t$ in T, let

$$\delta_{s,t}(u, v) = \begin{cases} 1 & \text{if} \quad u = s \quad \text{and} \quad v = t \\ 0 & \text{otherwise} \end{cases}$$

and for $t \in T$ let $e_t = \delta_{t,t}$. The e_t are *orthogonal idempotents*: $e_t^2 = e_t$ and $e_s e_t = 0$ if $s \neq t$.

The identity element of R is given by

$$1(s, t) = \begin{cases} 1 & \text{if} \quad s = t \\ 0 & \text{if} \quad s \neq t \end{cases},$$

and can be written as $1 = \sum_{t \in T} e_t$. (This kind of "formal" infinite sum makes sense if both sides are considered as functions. See also [DRS72, p. 271].)

The *zeta function* of T is given by $\zeta(s, t) = 1$ for all $s \leqslant t$. The Möbius function is $\mu = \zeta^{-1}$. The need for a general theory of Mobius inversion in combinatorics has motivated much of the research on incidence algebras. (For a survey, see (BG75].)

Let $\mathscr{K} = k^T = \{a: T \to k\}$ be the ring of k-valued time functions with the usual definitions of addition and multiplication. The embedding $K \subset \mathscr{A}_{\mathrm{f}}$ given by

$$a(s, t) = \begin{cases} a(t) & \text{if} \quad s = t \\ 0 & \text{otherwise} \end{cases}$$

makes $\mathscr{K}$ into a subring of $\mathscr{A}_{\mathrm{f}}$. The ring $\mathscr{K}$ is of course commutative, but it does not lie in the center of $\mathscr{A}_{\mathrm{f}}$ since $(\beta\alpha)(s, t) = \beta(s, t)\alpha(t)$, while $(\alpha\beta)(s, t) = \alpha(s)\beta(s, t) = \beta(s, t)\alpha(s)$.

In the next section we shall consider linear systems on posets as $R/\mathscr{K}$-systems. In particular, state spaces will be right R-modules, so it makes sense to consider carefully the structure of these modules.

Let $\mathscr{X}$ be any right R-module. Then for each $t \in T$ we define

$$\mathscr{X}_t = \mathscr{X} e_t = \{x e_t : x \in \mathscr{X}\}.$$

If $a \in \mathscr{K} \subset R$, then $a e_t = e_t a$ for all t, so that each X_t is a right $\mathscr{K}$-submodule of $\mathscr{X}$. We denote $x e_t$ by $x(t)$ and call it the "*t-component*" of x. The collection of mappings $x \to x(t)$ gives a $\mathscr{K}$-module map

$$\mathscr{X} \to \prod_{t \in T} X_t.$$

If T is a finite set, then $1 = \sum_{t \in T} e_t$ is a decomposition of the identity of $\mathscr{A}_{\mathrm{f}}$ into a finite sum of orthogonal idempotents. Using this, it is easy to see that $(*)$ is an isomorphism. For arbitrary T, we show in the next section that $(*)$ is bijective for modules occurring in system theory.

3.2. Definition. Let T be a poset, and let $\mathscr{A}_{\mathrm{f}}$ be the finite memory algebra of T. A right $\mathscr{A}_{\mathrm{f}}$-module $\mathscr{X}$ is called *dynamical* if the map $(*)$ is a $\mathscr{K}$-module isomorphism.

Dynamical modules can be given an attractive explicit description.

3.3. THEOREM. *Suppose given a poset T with finite memory algebra $\mathscr{A}_{\mathrm{f}}$ over a field k. Suppose that $\mathscr{X}$ is a dynamical $\mathscr{A}_{\mathrm{f}}$-module, so that $\mathscr{X} \cong \prod X_t$ with $X_t = \mathscr{X} e_t$. Then*

(a) *for each $s \leqslant t$, there exists a k-linear map $\Phi(s, t): X_s \to X_t$ such that for all $\alpha \in \mathscr{A}_{\mathrm{f}}$*

$$(x\alpha)(t) = \sum_{s \leqslant t} x(s)\Phi(s, t)\alpha(s, t),$$

(b) *the maps $\Phi(s, t)$ satisfy the following "coherence" assumptions:*

1. *$\Phi(t, t) = I$, the identity on X_t, for all $t \in T$,*
2. *$\Phi(s, r)\Phi(r, t) = \Phi(s, t)$ whenever $s \leqslant r \leqslant t$.*

Conversely, a collection $\{X_t\}$ of vector spaces together with k-linear mappings $\Phi(s, t)$ satisfying (b) *gives rise to a right $\mathscr{A}_{\mathrm{f}}$-module.*

Proof. We verify the converse part first. Suppose given $\{X_t\}$, and let $\mathscr{X} = \prod X_t$. We write $x \in \mathscr{X}$ as a "T-tuple" $x = (x(t))$, and define

$$(x\alpha)(t) = \sum_{s \leqslant t} x(s)\Phi(s, t)\alpha(s, t).$$

Note that the defining sum is finite since α has finite memory. We claim that this formula defines a right $\mathscr{A}_{\mathrm{f}}$-module structure on $\mathscr{X}$. The only difficulty lies in showing that $(x\alpha)\beta = x(\alpha\beta)$. To see this, write

$$\begin{aligned}
(x\alpha)\beta(t) &= \sum_{s \leqslant t} (x\alpha)(s)\Phi(s, t)\beta(s, t) \\
&= \sum_{s \leqslant t} \sum_{u \leqslant s} x(u)\Phi(u, s)\alpha(u, s)\Phi(s, t)\beta(s, t) \\
&= \sum_{u \leqslant t} x(u)\Phi(u, t) \sum_{s: u \leqslant s \leqslant t} \alpha(u, s)\beta(s, t) \\
&= \sum_{u \leqslant t} x(u)\Phi(u, t)(\alpha\beta)(u, t) \\
&= x(\alpha\beta)(t),
\end{aligned}$$

where we used the identity $\Phi(u, s)\Phi(s, t) = \Phi(u, t)$ for all s, $u \leqslant s \leqslant t$.

Now suppose given a right $\mathscr{A}_{\mathrm{f}}$-module $\mathscr{X}$ and consider $\mathscr{X} e_t = X_t \subset \mathscr{X}$. Since $e_s e_t = e_t$ if $s = t$ and $e_s e_t = 0$ if $s \neq t$, we see that $X_t = \{x \in \mathscr{X} : xe_t = x\}$ and $X_t e_s = 0$ for $s \neq t$. Another calculation in $\mathscr{A}_{\mathrm{f}}$ shows that $e_s \delta_{s,t} = \delta_{s,t} = \delta_{s,t} e_t$. Therefore, if $xe_s \in X_s$, then $(xe_s)\delta_{s,t} \in X_t$ and right multiplication by $\delta_{s,t}$ sends X_s to X_t. We denote this map, easily shown to be k-linear, by $\Phi(s, t)$. Since $e_t{}^2 = e_t$ and $\delta_{s,r}\delta_{r,t} = \delta_{s,t}$ for $s \leqslant r \leqslant t$, we conclude that $\Phi(t, t) = I$ on X_t and $\Phi(s, r)\Phi(r, t) = \Phi(s, t)$.

Finally, assume that X is dynamical and write $x = (x(t))$, where $x(t) = xe_t$ in X_t. Now, for $\alpha \in \mathscr{A}_{\mathrm{f}}$, we can write $\alpha = \sum_{s \leqslant t} \alpha(s, t) \cdot \delta_{st}$, where the sum is an infinite "formal" sum, but for any u, v, $\alpha(u, v) = \sum_{s \leqslant t} \alpha(s, t)\delta_{st}(u, v)$ is a valid equation involving finite sums. It follows that for $x \in \mathscr{X}$

$$\begin{aligned}(x\alpha)(t) &= \sum_{s \leqslant t} x\delta_{s,t} \cdot \alpha(s, t) \\ &= \sum_{s \leqslant t} xe_s\delta_{s,t}\alpha(s, t) \\ &= \sum_{s \leqslant t} x(s)\Phi(s, t)\alpha(s, t),\end{aligned}$$

which agress with the claim in (a) above. Q.E.D.

We anticipate the next section, and call the transformations $\Phi(s, t)$ the *state-transition matrix* for $\mathscr{X}$. In the special case that the spaces X_t have constant finite dimension, then $\Phi(s, t)$ can be thought of as a time-varying square matrix. For the special case of the line, see Section 7.

4. Linear Systems on Locally Finite Posets

Suppose given a locally finite poset T and a field k of scalars, giving rise to a ring $\mathscr{K} = k^T$ of k-valued time functions and a finite memory algebra $\mathscr{A}_{\mathrm{f}}$, which we denote by R in this section. As usual, there is a natural imbedding $\mathscr{K} \subset R$, and our first inclination is to define a *linear dynamical system on T* to be nothing but an $R/\mathscr{K}$ system in the sense of Section 2. This is essentially what we will do, but we put some restriction on the modules involved.

4.1. Definition. Suppose given T, k, $\mathscr{K}$, and R as above. Let U and Y be finite-dimensional vector spaces over k and define $t \in T$, $\mathscr{U} = U^T$, $\mathscr{Y} = Y^T$. Furthermore, let $\mathscr{X} \cong \prod_{t \in T} X_t$ be a dynamical R-module, and for each $t \in T$ assume given $G(t): U \to X_t$ and $H(t): X_t \to Y$. Define $\mathscr{K}$-module maps $G: \mathscr{U} \to \mathscr{X}$ by $G = \prod_{t \in T} G(t)$ and $H: \mathscr{X} \to \mathscr{Y}$ by $H = \prod_{t \in T} H(t)$. Then the $R/\mathscr{K}$ system $\Sigma = (\mathscr{X}, \mathscr{U}, \mathscr{Y}; G, H)$ will be called a *linear dynamical system on the poset T*.

The system-theoretic intuition goes as follows: We are given a space $\mathscr{U} = U^T$ of input functions $u(t): T \to U$ as well as an output space $\mathscr{Y} = Y^T$. The state-module $\mathscr{X}$ is dynamical, so that it is described by a family of "instantaneous" state-modules X_t together with "state-transition maps" $\Phi(s, t): X_s \to X_t$, when $s \leqslant t$. If $x(s) \in X_s$ is a state at time s, then $x(s)\Phi(s, t)$ is the state at time t resulting from $x(s)$ by "free evolution" of the system. In other words, $\{\Phi(s, t)\}$ is a "weighting pattern" (see [B70, p. 91]).

The maps $G(s)$ and $H(t)$ are in Kamen's terminology, "local in time": an input $u(s) = u(s)G(s)$ at time s sets up a state $x(s)$ at time s, and a state $x(t)$ yields output $x(t)H(t)$ at time t.

Just as in the abstract case, a system on T gives rise to an i/o-diagram of right R-modules

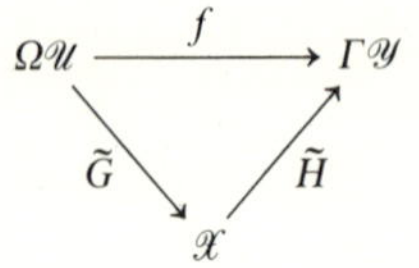

which we proceed to interpret in suggestive language.

We consider $\Omega\mathscr{U} = \mathscr{U} \otimes_{\mathscr{K}} R$ first. The map $G:\mathscr{U} \to \mathscr{X}$ is given by $(uG)(t) = u(t)G(t)$ for all $t \in T$. A simple tensor $u \otimes \alpha$ is though of as an input function u together with a "weight" α, and

$$(u \otimes \alpha)G(t) = (uG \cdot \alpha)(t) = \sum_{s \leqslant t} u(s)G(s)\alpha(s, t)\Phi(s, t).$$

Verbally, the input $u(s)$ at time s is multiplied by the weight $\alpha(s, t)$ and allowed to evolve freely to time t. The resulting "superposition" is finite, since α has finite memory.

The output module is given by $\Gamma\mathscr{Y} = \mathrm{Hom}_{\mathscr{K}}(R, \mathscr{Y})$. Suppose $x \in \mathscr{X}$ and $\beta \in R$; then

$$(x\tilde{H})(\beta)(t) = (x\beta)H(t) = \sum_{s \leqslant t} x(s)\Phi(s, t)\beta(s, t)H(t).$$

That is, the value of $x\tilde{H}$ at β depends on a weighted sum (determined by β) over part of the history of x.

In this context we can also speak of "abstract" input/output maps: these right R-module maps $f:\Omega\mathscr{U} \to \Gamma\mathscr{Y}$ arise in the following way. Given finite-dimensional vector spaces U and Y, a set of *input/output data* is a family of linear transformations $A(s, t): U \to Y$ for each pair $s \leqslant t$ in T. This is a "time-varying impulse response," indicating that an input $u(s)$ at time s produces the output $u(s)A(s, t)$ at time t.

Input/output data define an i/o-map on the module level. The map $f:\mathscr{U} \to \Gamma\mathscr{Y}$ is given by

$$(uf)(\theta)(t) = \sum_{s \leqslant t} u(s)\theta(s, t)A(s, t).$$

The natural extension (i/o-map) $f:\Omega\mathscr{U} \to \Gamma\mathscr{Y}$ is given by

$$(u \otimes \alpha)f(\theta)(t) = \sum_{s \leqslant t} u(s)(\alpha\theta)(s, t)A(s, t).$$

Furthermore, every i/o-map can be written in this form. Given such an f, define $A(s, t): U \to Y$ by

$$uA(s, t) = (u \otimes 1)f(\delta_{s,t})(t).$$

Then, writing any θ in R formally as $\theta = \sum \theta(s, t)\delta_{s,t}$ (cf. the proof of Theorem 3.3) we see that

$$(u \otimes 1)f(\theta)(t) = \sum_{s \leqslant t} u(s)\theta(s, t)A(s, t),$$

as required.

Suppose given, then, an abstract i/o-map $f: \Omega\mathscr{U} \to \Gamma\mathscr{Y}$, or (equivalently) i/o-data $A(s, t)$. According to Theorem 2.7, f has a unique canonical realization as an $R/\mathscr{K}$-system. The next theorem shows that this canonical realization satisfies Definition 4.1.

4.2. THEOREM. *Suppose $\mathscr{U} = U^T$ and $\mathscr{Y} = Y^T$, where U and Y are finite-dimensional vector spaces over k. If $f: \Omega\mathscr{U} \to \Gamma\mathscr{Y}$ is an i/o-map, then the canonical realization of f as an $R/\mathscr{K}$-system is a linear dynamical system on the poset T in the sense of Definition* 4.1. *In particular, the canonical state module $\mathscr{X}$ is dynamical.*

Proof. We are given a commutative diagram of right R-modules

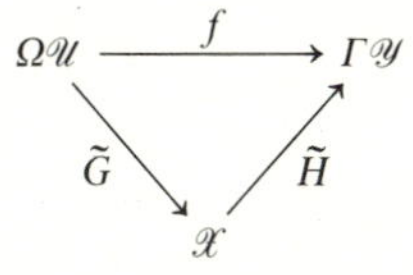

where $\tilde{G}$ is surjective and $\tilde{H}$ is injective. We show first that the state module $\mathscr{X}$ is dynamical: that is, the map $(*)$ $\mathscr{X} \to \prod X_t$ is bijective.

The proof proceeds in several stages. First, we show that $\Omega\mathscr{U}$ is dynamical and conclude from the surjectivity of $\tilde{G}$ that $(*)$ is surjective. Then we show that $\Gamma\mathscr{Y}$ injects into $\prod(\Gamma\mathscr{Y})e_t$ and conclude that $(*)$ is injective.

To show that $\Omega\mathscr{U}$ is dynamical, assume first that $U = k$ is one dimensional, so that $\mathscr{U} = \mathscr{K}$ and $\Omega\mathscr{U} \cong R$. What is Re_t? If $\alpha \in R$ and $t \in R$ and $t \in T$, then $\alpha e_t(u, v) = \alpha(u, t)$ if $v = t$; otherwise, $\alpha e_t(u, v) = 0$. This shows immediately that $R \to \prod Re_t$ is injective. Now, giving an element of $\prod Re_t$ amounts to giving functions $\alpha_t(u, t)$ for each t in T, so we can define α by $\alpha(u, t) = \alpha_t(u, t)$ for all u and t. The function α has finite memory if all the α_t do, showing that $R \to \prod Re_t$ is surjective. This argument extends easily to $U = k^n$, since then $\Omega\mathscr{U} = R^n$ and $R^n e_t \cong (Re_t)^n$. Therefore, $\Omega\mathscr{U}$ is dynamical.

Consider the diagram

$$\begin{array}{ccc} \Omega\mathscr{U} & \longrightarrow & \prod(\Omega\mathscr{U})e_t \\ \Big\downarrow \tilde{G} & & \Big\downarrow \prod \tilde{G}e_t \\ \mathscr{X} & \longrightarrow & \prod \mathscr{X}e_t \end{array}$$

Since $\tilde{G}$ is a surjective $\mathscr{K}$-linear map, we can write $\tilde{G} = \prod \tilde{G}e_t$, where each $\tilde{G}e_t:(\Omega\mathscr{U})e_t \to \mathscr{X}e_t$ must be a surjective k-linear map. Therefore $\prod \tilde{G}e_t$ is surjective, and we conclude that the bottom horizontal arrow is surjective since the top one is.

Next consider $\Gamma\mathscr{Y}$. Again, it suffices to assume $Y = k$, so that $\mathscr{Y} = \mathscr{K}$ and $\Gamma\mathscr{Y} = \Gamma\mathscr{K} = \mathrm{Hom}_{\mathscr{K}}(R, \mathscr{K})$. For h in $\Gamma\mathscr{K}$ and t in T we have $(he_t)(\beta) = h(e_t\beta)$. The crucial statement here is that $\Gamma\mathscr{K} \to \prod(\Gamma\mathscr{K})e_t$ is *injective*. To see this, assume that $he_t = 0$ for all t in T. If γ in R is arbitrary, how can we conclude that $h(\gamma) = 0$? Since h is $\mathscr{K}$-linear, $h(\gamma) = 0$ if and only if $h(\gamma e_u) = 0$ for all u in T. But γ has finite memory, so that γe_u is a finite sum $\sum \gamma(r, u)\delta_{r,u}$ for $r \leqslant u$, r in a finite set S. That is, $\gamma e_u = \sum_{r \in S} e_r\gamma e_u$, so that $h(\gamma e_u) = \sum_{r \in S} h(e_r\gamma e_u) = \sum_{r \in S} (he_r)(\gamma e_u) = 0$, since each $he_r = 0$. Therefore, $\Gamma\mathscr{K} \to \prod(\Gamma\mathscr{K})e_t$ is injective, and the argument extends to finite-dimensional Y. Now consider

$$\begin{array}{ccc} \mathscr{X} & \xrightarrow{\ \tilde{H}\ } & \Gamma\mathscr{Y} \\ \Big\downarrow & & \Big\downarrow \\ \prod \mathscr{X}e_t & \xrightarrow{\ \prod \tilde{H}e_t\ } & \prod(\Gamma\mathscr{Y})e_t \end{array}$$

Since $\tilde{H}$ and the right-hand vertical arrow are both injective, we conclude that the left-hand arrow is injective, completing the proof that $\mathscr{X}$ is dynamical.

To conclude that we have a system on T, we must show that $G = \prod G(t)$ and $H = \prod H(t)$, but this is just the fact that G and H are $\mathscr{K}$-linear. Q.E.D.

4.3. COROLLARY. *Suppose given input/output data $A(s, t): U \to Y$ for each $s \leqslant t$ as discussed before Theorem* 4.1 *above. Then there exists a family $\{X_t : t \in T\}$ of vector spaces, and k-linear maps $\Phi(s, t): X_s \to X_t$ satisfying $\Phi(t, t) = \mathrm{id}$ and $\Phi(s, u)\Phi(u, t) = \Phi(s, t)$ when $s \leqslant u \leqslant t$, together with k-linear maps $G(s): U \to X_s$ and $H(t): X_t \to Y$ such that for all s, t, $s \leqslant t$:*

$$A(s, t) = G(s)\Phi(s, t)H(t).$$

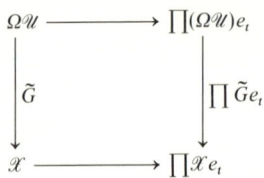

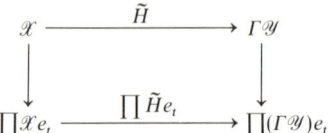

Proof. As above, define an i/o-map $f: \Omega\mathscr{U} \to \Gamma\mathscr{Y}$ by $(uf)(\theta)(t) = \sum_{s \leqslant t} u(s)\theta(s, t)A(s, t)$, and let $\sum = (\mathscr{X}, \mathscr{U}, \mathscr{Y}; G, H)$ be a canonical realization of f. By the theorem, $\mathscr{X}$ is dynamical, so we get $\Phi(s, t)$ from Theorem 3.3. Factor $f = \tilde{G}\tilde{H}$, so that for u in $\mathscr{U}$, α in R, and t in T, we have $(uf)(\alpha)(t) = (u\tilde{G}\tilde{H})(\alpha)(t)$. That is,

$$\sum_{s \leqslant t} u(s)A(s, t)\alpha(s, t) = \sum_{s \leqslant t} u(s)G(s)\Phi(s, t)\alpha(s, t)H(t).$$

Now set $\alpha = \delta_{s,t}$ and conclude that $A(s) = G(s)\Phi(s, t)H(t)$. Q.E.D.

We can also say something about uniqueness.

4.4. COROLLARY. *Suppose given i/o-data $A(s, t)$ with*

$$A(s, t) = G_1(s)\Phi_1(s, t)H_1(t) = G_2(s)\Phi_2(s, t)H_2(t)$$

for $s \leqslant t$, where (G_i, Φ_i, H_i) both come from canonical realizations as in Corollary 4.3. Then there exist k-linear isomorphisms $P(t):(X_1)_t \to (X_2)_t$ such that for all s, t in T:

$$G_1(s) = G_2(s)P(s),$$
$$P(t)H_1(t) = H_2(t),$$
$$P(s)\Phi_1(s, t) = \Phi_2(s, t)P(t).$$

Proof. This is a concrete version of Theorem (2.7). The reader should check that a system isomorphism gives the required $P(s)$ maps.

For this corollary to be useful, we need to know when a concrete realization (G, H, Φ) comes from a canonical system. The next theorem says that a canonical system is "locally of minimum dimension."

4.5. THEOREM. *Suppose given an i/o-map $f:\Omega\mathscr{U} \to \Gamma\mathscr{Y}$. For each t in T, let $n_t = \min\{\dim_k X_t : X_t$ occurs as the t-component of some realization of $f.\}$ Assume that n_t is finite for all t. Then a realization $\sum_0 = (\mathscr{X}_0, G_0, H_0)$ of f is canonical if and only if $\dim(X_0)_t = n_t$ for all t in T, where $(X_0)_t = (\mathscr{X}_0)e_t$.*

Proof. Choose t in T. Let $\sum_0 = (\mathscr{X}_0; G_0, H_0)$ be a canonical realization of f, and let $\sum_1 = (\mathscr{X}_1; G_1, H_1)$ be a realization with $\dim(X_1)_t = n_t$. It is implicit in the proof of Theorem 2.7 that $\mathscr{X}_0$ is isomorphic to a subquotient of $\mathscr{X}_1$, so that $\dim(X_0)_t \leqslant n_t$, hence equal since n_t is minimal. Q.E.D.

5. DIFFERENCE EQUATIONS ON LOCALLY FINITE POSETS

In this section we define difference equations on posets in a way suggested by work of Elliott and Mullans [EM73]. The same idea occurs in a different context in Hafiz [H75, p. 25] and in the combinatorial literature [C68]. These difference equations can be used to construct i/o-maps as in the last section, and the canonical realizations of these maps can be studied.

Classically, realization theory can be viewed as a device for replacing a high-order scalar difference equation with a first-order vector difference equation. This view survives in the present context, but the proper definition of "first-order" involves the Möbius function of the poset T.

Suppose given a locally finite poset $(T, \leqslant)$, a field k of scalars, and a finite memory algebra $R = \mathscr{A}_f(T)$ with coefficients in k. Set $\mathscr{K}^T$ as usual and identify $\mathscr{K} \subset R$. In this section we fix finite-dimensional vector spaces U and Y over k and set

$$\mathscr{U} = U^T = U \otimes_k \mathscr{K}, \qquad \mathscr{Y} = Y^T = Y \otimes_k \mathscr{K}.$$

These are right $\mathscr{K}$-modules, and we consider them as "dynamically trivial" right R-modules by setting for $u \in \mathscr{U}$ and $\alpha \in R$

$$(u\alpha)(t) = \sum_{s \leqslant t} u(s)\alpha(s, t).$$

The action on $\mathscr{Y}$ is similar.

Suppose now that U and Y are one dimensional, so that $\mathscr{U} \cong \mathscr{Y} \cong \mathscr{K}$. In this context, a *scalar difference* equation is given (by definition) by

$$y\alpha = u\beta \tag{5.1}$$

Next, suppose given the difference equation $y\alpha = u\beta$ with associated i/o-map $f: \Omega\mathscr{U} \to \Gamma\mathscr{Y}$ as above. Let $\Sigma = (\mathscr{X}, \mathscr{U}, \mathscr{Y}; G, H)$ be any realization of f. If the Möbius function $\mu = \zeta^{-1}$ of $(T, \leqslant)$ has finite memory, we can define the *first-order vector difference system* on Σ by the equations

$$x\mu = uG, \qquad y = xH. \tag{5.2}$$

Here, $x\mu$ is computed using the right R-module structure on $\mathscr{X}$. If $\mathscr{X}$ is dynamical with state-transition maps $\Phi(s, t)$, then (5.2) becomes, for each t in T,

$$\sum_{s \leqslant t} x(s)\Phi(s, t)\mu(s, t) = u(t)G(t), \qquad y(t) = x(t)H(t). \tag{5.2$'$}$$

If ζ has finite memory, we can solve for y using algebraic manipulations described in Section 4. Explicitly

$$\begin{aligned} y &= (uG)\zeta H \\ &= ((u\tilde{G})\zeta\tilde{H})(1) \\ &= (u\tilde{G}\tilde{H})(\zeta) \\ &= (uf)(\zeta). \end{aligned}$$

That is, if μ and ζ have finite memory, then the first-order vector difference equations (5.2) yield the solution of the high-order scalar difference equation $y\alpha = u\beta$. The same technique works even if ζ does not have finite memory, as long as $(uG)\zeta$ is defined.

This shows that the "correct" approach to state space equations on a poset involves the calculation of a Möbius function. Such a calculation is also a crucial step for many combinatorial problems [R64, DRS72]. For systems on the line, treated in succeeding sections, these equations have the usual shape. More complicated posets will be treated in later work.

6. The Incidence Algebra of the Discrete Line

This section contains some of the technical background for our study of systems and difference equations on the discrete line $T = (\mathbb{Z}, \leqslant)$.

Let $\mathscr{A}$ be the full incidence algebra of $T = (\mathbb{Z}, \leqslant)$ with coefficients in the field k. Let $\mathscr{K} = k^T$, and identify $\mathscr{K} \subset \mathscr{A}$ as before: for a in $\mathscr{K}$,

$$a(s, t) = \begin{cases} a(t) & \text{if} \quad s = t \\ 0 & \text{otherwise.} \end{cases}$$

We also introduce a special function z in $\mathscr{A}$ by

$$z(s, t) = \begin{cases} 1 & \text{if} \quad s = t - 1 \\ 0 & \text{otherwise.} \end{cases}$$

Now suppose α is any function in $\mathscr{A}$. Then easy calculations show that

$$(\alpha z)(s, t) = \alpha(s, t - 1), \qquad (z\alpha)(s, t) = \alpha(s + 1, t).$$

In particular, for $a \in \mathscr{K}$, we have

$$az(s, t) = za^{\sigma}(s, t) = \begin{cases} a(s), & s = t - 1 \\ 0 & \text{otherwise,} \end{cases}$$

where $a^{\sigma}(t) = a(t - 1)$ by definition. This formula generalizes immediately to $az^n = z^n a^{\sigma^n}$, where $a^{\sigma^n}(t) = a(t - n)$.

Next, we consider the smallest subring of $\mathscr{A}$ containing $\mathscr{K}$ and z. This ring, called $\mathscr{K}_{\sigma}[z]$, consists of all polynomials $\alpha = a_0 + za_1 + z^2 a_2 + \cdots + z^n a_n$ with $a_i \in K$. By direct computation, we verify that

$$\alpha(t - i, t) = \begin{cases} a_i(t), & 0 \leqslant i \leqslant n \\ 0 & n < i. \end{cases}$$

From the point of view of classical difference algebra, the ring $\mathscr{K}_{\sigma}[z]$ is the *skew-polynomial ring* with coefficients in the difference ring $(\mathscr{K}, \sigma)$, since $az = za^{\sigma}$. (See [H75] or [CP71].)

If $\alpha = a_0 + za_1 + \cdots + z^n a_n$ and a_n is not (identically) zero, we say that α has *degree n*. More generally, for any $\beta \in \mathscr{A}$, the degree of β is defined by

$$\deg(\beta) = \max\{i : \beta(t - i, t) \neq 0 \text{ for some } t\}.$$

Of course, $\deg(\beta)$ may be infinite. In fact, $\deg(\beta) < \infty$ if, and only if, $\beta \in \mathscr{K}_\sigma[z]$. To handle all incidence functions, we can introduce the ring $\mathscr{K}_\sigma[[z]]$ of all formal power series

$$\alpha = a_0 + za_1 + z^2 a_2 + \cdots + z^n a_n + \cdots,$$

so that $\alpha(t - i, t) = a_i(t)$ for all $i \geqslant 0$. Since every incidence function has such a representation, we conclude that

$$\mathscr{A} = \mathscr{K}_\sigma[[z]].$$

If $\mathscr{A}_f$ is the finite memory algebra, we have

$$\mathscr{K}_\sigma[z] \subset \mathscr{A}_f \subset \mathscr{A} \cong \mathscr{K}_\sigma[[z]],$$

and both inclusions are proper. Kamen and Hafiz consider difference equations of finite degree using the ring $\mathscr{K}_\sigma[z]$; we shall see that $\mathscr{A}_f$ is technically a more convenient ring for use in this paper, but Theorem 6.1 below allows us to apply their results [K74, H75].

From now on, write $R = \mathscr{A}_f$. As usual, we have a forgetful functor $((\text{Mod-R})) \to ((\text{Mod-}\mathscr{K}))$ with adjoints Ω and Γ. The left adjoint Ω can be written $\Omega\mathscr{U} = \mathscr{U} \otimes_{\mathscr{K}} R$, and, in particular, $\Omega\mathscr{K} = R$. The right adjoint is given by $\Gamma\mathscr{Y} = \text{Hom}_{\mathscr{K}}(R, \mathscr{Y})$, which is very difficult to compute. However, the inclusion $\mathscr{K}_\sigma[z] \subset R$ induces a restriction map,

$$\text{Hom}_{\mathscr{K}}(R, \mathscr{Y}) \to \text{Hom}_{\mathscr{K}}(\mathscr{K}_\sigma[z], \mathscr{K}),$$

which will allow us to replace R by $\mathscr{K}_\sigma[z]$ while studying observability in this paper. Since $\mathscr{K}_\sigma[z]$ is a free $\mathscr{K}$-module with basis $1, z, z^2, \ldots$, a map $h : \mathscr{K}_\sigma[z] \to \mathscr{Y}$ is exactly determined by the vectors $h(z^i) \in \mathscr{Y}$, $i = 0, 1, 2, \ldots$. In a later paper, we shall discuss $\text{Hom}_{\mathscr{K}}(\mathscr{K}_\sigma[z], \mathscr{Y})$ from the point of view of "skew-formal Laurent series," writing

$$h \to \sum_{i=0}^{\infty} h(z^i) z^{-i}.$$

We conclude this section with a "division algorithm" which leads to an explicit description of a special kind of R-module and ultimately (in Section 8) to some "canonical forms" for time-varying systems. This part is directly inspired by [H75].

6.1. THEOREM. *Suppose $\beta = z^n + z^{n-1}\beta_1 + \cdots + \beta_n$ is a monic polynomial in $\mathscr{K}_\sigma[z] \subset R$. Then for any θ in R, there exists a unique λ in R such that*

$(\beta\lambda)(s, t) = \theta(s, t)$ whenever $t - s \geqslant n$. That is, $\theta - \beta\lambda$ is a polynomial in $\mathscr{K}_\sigma[z]$ of degree $\leqslant n - 1$.

Proof. Given t, we construct $\lambda(s, t)$ as a function of s. Suppose that $t - s \geqslant n$, and try to solve $(\beta\lambda)(s, t) = \theta(s, t)$ for λ. Since β is monic of degree n, this gives

$$\lambda(s + n, t) + \sum_{r=s}^{s+n-1} \beta(s, r)\lambda(r, t) = \theta(s, t). \tag{$*$}$$

If $\theta = 0$, we can choose $\lambda = 0$, so assume $\theta \neq 0$. Since θ has finite memory, we can find $l \leqslant t$ such that $\theta(l, t) \neq 0$, but $\theta(u, t) = 0$ for all $u \leqslant l$. Set $\lambda(u, t) = 0$ for $u < l + n$, so that $(*)$ holds with $0 = 0$ whenever $s < l$. For $s = l$, $(*)$ becomes $\lambda(l + n, t) = \theta(l, t)$. For $s = l + 1, \ldots, t - n$, $(*)$ gives a formula for $\lambda(s + n, t)$ in terms of previously defined quantities, completing the definition of $\lambda(r, t)$ for all $r \leqslant t$. This shows existence of λ.

For uniqueness, suppose λ_1 and λ_2 are two solutions, so that

$$\beta(\lambda_1 - \lambda_2)(s, t) = 0$$

whenever $s \leqslant t - n$. Claim $\lambda_1 - \lambda_2 = 0$: if not, there exists l, t with $l \leqslant t$ such that $(\lambda_1 - \lambda_2)(l, t) \neq 0$ but $(\lambda_1 - \lambda_2)(u, t) = 0$ for all $u < l$, since $\lambda_1 - \lambda_2$ has finite memory. But then

$$\beta(\lambda_1 - \lambda_2)(l - n, t) = \beta(l - n, l) \cdot (\lambda_1 - \lambda_2)(l, t) \neq 0,$$

a contradiction. Q.E.D.

6.2. COROLLARY. *Suppose $\beta = z^n + z^{n-1}\beta_1 + \cdots + \beta_n$ is a monic polynomial in $\mathscr{K}_\sigma[z] \subset R$. Then $R/\beta R \cong \mathscr{K}^n$ is a free right $\mathscr{K}$-module of rank n. One basis is given by $1, z, z^2, \ldots, z^{n-1}$ (mod βR).*

Proof. Let $\mathscr{U}$ be the $\mathscr{K}$-module of all polynomials of degree $\leqslant n - 1$. Then $\mathscr{U}$ is free of rank n with bases $1, z, \ldots, z^{n-1}$. For each θ in R, let λ_θ be the unique function of Theorem 6.1 such that $\theta - \beta\lambda_\theta = \rho_\theta$ (say) is a polynomial of degree $\leqslant n - 1$. Then $\theta \to \rho_\theta$ obviously gives a $\mathscr{K}$-linear surjective map from $R/\beta R$ to $\mathscr{U}$, which is injective since θ determines ρ_θ uniquely. Q.E.D.

7. LINEAR SYSTEMS ON THE DISCRETE LINE

In this short section, we only recall definitions and list formulas for the special case $T = (\mathbb{Z}, \leqslant)$. We consider time-varying linear systems on T as $R/\mathscr{K}$-systems, where $R = \mathscr{A}_f(T)$ is the finite-memory algebra. Our systems must have "dynamical" state modules as in Definition 4.1. Recall that a system $\sum = (\mathscr{X}; \mathscr{U}, \mathscr{Y}; G, H)$ consists of input/output modules $\mathscr{U} = U^T$, $\mathscr{Y} = T^T$, and state module $\mathscr{X} = \prod X_t$. We take U and Y finite dimensional over k, and in interesting cases each X_t will also be finite dimensional. In this case $G = (G(s))$ and $H = (H(t))$ can be considered as time-varying rectangular matrices.

The right R-module action or "dynamical structure" on $\mathscr{X}$ is completely determined by

$$(xz)(t) = x(t-1)F(t),$$

where $F(t): X_{t-1} \to X_t$ is k-linear. Alternatively, we have state-transition matrices

$$\Phi(s, t) = \begin{cases} I & \text{if} \quad s = t \\ F(s)F(s+1)\cdots F(t) & \text{if} \quad s < t \end{cases}.$$

The input map $\tilde{G}: \mathscr{U} \otimes R \to \mathscr{X}$ is given by

$$(u \otimes \alpha)\tilde{G}(t) = \sum_{s \leqslant t} u(s)G(s)\Phi(s, t)\alpha(s, t)$$

and, in particular,

$$(uz^i)\tilde{G}(t) = u(t-i)G(t-i)\Phi(t-i, t).$$

Intuitively, uz^i corresponds to an input $u(t-i)$ at time $t-i$, followed by free evolution to time t. Thus a "local state" $x(t)$ is reachable if there exist finitely many inputs $u(t-i)$ which set up $c(t)$ at time t. "Global reachability," that is, $\tilde{G}$ surjective, means simultaneous reachability at all times t. This is stronger than the assumption that X_t is reachable for all t. (See Example (9.1).)

Outputs go to $\Gamma\mathscr{Y} = \operatorname{Hom}_{\mathscr{K}}(R, \mathscr{X})$, and for θ in R we have

$$(x\tilde{H})(\theta)(t) = \sum_{s \leqslant t} x(s)\Phi(s, t)\theta(s, t)H(t).$$

Also, $\mathscr{K}_\sigma[z] \subset R$ is dense, so that $\tilde{H}: \mathscr{X} \to \Gamma\mathscr{Y}$ is determined by

$$(x\tilde{H})(z^i)(t) = x(t-i)\Phi(t-i, t)H(t).$$

Perhaps more intuitively, for each $i > 0$,

$$(x\tilde{H})(z^i)(t+i) = x(t)\Phi(t, t+i)H(t+i),$$

which is the output at time $t+i$ resulting solely from the local state $x(t)$. We can say that the system is *locally observable at time* t if $x(t)\Phi(t, t+i)H(t+i) = 0$ (for all $i \geqslant 0$) implies $x(t) = 0$. It is not too hard to see that observability is equivalent to local observability at all times t.

The input/output map $f_\Sigma = \tilde{G}\tilde{H}$ is given by

$$(uf_\Sigma)(z^i)(t) = u(t-i)G(t-i)\Phi(t-i, t)H(t).$$

That is, f_Σ is equivalent is the time-varying impulse response

$$A(t-i, t) = G(t-i)\Phi(t-i, t)H(t).$$

Finally, the Möbius function for $(\mathbb{Z}, \leqslant)$ is $1 - z$, so that the first-order vector difference equation corresponding to the system Σ is

$$x(1-z) = uG, \qquad y = xH,$$

or

$$x(t) = x(t-1)F(t) + u(t)G(t), \qquad y(t) = x(t)H(t).$$

8. Difference Equations, Adjoints, and Realizations

In this section we consider in some detail a fairly general "autoregressive" time-varying difference equation on $T = (\mathbb{Z}, \leqslant)$:

$$y(t) + \alpha_1(t)y(t-1) + \cdots + a_n(t)y(t-n) = u(t),$$

where $\alpha_1, \alpha_2, \ldots, \alpha_n$ in $\mathcal{K}$ are arbitrary.

We can also write this as

$$y\alpha = u$$

where $\alpha = 1 + z\alpha_1 + z^2\alpha_2 + \cdots + z^n\alpha_n$ in $\mathcal{K}_\sigma[z] \subset R$.

Since $\alpha(t, t) = 1$ for all $t \in \mathbb{Z}$, there exists an inverse α^{-1} in the full incidence algebra of $\mathbb{Z}$. We use α^{-1} to define a scalar i/o-map $f: \Omega\mathcal{K} \to \Gamma\mathcal{K}$ as follows: For β in $\Omega\mathcal{K} = R$, define $\beta f: R \to \mathcal{K}$ in $\Gamma\mathcal{K}$ by

$$(\beta f)(\theta)(t) = \sum_{s \leqslant t} (\beta\theta)(s, t)\alpha^{-1}(s, t).$$

As a special case, we compute

$$(1f)(z^i)(t) = \alpha^{-1}(t-i, t).$$

The function $A(s, t) = \alpha^{-1}(s, t)$ is essentially the same as the classical "one-sided Green's function." See [MK68, pp. 14–16, 39 ff.].

Our main goal is to construct a canonical realization

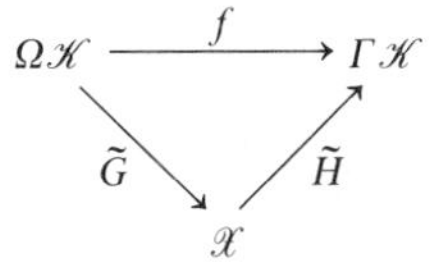

so that, if the dynamics of $\mathscr{X}$ are determined by $\{F(t)\}$, we have

$$\alpha^{-1}(s, t) = G(s)F(s+1)\cdots F(t)H(t),$$

with $\tilde{G}$ surjective and $\tilde{H}$ injective.

Since $\Omega\mathscr{K} \cong R$, we can take $\mathscr{X} \cong R/\ker f$. We begin by presenting an important element of $\ker f$.

8.1. Definition. Suppose given β in $\mathscr{K}_\sigma[z]$ of degree n. Then the (classical scalar) *adjoint* of β is the function $\beta^\#$ given by

$$\beta^\#(s, t) = \begin{cases} \beta(t-n, s) & \text{if } t-n \leqslant s \\ 0 & \text{otherwise.} \end{cases}$$

First we verify that this adjoint is a familiar object in the theory of difference equations (cf. [MK68, p. 46).

8.2. Theorem. *If* $\alpha = \alpha_0 + z\alpha_1 + \cdots + z^n\alpha_n$, *then*

$$\begin{aligned} \alpha^\# &= \alpha_0 z^n + \alpha_1 z^{n-1} + \cdots + \alpha_{n-1}z + \alpha^n \\ &= z^n\alpha_0^{\sigma^n} + z^{n-1}\alpha_1^{\sigma^{n-1}} + \cdots + z\alpha_{n-1}^\sigma + \alpha_n. \end{aligned}$$

Proof. Set $\gamma = \alpha_0 z^n + \alpha_1 z^{n-1} + \cdots + \alpha_{n-1}z + \alpha_n$. If $t - s > n$, then $\alpha^\#(s, t) = \gamma(s, t) = 0$. If $t - s = i \leqslant n$, then $\alpha^\#(s, t) = \alpha(t-n, s) = \alpha(t-n, t-i) = \alpha_{n-i}(t-i)$; also, $\gamma(s, t) = \gamma(t-i, t) = (\alpha_{n-i}z^i)(t-i, t) = \alpha_{n-i}(t-i)$. Q.E.D.

The main point is that if f is defined by the weighting function $\alpha^{-1}(s, t)$, then $\alpha^\# \in \ker f$. This follows from the following slightly more general statement.

8.3. Theorem. *Suppose given a scalar i/o-map* $f: \Omega\mathscr{K} \to \Gamma\mathscr{K}$ *corresponding to a weighting function* $A(s, t)$. *Then* ker f *contains the monic polynomial*

$$\alpha^\# = z^n + \alpha_1 z^{n-1} + \cdots + \alpha_{n-1}z + \alpha_n$$

if, and only if, $A(s, t)$ *satisfies the recurrence equation*

$$(\alpha A)(s, t) = \sum_{l: s \leqslant l \leqslant t} \alpha(s, l)A(l, t) = 0$$

for $t > s + n$. *Here* $\alpha = 1 + z\alpha_1 + \cdots + z^n\alpha_n$.

Proof. Choose θ in $\mathscr{K}_\sigma[z]$. Then for t in T,

$$\begin{aligned}(\alpha^\# f)(\theta)(t) &= \sum_{w \leqslant t} (\alpha^\# \theta)(w, t)A(w, t) \\ &= \sum_{w \leqslant t} \sum_{l:w \leqslant l \leqslant t} \alpha^\#(w, l)\theta(l, t)A(w, t) \\ &= \sum_{l \leqslant t} \left(\sum_{w \leqslant l} \alpha^\#(w, l)A(w, t) \right) \theta(l, t).\end{aligned}$$

Now, for $w \leqslant l$, $\alpha^\#(w, l) = \alpha(l - n, w)$, and for $w > l$, $\alpha(l - n, w) = 0$, since α has degree n. Therefore, the inner sum is just $\sum_{w:l-n \leqslant w \leqslant t} \alpha(l - n, w)A(w, t) = (\alpha A)(l - n, t)$. Since θ is arbitrary, this shows that $\alpha^\# \in \ker f$ if, and only if, $(\alpha A)(l - n, t) = 0$ for all $l \leqslant t$. Setting $l - n = s$ gives the theorem. Q.E.D.

8.4. Corollary. *If $A(s, t) = \alpha^{-1}(s, t)$, then $\alpha^\# \in \ker f$, and $\alpha^\# R \subseteq \ker f$.*

Proof. We have $(\alpha \cdot \alpha^{-1})(s, t) = 1(s, t) = 0$ for $t - s > 1$. Therefore $\alpha^\# \in \ker f$ by the theorem.

Next, we define $\mathscr{X} = R/\alpha^\# R$ and consider the reachable realization

$$\Sigma: \quad \begin{array}{ccc} R & \longrightarrow & \Gamma\mathscr{K} \\ & {\scriptstyle \tilde{G}} \searrow \quad \nearrow {\scriptstyle \tilde{H}} & \\ & \mathscr{X} & \end{array} \tag{8.5}$$

where $\tilde{G}: R \to \mathscr{X}$ is the natural projection and $\tilde{H}: \mathscr{X} \to \Gamma\mathscr{K}$ comes from a standard isomorphism theorem. Since $\alpha^\# = z^n + z^{n-1}\alpha^{\sigma^{n-1}} + \alpha_n$ is monic, Corollary 6.2 shows that $\mathscr{X}$ is a free $\mathscr{K}$-module of rank n, with basis given by appropriate equivalence classes of $1, z, z^2, \ldots, z^{n-1}$.

The next corollary, trivial in the present context, answers an important system theoretic question. We will return to this topic in Section 9.

8.6. Corollary. *The system Σ is reachable in time n. That is, any x in $\mathscr{X}$ is $\omega\tilde{G}$, where ω is a polynomial in $\mathscr{K}_\sigma[z] \subset R$ of degree $\leqslant n - 1$.*

Now we are ready to present our first "canonical form." The action of z is given by

$$\begin{aligned} z^i \cdot z &= z^{i+1} \qquad \text{for} \quad i = 0, \ldots, n - 2, \\ z^{n-1} \cdot z &= z^n \equiv -z^{n-1}\alpha_1^{\sigma^{n-1}} - \cdots - z\alpha_{n-1}^{\sigma} - \alpha_n, \end{aligned}$$

where $\equiv$ denotes congruence mod $\alpha^\# R$.

Write a general state vector as

$$x = x_0 + zx_1 + \cdots + z^{n-1}x_{n-1}$$
$$= (x_0, x_1, \ldots, x_n).$$

Then we have

$$xz = zx_0^{\sigma} + z^2 x_k^{\sigma} + \cdots + z^n x_n^{\sigma}.$$

The action of $\mathscr{K}_{\sigma}[z]$ on $\mathscr{K}^n$ can be described by a time-varying matrix $F(t)$ such that $xz = x^{\sigma}F$, or $(xz)(t) = x(t-1)F(t)$ for all t. The calculation above shows that

$$[F] = \begin{bmatrix} 0 & \cdots & 0 & \cdots & 0 & \cdots & 0 \\ 0 & \cdots & 0 & \cdots & \vdots & \cdots & \vdots \\ 0 & \cdots & 0 & \cdots & 1 & \cdots & 0 \\ \vdots & & & & 0 & & 1 \\ & & \vdots & & \vdots & & \vdots \\ -\alpha_n & \cdots & -\alpha_{n-1}^{\sigma} & \cdots & -\alpha_2^{\sigma^{n-2}} & \cdots & -\alpha_1^{\sigma^{n-1}} \end{bmatrix}$$

Next, the map $G: \mathscr{K}_{\sigma}[z] \to \mathscr{X}$ sends 1 in $\mathscr{K}_{\sigma}[z]$ to $1 = (1, 0, 0, \ldots, 0)$ in $\mathscr{X}$, so that

$$[G] = (1, 0, 0, \ldots, 0).$$

Finally, the output H is defined by $(1H)(z^i) = (1f)(z^i)$, and $(1f)(z^i)(t) = \alpha^{-1}(t-i, t)$ by direct calculation. Hence,

$$[H] = \begin{pmatrix} \alpha^{-1}(t, t) \\ \alpha^{-1}(t-1, t) \\ \alpha^{-1}(t-n+1) \end{pmatrix}$$

To summarize:

8.7. THEOREM. *The difference equation*

$$y(t) + \alpha_1(t)y(t-1) + \cdots + \alpha_n(t)y(t-n) = u(t)$$

has a free reachable realization of rank n over $\mathscr{K}$ *given by the matrices* $F(t)$, $G(s)$, $H(t)$ *described above.*

This particular matrix realization might be called the "algebraists' realization," since the basis $\{1, z, z^2, \ldots, z^{n-1}\}$ would occur immediately to an algebraist. A more useful realization for system theory is the *control canonical realization* discussed by Hafiz in [H75, p. 34].

In our notation, the control canonical basis is given by

$$\begin{aligned} c_n &= 1, \\ c_{n-1} &= z + \alpha_1, \\ c_{n-2} &= z^2 + \alpha_1 z + \alpha_2, \\ &\vdots \\ c_1 &= z^{n-1} + \alpha_1 z^{n-2} + \cdots + \alpha_{n-2} z + \alpha_{n-1}, \end{aligned}$$

and yields matrices

$$\begin{aligned} G_{\mathrm{c}}(s) &= (0, 0, \ldots, 1), \\ H_{\mathrm{c}}(t) &= (0, 0, \ldots, 1), \\ F_{\mathrm{c}}(t) &= \begin{bmatrix} 0 & 0 & \cdots & 0 & -\alpha_n(t) \\ 1 & 0 & \cdots & 0 & -\alpha_{n-1}(t) \\ 0 & 1 & \cdots & & -\alpha_{n-1}(t) \\ \vdots & 0 & & \vdots & \vdots \\ & \vdots & & & \\ 0 & 0 & \cdots & 1 & -\alpha_1(t) \end{bmatrix} \end{aligned}$$

Next, we present an example to show that our reachable realization (8.5) with $\mathscr{X} = R/\alpha^{\#}R$ is not necessarily observable, followed by a theorem about when it is observable.

8.8. Example. Consider the equation $y(t) - \alpha_1(t)y(t-1) = u(t)$, or $y\alpha = u$ with $\alpha = 1 - z\alpha_1$. This has a realization with $\mathscr{X} = R/(z-\alpha)R \cong \mathscr{X}$, and $G(s) = 1$, $H(t) = 1$, $F(t) = \alpha(t)$. This realization is always observable, since $(x\tilde{H})(1)(t) = x(t)$.

On the other hand, for any larger dimension the reachable systems we have constructed sometimes fail to be observable. The situation is described completely in the following theorem.

8.9. Theorem. *Suppose given a difference equation* $y(t) + \alpha_1(t)y(t-1) + \cdots + \alpha_n(t)y(t-n) = u(t)$ *with* $n > 1$. *Then the reachable realization* Σ *of Theorem* 8.7 *above is observable if, and only if,* $\alpha_n(t) \neq 0$ *for all* t *in* T.

Proof. We consider $x = (x_1, \ldots, x_n)$ in $\mathscr{X} = \mathscr{K}^n$, written in the control canonical basis. If $x\tilde{H} = 0$ then, for each t, $(x\tilde{H})(z^i)(t+i) = 0$. We must show that $x(t) = 0$ (under the assumption that $\alpha_n(t) \neq 0$ for all t.)

Explicit computation of

$$(x\tilde{H})(z^i)(t) = x(t)F_c(t)F_c(t+1)\cdots F_c(t+i)H_c(t+i) = 0$$

for $i = 0, \ldots, n-1$ yields

$$\begin{aligned}
&x_n(t) = 0,\\
&\alpha_n(t)x_1(t) + \alpha_{n-1}(t)x_2(t) + \cdots + \alpha_1(t)x_n(t) = 0,\\
&\alpha_n(t+1)x_2(t) + \cdots + \alpha_2(t+1)x_n(t) = 0,\\
&\quad\vdots\\
&\alpha_n(t+n-1)x_{n-1}(t) + \alpha_{n-1}(t+n-1)x_n(t) = 0.
\end{aligned}$$

Solving this triangular system backward, using $\alpha_n(s) \neq 0$ for all s, shows that $x_i(t) = 0$ for all i.

For the converse, assume that $\alpha_n(t) = 0$ for some t. Consider a state $x = (x_1, 0, 0, \ldots, 0)$ such that $x_1(t) \neq 0$ but $x_1(s) = 0$ for all $s \neq t$. The formulas above imply easily that $x\tilde{H} = 0$. Q.E.D.

We conclude this section with an analogue of Corollary 8.6.

8.10. Corollary. *The reachable realization* $\Sigma = (F, G, H)$ *of the equation* $y\alpha = u$ *(with* α *of degree* $n \geqslant 1$*) is observable if, and only if, it is observable in time n, that is, if and only if*

$$\operatorname{rank}[H(t); F(t)H(t+1); \ldots; F(t+n-1)H(t+n)] = \dim X_t$$

for all t in T.

Proof. The case $n = 1$ is contained in Example 8.9, and for $n \geqslant 2$ this follows from the proof of the last proposition.

We shall see in the next section that the finiteness results (8.6) and (8.10) are rather special.

9. Some Examples

In this section we present examples which show some of the limitations of the present theory. In particular, we shall see that the nice finiteness results (8.6) and (8.10) are rather special: General equations of the form $y\alpha = u\beta$ present serious difficulties. We shall also motivate further the use of the finite memory algebra instead of $\mathcal{K}_\sigma[z]$ for systems on the line.

The first two examples start by considering the difference equation

$$y(t) - y(t-1) = u(t)b(t)$$

for given $b(t)$ in $\mathscr{K}$. One realization is given by $\Sigma = (\mathscr{X}, F, G, H)$, with $\mathscr{X} = \mathscr{K}$, $F(t) \equiv 1$, $H(t) \equiv 1$, and $G(s) = b(s)$ for all s in T. For α in the finite memory algebra $\mathscr{A}_{\mathrm{f}}$, we have

$$(\alpha\tilde{G})(t) = \sum_{s \leqslant t} b(s)\alpha(s, t).$$

9.1. Example. With notation as above, choose $k \geqslant 0$ and set $b(t) = 0$ for $t = 1, 2, \ldots, k$, ($b(t) = 1$ otherwise. For x in $\mathscr{X}$, define

$$\alpha(s, t) = \begin{cases} x(t), & s = t \\ 0, & s < t \end{cases}; \qquad t \leqslant 0 \qquad \text{or} \qquad t > k,$$

$$\alpha(s, t) = \begin{cases} x(t), & s = 0 \\ 0, & s \neq 0 \end{cases}; \qquad t = 1, 2, \ldots, k.$$

Then α has degree k (if $x(k) \neq 0$) and $\alpha\tilde{G} = x$. Also, there is no function β of smaller degree with $\beta\tilde{G} = x$. That is, with this b, the system Σ is reachable, but only in time k, which can be arbitrarily large. (Compare Corollary 8.6.)

9.2. Example. We change Example 9.1 slightly by setting $b(s) = 1$ for $s \leqslant 0$, $b(s) = 0$ for $s > 0$. For x in $\mathscr{X}$ the corresponding α has $\alpha(0, t) = x(t)$ for all $t \geqslant 0$, and therefore it has finite memory but infinite degree. Therefore, Σ is not reachable when considered as a system over $\mathscr{K}_\sigma[z]$. On the other hand, Σ must have a canonical subsystem, whose state module must therefore be a proper submodule of $\mathscr{K}$. The original $\mathscr{X}$ is, however, locally reachable and (in the notation of Theorem 4.5) $n_t = 1$ for all t in T. If $\mathscr{X}_0$ is the $\mathscr{K}_\sigma[z]$-canonical realization, then $\dim(\mathscr{X}_0)_t = 1$; it follows that $\mathscr{X}_0 \subsetneqq \prod(\mathscr{X}_0)_t = \mathscr{K}$, so Theorem 4.2 is false over $\mathscr{K}_\sigma[z]$. This is an important reason for the use of the finite memory algebra.

9.3. Example. This example treats observability. Consider a scalar system $\Sigma_1 = (\mathscr{X}; F, G, H)$ with $\mathscr{X} = \mathscr{K}$, $G(s) \equiv 1$, $F(t) \equiv 1$, and $H(t) = b(t)$, where $b(t) = 1$ for $t = 1, 2, 4, \ldots, 2^k, \ldots$; $b(t) = 0$ otherwise. We have

$$(x\tilde{H})(z^i)(t + i) = x(t)b(t + i)$$

for all t in T and $i \geqslant 0$. This shows that Σ_1 is observable, but not in finite time. Compare Corollary 8.10.

References

AM75 M. A. Arbib and E. G. Manes, Basic concepts of category theory, *in* "Category Theory Applied to Computation and Control" (E. G. Manes, ed.), Lecture Notes in Computer Science No. 25, Springer–Verlag, Berlin and New York, 1975.

B70 R. W. BROCKETT, "Finite-dimensional Linear Systems," Wiley, New York, 1970.

BG75 E. A. BENDER AND J. GOLDMAN, On the applications of Möbius inversion on combinatorial analysis, *Amer. Math. Monthly* **82** (1975), 789–802.

C68 H., CRAPO, "Möbius inversion in Lattices," *Arch. Math.* (*Basel*) **19** (1968), 595–607.

CE56 H. CARTAN AND S. EILENBERG, "Homological Algebra," Princeton Univ. Press, Princeton, New Jersey, 1956.

CP71 P. M. COHN, "Free Rings and their Relations," Academic Press, New York, 1971.

DRS72 P. DOUBILET, G.-C. ROTA, AND R. STANLEY, On the foundations of combinatorial theory VI: the idea of generating function, *Proc. Sixth Berkeley Symp. Math. Stat. and Prob., Berkeley, 1972*, Vol. II, pp. 267–318. Univ. of California Press, Berkeley, 1972.

EM73 D. ELLIOTT AND R. MULLANS, Linear systems on partially ordered time sets, *IEEE Decision and Control Conf., Dec. 1973.*

H75 K. M. HAFIZ, New results on discrete-time-varying linear systems. Ph. D. Thesis, Georgia Institute of Technology, 1975.

K74 E. W. KAMEN, A new algebraic approach to linear time-varying systems, preprint March, 1974 (to appear in *J. Comp. Sys. Sci.*).

M71 S. MAC LANE, "Categories for the Working Mathematician," Springer–Verlag, Berlin and New York, 1971.

ME75 E. G. MANES (ed.), "Category Theory Applied to Computation and Control," Lecture Notes in Computer Science No. 25, Springer–Verlag, Berlin and New York, 1975.

MK68 K. S. MILLER, "Linear Difference Equations," Benjamin, New York, 1968.

R64 G.-C. ROTA, On the foundations of combinatorial theory I: theory of Möbius functions, *Z. Wahrscheinlichkeitstheorie and Verw. Gebiete* **2** (1964), 340–368.

W72 B. F. WYMAN, Linear systems over commutative rings, unpublished lecture notes.

W75 B. F. WYMAN, Linear systems over rings of operators, *in* "Category Theory Applied to Computation and Control" (E.G. Manes, ed.). pp. 218–223, Lecture Notes in Computer Science No. 25, Springer–Verlag, Berlin and New York, 1975.

W76 B. F. WYMAN, Linear difference systems on partially ordered sets, *in* "Mathematical Systems Theory," Lecture Notes in Economic and Mathematical Systems No. 131, Springer–Verlag, Berlin and New York, 1976.

W76b B. F. WYMAN, Time-varying linear difference systems, *Proc. IEEE Conf. Decision and Control, Dec. 1–3, 1976.*

AMS (MOS) subject classifications: 93C50, 93C55, 06A10.

STUDIES IN FOUNDATIONS AND COMBINATORICS
ADVANCES IN MATHEMATICS SUPPLEMENTARY STUDIES, VOL. 1

Combinatorial Problems on Subsets and Their Intersections

M. DEZA

Centre National des Recherches Scientifiques, Paris, France

P. ERDOS

The Hungarian Academy of Sciences, Budapest, Hungary

AND

N. M. SINGHI†

School of Mathematics, Tata Institute of Fundamental Research
Colaba, Bombay, India

Let $|S| = n$, $m(n, l_1, l_2, k)$ (respectively, $m'(n, l_1, l_2, k)$) denote the cardinality of the largest family of subsets $A_i \subset S$ satisfying $|A_i| = k$ (respectively, $|A_i| \leqslant k$) and $|A_{i_1} \cap A_{i_2}| = l_1$ or l_2. In this paper we prove

(a) $m(n, 0, l_2, k) \leqslant \binom{n}{2}$, $m'(n, 0, l_2, k) \leqslant \binom{n}{2} + n + 1$; equality iff $k = 2$;
(b) $m(n, 0, l_2, k) \leqslant n$, if $l_2 \nmid k$, with equality for an infinity of n.

For $n \geqslant n_0(k)$ we show that

(a)
$$m(n, l_1, l_2, k) \leqslant \binom{n - l_1}{2}, \qquad m'(n, l_1, l_2, k) \leqslant \binom{n - l_1}{2} + (n - l_1) + 1;$$

(b) more exactly,
$$m(n, l_1, l_2, k) \leqslant \left[\frac{n - l_1}{k - l_1}\left[\frac{n - l_2}{k - l_2}\right]\right],$$

with equality for an infinity of n.

Let integers $0 \leqslant l_1 \leqslant l_2 < k < n$ be given. Let $M(n, l_1, l_2, k)$ denote any maximal system $\alpha = \{A_i\}$ of different sets such that

$$\left|\bigcup_{A_i \in \alpha} A_i\right| \leqslant n, \quad |A_i| = k(A_i \in \alpha), \quad |A_i \cap A_j| = l_1, l_2, \quad A_i, A_j \in \alpha, \quad i \neq j; \quad (1)$$

† This paper was written when the author was visiting Colorado State University, Fort Collins, Colorado, 1975.

ISBN 0-12-599101-0

let

$$m(n, l_1, l_2, k) = |M(n, l_1, l_2, k)|; \tag{2}$$

let $M'(n, l_1, l_2, k)$ denote any maximal system $\alpha = \{A_i\}$ such that

$$\left|\bigcup_{A_i \in \alpha} A_i\right| \leqslant n, \quad |A_i| \leqslant k(A_i \in \alpha), \quad |A_i \cap A_j| = l_1, l_2, \quad A_i, A_j \in \alpha, \quad i \neq j; \tag{1'}$$

and let

$$m'(n, l_1, l_2, k) = |M'(n, l_1, l_2, k)|. \tag{2'}$$

Let $l > 0$ be a given integer. The *kernel* of the system $\alpha = \{A_i\}$ is the intersection

$$K(\alpha) = \bigcap_{A_i \in \alpha} A_i. \tag{3}$$

System α is an *l-star* if

$$|K(\alpha)| \geqslant l. \tag{4}$$

System α is a *Δ-system* if all sets

$$A_i \backslash K(\alpha) \tag{5}$$

are disjoint, $|\alpha| > 2$.

Assume first $l_1 = l_2 = l$. Then Ryser proved the following (in other terms):

THEOREM 1. [10]

$$m(n, l, l, k) \leqslant n, \tag{6}$$

$$m'(n, l, l, k) \leqslant n + 1, \tag{6'}$$

and equality holds if there exists an (n, k, l)-design.

In fact, it was also shown in Theorem 1 of [10] that if $\alpha = \{A_1, A_2, \ldots, A_n\}$ satisfies $|\bigcup_{i=1}^n A_i| = n$, $|A_i \cap A_j| = l$ $(\forall 1 \leqslant i < j \leqslant n)$, then it is either an (n, k, l)-design or a λ-design, $\lambda = l$. Theorem 1 is a generalization of the de Bruijn–Erods's theorem (case $l = 1$), which in turn is a generalization of Fisher's inequality for a (b, v, r, k, λ)-design. Deza proved (in other terms):

THEOREM 2. [3] *There is an $r(k, l)$ such that*

$$r(k, l) \leqslant k^2 - k + 1, \tag{7}$$

$$\begin{aligned} n > l + r(k, l)(k - l) &\Rightarrow m(n, l, l, k) > r(k, l) \\ &\Rightarrow \text{any } M(n, l, l, k) \text{ is a } \Delta\text{-system} \\ &\Rightarrow m(n, l, l, k) = \left[\frac{n-l}{k-l}\right], \end{aligned} \tag{8}$$

$$\begin{aligned} n > l + r(k, l) - 1 &\Rightarrow m'(n, l, l, k) > r(k, l) \\ &\Rightarrow \textit{any } M'(n, l, l, k) \textit{ is a } \Delta\textit{-system} \\ &\Rightarrow m'(n, l, l, k) = n - l + 1. \end{aligned} \tag{8'}$$

For $l = 1$ and infinitely many k, (7) is best possible. We obtain from [2, 3, 8] that

$$\begin{aligned} k^2 - k + 1 &\geqslant \max(l + 2, (k - l)^2 + k - l + 1) \\ &\geqslant r(k, l) \geqslant \max(l + 2, q^2 + q + 1), \end{aligned} \tag{9}$$

where $q = \max q^*$, such that $q^* \leqslant k - l$ and $PG(2, q^*)$ exists.

In this paper we consider the case $l_1 < l_2$. From now on we assume $l_1 < l_2$. It is evident that

$$m(n, l_1, l_2, k) \geqslant m(n - l_1, 0, l_2 - l_1, k - l_1), \tag{10}$$

$$m'(n_1, l_1, l_2, k) \geqslant m'(n - l_1, 0, l_2 - l_1, k - l_1), \tag{10'}$$

since, for example, if $\alpha = \{A_i\} = M(n - l_1, 0, l_2 - l_1, k - l_1\}$ and $|A| = l_1$, $A \cap (\bigcup_{A_i \in \alpha} A_i) = \varnothing$, then

$$|\{A_i \cup A\}| \leqslant m(n, l_1, l_2, k).$$

Deza and Erdos proved the following (this is the inversion of (10), (10′) and generalization of Theorem 2):

THEOREM 3. [4] *Let* $0 < l_1 < l_2 < k < n$. *Let* $c(l_1, l_2, k) = \max(k - l_1 + 1, l_2{}^2 - l_2 + 1)$.

$$\begin{aligned} &m(n, l_1, l_2, k) > c(l_1, l_2, k)\frac{n - l_2}{k - l_2} \\ &\qquad \Rightarrow \textit{any } M(n, l_1, l_2, k) \textit{ is an } l_1\textit{-star} \\ &\qquad \Rightarrow m(n, l_1, l_2, k) = m(n - l_1, 0, l_2 - l_1, k - l_1), \end{aligned} \tag{11}$$

$$m(n, l_1, l_2, k) > 2k^2 n \Rightarrow l_2 - l_1 | k - l_2;$$

$$\begin{aligned} &m'(n, l_1, l_2, k) > kc(l_1, l_2, k)\frac{n - l_2}{k - l_2} \\ &\qquad \Rightarrow \textit{any } M'(n, l_1, l_2, k) \textit{ is an } l_1\textit{-star} \\ &\qquad \Rightarrow m'(n, l_1, l_2, k) = m'(n - l_1, 0, l_2 - l_1, k - l_1), \end{aligned} \tag{11'}$$

$$\begin{aligned} &m'(n, l_1, l_2, k) > 2k^3 n \\ &\qquad \Rightarrow l_2 - l_1 | \tilde{k} - l_2 \qquad \textit{for some} \quad \tilde{k}, \quad l_2 < \tilde{k} \leqslant k. \end{aligned}$$

Assume now $l_1 = 0$, $l_2 = l > 0$.

THEOREM 4. *Let $0 < l < k < n$. Then*

$$m(n, 0, l, k) = \binom{n}{2} \qquad \text{for} \quad k = 2, \tag{12}$$

$$m(n, 0, l, k) \leqslant \left[\frac{n(n-1)}{k}\right] \quad \text{for} \quad k > 2,$$

$$m(n, 0, l, k) \leqslant \left[\frac{n}{k}\left[\frac{n-l}{k-l}\right]\right] \qquad \text{for} \quad n > l + r(k, l)(k-l), \tag{13}$$

$$m(n, 0, l, k) = \frac{n(n-l)}{k(k-l)} \qquad \text{for the case } l \mid k \text{ and}$$

$$n > f_0(k, l), \qquad l \mid n, \qquad \frac{k}{l} - 1 \,\Big|\, \frac{n}{l} - 1, \qquad \frac{k}{l}\left(\frac{k}{l} - 1\right) \Big|\, \frac{n}{l}\left(\frac{n}{l} - 1\right),$$

$$m(n, 0, l, k) \leqslant n \qquad \text{for} \qquad l \nmid k, \tag{14}$$

$m(n, 0, l, k) = n$ for $v \nmid n$, where v is an integer such that there exists a(v, k, l)-design.

Proof. In fact, equality (12) is trivial, because $m(n, 0, l, 2) = m(n, 0, 1, 2) \leqslant |\{A_i : |A_i| = 2\}| = \binom{n}{2}$. It is easy to see that $M(n^*, 0, 1, k^*)$ is a *pairwise balanced design* PBD$[k^*, n^*]$. R. M. Wilson proved in [11] that a PBD$[k^*, n^*]$ exists if $n^* > f_0(k^*)$, $k^* \mid n^*$, $k^*(k^* - 1) \mid n^*(n^* - 1)$. In this case, we have

$$m(n^*, 0, 1, k^*) = \frac{n^*(n^* - 1)}{k^*(k^* - 1)}.$$

Now we take an l-multiple of PBD$[k^*, n^*]$ and put $n = ln^*$, $k = lk^*$. We obtain

$$m(n, 0, l, k) \geqslant m(n^*, 0, 1, k^*) = \frac{\dfrac{n}{l}\left(\dfrac{n}{l} - 1\right)}{\dfrac{k}{l}\left(\dfrac{k}{l} - 1\right)} = \frac{n(n-l)}{k(k-l)}$$

for $n^* = n/l > f_0(k^*)$, i.e., $n > lf_0(k/l)$. If also $n > l + r(k, l)(k - l)$, then we have equality in (13). We obtain the second equality, (14), by taking n/v (v, k, l)-designs $\alpha_j = \{A_{ij}\}$, $1 \leqslant j \leqslant n/v$, such that

$$\left(\bigcup_{A_{ij_1} \in \alpha_{j_1}} A_{ij_1}\right) \cap \left(\bigcup_{A_{ij_2} \in \alpha_{j_2}} A_{ij_2}\right) = \varnothing \qquad \text{for} \quad 1 \leqslant j_1 < j_2 \leqslant (n/v);$$

It is evident that $m(n, 0, l, k) \geqslant |\alpha_1| \cdot n/v = n$. Now we will prove upper bounds (12), (13), and (14). Let any $M(n, 0, l, k) = \alpha = \{A_i\}$ be given. We have

$$|\alpha|k \leqslant nm(n-1, l-1, l-1, k-1)$$

and so

$$|\alpha| \leqslant [m(n-1, l-1, l-1, k-1)n/k]. \tag{15}$$

Now inequality (12) follows from (15) and (6) of Theorem 1; inequality (13) follows from (15) and (8) of Theorem 2.

To prove (14), assume that there exists $M(n, 0, l, k) = \{A_1, A_2, \ldots, A_b\}$, $b > n$. Let $\bigcup_{i=1}^{b} A_i = \{x_1, x_2, \ldots, x_n\}$. Define the $n \times b$ incidence matrix N as follows:

$$N = (n_{ij}) \qquad \text{where} \quad ij = \begin{cases} 1 & \text{if} \quad x_i \in A_j \\ 0 & \text{if} \quad x_i \notin A_j. \end{cases}$$

Clearly, $N^T N = (b_{ij})$, where

$$b_{ij} = \begin{cases} k & \text{if} \quad i = j \\ 0 & \text{if} \quad |A_i \cap A_j| = 0 \\ l & \text{if} \quad |A_i \cap A_j| = l. \end{cases}$$

Since N is an $n \times b$ matrix and $b > n$, $N^T N$ is singular. Hence, there exists a rational vector $(y_1, y_2, \ldots, y_b)^T$ such that

$$N^T N(y_1, y_2, \ldots, y_b)^T = 0. \tag{16}$$

Now, by choosing $(y_1, y_2, \ldots, y_b)$ suitably, we can assume that $y_1, y_2, \ldots, y_b$ are integers, and if $y_{i_1}, y_{i_2}, \ldots, y_{i_2}$ are the nonzero integers among these, then

$$\text{g.c.d.}(y_{i_1}, y_{i_2}, \ldots, y_{i_r}) = 1.$$

From (16) we have

$$ky_i + l(\textstyle\sum y_j) = 0, \qquad i = 1, 2, \ldots, b, \tag{17}$$

where terms in the sum $\sum y_j$ are those for which $b_{ij} = l$.

Hence from (17), $l|ky_i$ for each i; in particular, $l|ky_{i_j}, j = 1, 2, \ldots, r$. Since g.c.d.$(y_{i_1}, y_{i_2}, \ldots, y_{i_2}) = 1$, we have a contradiction and so $l|k$.

An interesting example of a system $\alpha = \{A_i\}$ are the k-subsets of a given n-set such that $|A_i \cap A_j| = 0, 1$ and $|\alpha| > n$ is of locally symmetric design [1], but which is not symmetric.

THEOREM 5. *Let $0 < l < k < n$. Then*

$$m'(n, 0, l, k) \begin{cases} = \binom{n}{2} + n + 1 & \text{for } \; l = 1 \\ = 9 < \binom{n}{2} + n + 1 & \text{for } \; n = 4, \; k = 3, \; l = 2 \\ \leqslant \dfrac{n^2}{l+1} + n + 1 < \binom{n}{2} + n + 1 & \text{otherwise}; \end{cases} \tag{18}$$

$$m'(n, 0, l, k) \leqslant \left[\frac{n(n-l+1)}{l+1}\right] + n + 1 \qquad \text{for} \quad n > l + r(k, l) - 1. \tag{19}$$

In fact, the proof is analogous to the proof of Theorem 4. But instead of (15) we have

$$|\alpha| \leqslant [m'(n-1, l-1, l-1, k-1)n/(l+1)] + n + 1 \tag{15'}$$

for $M'(n, 0, l, k) = \alpha = \{A_i\}$, because, putting $\alpha^* = \{A_i \in \alpha : |A_i| \geqslant l + 1\}$, we obtain

$$|\alpha^*|(l+1) \leqslant nm'(n-1, l-1, l-1, k-1),$$
$$|\alpha^*| \geqslant |\alpha| - m'(n, 0, 0, l).$$

Now we return to the general case.

THEOREM 6. *Let $0 \leqslant l_1 < l_2 < k \leqslant n$. Then*

$$m(n, l_1, l_2, k) \leqslant \left[\frac{(n-l_1)}{(k-l_1)}\left[\frac{(n-l_2)}{(k-l_2)}\right]\right] \qquad \text{for} \quad n \geqslant n_0(k, l), \tag{20}$$

$$m'(n, l_1, l_2, k) \leqslant \left[\frac{(n-l_1)(n-l_2)}{l_2 - l_1 + 1}\right] + (n - l_1) + 1 \qquad \text{for} \quad n \geqslant n_0'(k, l). \tag{21}$$

Proof. In fact, (20), (21) follow from Theorems 3 and 4 applied to the case $m(n - l_1, l_1 - l_1, l_2 - l_1, k - l_1)$.

This paper was initiated by the following problem of R. Lemmon communicated to P. Erdos by A. Stone: Estimate $f(m, l, k) = \min|\bigcup_{i=1}^m A_i|$ if there exists a family $A_1, A_2, \ldots, A_m$ such that $|A_i| = k$ $(1 \leqslant i \leqslant m)$, $|A_i \cap A_j| = 0, l$ $(1 \leqslant i < j \leqslant m)$. A. Stone and R. Lemmon considered $f(m, l, k)$ for small m; it is easy to show that $f(m, l, k) \geqslant mk - l\binom{m}{2}$ (with equality for $m = k/l + 1$ if $l|k$).

Using elementary methods, Lemmon and Stone proved the following results (communicated privately [7]), assuming of course that $l \leqslant k$:

$$f(3, l, k) \begin{cases} = 2k - 1 & \text{if} \quad k \leqslant 2l \\ = 3k - 3l & \text{if} \quad k \geqslant 2l; \end{cases}$$

$$f(4, l, k) \begin{cases} = 2k - l & \text{if} \quad 2k \leqslant 3l \\ = -[(6l - 8k)/3] & \text{if} \quad k \leqslant 3l \leqslant 2k \\ = 4k - 6l & \text{if} \quad k \geqslant 3l. \end{cases}$$

Also, $f(m, 1, 3) \geqslant [1 + (24m + 1)^{1/2}]/2$.

Let $|S| = n$, $M(n, L, k)$ denote the largest family of subsets $A_i \subset S$ satisfying $|A_i| = k$ and $|A_{i_1} \cap A_{i_2}| \in L$, where $L = \{l_1, \ldots, l_r\}$, $0 \leqslant l_1 < \cdots < l_r < k$. Ray-Chaudhuri and Wilson proved in Theorem 3 of [9] (using, in particular, the nonsingularity of one matrix) that $|M(n, L, k)| \leqslant \binom{n}{r}$. In [4] it was shown that

$$|M(n, L, k)| \leqslant \left[\frac{n - l_1}{k - l_1}\left[\frac{n - l_2}{k - l_2}\left[\cdots \frac{n - l_r}{k - l_r}\right]\cdots\right]\right],$$

but only for the case $n > n_0(L, k)$. Also in [4] it was shown that $|M(n, L, k)| = O(n) \Rightarrow l_2 - l_1 | l_3 - l_2 | \cdots | l_r - l_{r-1} | k - l_r$ and that each $M(n, L, k)$ is an l_1-star (for $n > n_0(L, k)$); these results were generalized in [4] for the case $|A_i| \in K = \{k_1, \ldots, k_s\} \subset N$.

References

1. P. J. Cameron, Locally symmetric designs, *Geometriae Dedicata* **3** (1974) 65–76.
2. M. Deza, Une proprieté extrémale des plans projectifs finis dans une classe de codes équidistants, *Discrete Math.* **6** (1973) 343–352.
3. M. Deza, Solution d'un problemé de Erdos–Lovasz, *J. Combinatorial Theory Ser. B* **16**(2) (1974), 166.
4. M. Deza and P. Erdos, On intersection properties of the systems of finite sets, *Notices Amer. Math. Soc.* **22**(6) (1975), 657.
5. P. Erdos and R. Rado, Intersection theorems for systems of sets, *J. London Math. Soc.* **35** (1960), 85–90.
6. P. Erdos, Chao Ko, and R. Rado, Intersections theorems for systems of finite sets, *Quart. J. Math. Oxford* **12**(12) (1961), 313–320.
7. R. Lemmon and A. Stone, private communication.
8. R. C. Mullin, An asymptotic property of (r, λ)-systems, *Utilitas Math.* **3** (1973) 139–152.
9. D. K. Ray-Chaudhuri and R. M. Wilson, On t-designs, *Osaka J. Math.* **12** (1975), 737–744.
10. M. J. Ryser, An extension of a theorem of de Bruijn and Erdos on combinatorial designs, *J. Combinatorial Theory Ser. A* **10**(2) (1968), 246–259.
11. R. M. Wilson, An existence theorem for pairwise balanced designs II, *J. Combinatorial Theory Ser. A* **13** (1972), 246–273.

A B C 8 D 9 E 0 F 1 G 2 H 3 I 4 J 5